AF831281

INSTRUMENTATION FOR HEAVY ION NUCLEAR RESEARCH

NUCLEAR SCIENCE RESEARCH CONFERENCE SERIES

A series of monographs edited by Alexander Zucker,
Oak Ridge National Laboratory

VOLUME 1 **GIANT MULTIPOLE RESONANCES**
Edited by Fred E. Bertrand

VOLUME 2 **CONTINUUM SPECTRA OF HEAVY ION REACTIONS**
Edited by T. Tamura, J. B. Natowitz and D. H. Youngblood

VOLUME 3 **LASERS IN NUCLEAR PHYSICS**
Edited by C. E. Bemis, Jr. and H. K. Carter

VOLUME 4 **HIGH ANGULAR MOMENTUM PROPERTIES OF NUCLEI**
Edited by Noah R. Johnson

VOLUME 5 **HEAVY IONS AND NUCLEAR STRUCTURE**
Edited by B. Sikora and Z. Wilhelmi

VOLUME 6 **NUCLEAR PHYSICS WITH HEAVY IONS**
Prospects at Energies Below 20 MeV/amu
Edited by Peter Braun-Munzinger

VOLUME 7 **INSTRUMENTATION FOR HEAVY ION NUCLEAR RESEARCH**
Edited by Dan Shapira

Additional volumes in preparation

ISSN: 0250-4375

INSTRUMENTATION FOR HEAVY ION NUCLEAR RESEARCH

Proceedings of the International Conference
on Instrumentation for Heavy Ion Nuclear Research
held at Oak Ridge National Laboratory,
Oak Ridge, Tennessee
October 22-25, 1984

Edited by

Dan Shapira
Oak Ridge National Laboratory
Oak Ridge, TN

CRC Press
Taylor & Francis Group
Boca Raton London New York

CRC Press is an imprint of the
Taylor & Francis Group, an informa business

First published 1985 by Harwood Academic Publishers

Published 2019 by CRC Press
Taylor & Francis Group
6000 Broken Sound Parkway NW, Suite 300
Boca Raton, FL 33487-2742

© 1985 by Taylor & Francis Group, LLC
CRC Press is an imprint of Taylor & Francis Group, an Informa business

First issued in hardback 2019

No claim to original U.S. Government works

ISBN 13: 978-1-138-45988-5 (hbk)
ISBN 13: 978-3-7186-0240-7 (pbk)

This book contains information obtained from authentic and highly regarded sources. Reasonable efforts have been made to publish reliable data and information, but the author and publisher cannot assume responsibility for the validity of all materials or the consequences of their use. The authors and publishers have attempted to trace the copyright holders of all material reproduced in this publication and apologize to copyright holders if permission to publish in this form has not been obtained. If any copyright material has not been acknowledged please write and let us know so we may rectify in any future reprint.

Except as permitted under U.S. Copyright Law, no part of this book may be reprinted, reproduced, transmitted, or utilized in any form by any electronic, mechanical, or other means, now known or hereafter invented, including photocopying, microfilming, and recording, or in any information storage or retrieval system, without written permission from the publishers.

For permission to photocopy or use material electronically from this work, please access www.copyright.com (http://www.copyright.com/) or contact the Copyright Clearance Center, Inc. (CCC), 222 Rosewood Drive, Danvers, MA 01923, 978-750-8400. CCC is a not-for-profit organization that provides licenses and registration for a variety of users. For organizations that have been granted a photocopy license by the CCC, a separate system of payment has been arranged.

Trademark Notice: Product or corporate names may be trademarks or registered trademarks, and are used only for identification and explanation without intent to infringe.

Visit the Taylor & Francis Web site at
http://www.taylorandfrancis.com

and the CRC Press Web site at
http://www.crcpress.com

Library of Congress Cataloging in Publication Data
Main entry under title:

Instrumentation for heavy ion nuclear research.

(Nuclear science research conference series, ISSN 0250-4375 ; v. 7)
Papers presented at a conference dedicated to the memory of James L. C. Ford, Jr.
Includes index.
1. Heavy ion collisions—Instruments—Congresses.
2. Ford, James L. C., 1933-1984. I. Shapira, Dan, 1943- . II. Ford, James L. C., 1933-1984.
III. Series.
QC794.6.C6I558 1985 539.7 85-8707

ISSN 0250-4375.

CONTENTS

*Denotes Speaker

*Denotes Speaker

IN MEMORIAM

James L. C. Ford, Jr. (1933-1984)

Jim Ford was born in 1933 in New York City. He began his education in physics at Montana State University and continued in graduate school at the California Institute of Technology. Upon completion of the requirements for a doctoral degree in 1962, he began his professional career at the Physics Division of the Oak Ridge National Laboratory. It was there that he was stricken by a fatal heart attack on January 16, 1984. Earlier, he had taken a leave of absence from ORNL during 1973-74, when he was a recipient of the Alexander von Humboldt Fellowship at the Max Planck Institut für Kernphysik in Heidelberg, Germany.

Over the years, his research in nuclear physics led to the authorship of over a hundred articles in Coulomb excitation, elastic and inelastic scattering, heavy-ion reaction mechanisms, and instrumentation. It is through these papers that he is best known to the nuclear physics community. Among the highlights of his research were the precision measurements of the Coulomb excitation of the actinide nuclei, the use of heavy-ion reactions to study high-spin states in nuclei, and the measurement of cross sections for single-nucleon transfer in heavy-ion reactions.

Until the early 1970s, his research was performed primarily at the EN tandem accelerator laboratory. A major achievement of his during this period was to bring the split-pole magnetic spectrometer, including the adaptation of one of the earliest position-sensitive proportional detectors, into full operation. This system became the major experimental device at the EN tandem facility. In 1979, Jim Ford supervised the relocation of the spectrometer to the Holifield Heavy Ion Research Facility, where it continues to serve as a significant research tool.

In the seventies, his research program was gradually shifted from the EN tandem to the Oak Ridge Isochronous Cyclotron and, later, to the Holifield Facility. His interest continued in the study of heavy-ion induced reactions. Because of his skills in experimental hardware, he made many major contributions to the Holifield Facility. At the beginning of the Holifield Project in 1974, Jim developed the criteria for the beam transport lines between the tandem accelerator and experimental stations. His ideas for implementation of these lines continue to be used for new beam lines and to upgrade existing beam lines. During 1975–77, Jim designed and supervised the fabrication of a time-of-flight facility at Holifield. Another pioneering project was the development of a gas-jet target for use with heavy-ion beams that was both very thin and indestructible. This emphasis on instrumentation was partly due to his natural bent but was also the result of a highly developed sense of good citizenship that made him respond to the needs of the new Holifield Heavy-Ion Research Facility.

Working with Jim Ford on an experiment was a pleasure because of his enthusiasm and his sense of adventure in testing new or improved instrumentation. His research method was characterized by a strong interaction with the experimental equipment—if it was doing the job, he was a proud parent, but if it was not, he clearly showed mounting frustration. The next experiment would usually then have modified and improved equipment which he had developed in the meantime.

At the time of his death, Jim Ford was performing the important jobs of leading the Experimental Support Group and serving as Liaison Officer to the Holifield Users Group.

A man of many interests, he enjoyed camping, biking, and music. He was an avid reader, with a particular interest in history. Those who knew Jim well appreciated his enthusiasm, his support, his genuineness, and, especially, his concern for people.

Russell L. Robinson
Paul H. Stelson

PREFACE TO THE SERIES

Nuclear science is now in full flower. From obscure origins in a few university laboratories it has grown to be one of the richest fields of scientific endeavor in the history of science. It has spun off two virtually unlimited sources of energy for mankind, fission and fusion. It is used broadly in many applied areas that range from medicine to agriculture to mining to metallurgy, and withal it continues to be a happy and productive basic science. New approaches, new particles, new energy regimes, and above all new ideas continue to provide astonishing theoretical insights and frequently unexpected experimental results.

Burgeoning growth is a healthy sign, but it carries with it a concomitant information explosion: as the science expands, the number of scientists increases, and specialized areas appear where once there existed but one unified field. To be sure, the technical literature performs its task adequately, but if a practitioner in a narrow field of nuclear science is to be aware of everything that he needs to know, he must read a dozen or more journals. Frequently a year or more elapses before a physicist's results appear in the journal literature. Because of such formidable information barriers, rapid communication in science is now principally oral rather than written. With this realization, specialized scientific conferences become a prime arena of scientific exchange and scientific publication. Spectacular results are now nearly always first communicated at conferences where they also receive their first baptism under fire. Sometimes new ideas may die at the same conference where they were first revealed. Conferences provide the stage where experiment and theory are joined, where results of different groups are compared, and new syntheses are evolved.

A conference will frequently broaden its program to include a paper or two on ancillary but essential developments. For example, a meeting on heavy-ion nuclear reactions might well include a session on new accelerator ideas or ion source developments. Finally, a conference ends with a summary by a noted practitioner in the field, and this summary can have great and lasting value for the reader.

It is for all these reasons that we have undertaken to publish a conference series in nuclear science, NUCLEAR SCIENCE RESEARCH. We expect this series, while it specializes in the areas of nuclear physics, to cover topics both basic and applied. We expect each conference to have its specialized nature so well defined that the proceedings will serve as a reference volume of lasting importance, and we expect that the resource provided to the working physicist by this series will enable those who were not participants to get both the data and the flavor of the conference. Speed of publication is essential for such proceedings. We are told that with modern publishing methods the book can appear 3 to 4 months after receipt of the last manuscript. It is our intent that the series NUCLEAR SCIENCE RESEARCH publish proceedings of important conferences in the field, and our hope that it will provide a record of advances in a way that is most useful to the working scientist.

Alexander Zucker

PREFACE

Since its beginnings more than three decades ago, the study of nuclear reactions induced by heavy ions has become an important subfield of nuclear physics; experiments are performed at many dedicated facilities using instrumentation that has become increasingly specialized and complex. This field of nuclear science is evolving very quickly; since the last conference dedicated to heavy ion detectors held in 1982,[1] a large number of heavy ion detection systems have become available. Some of the papers published in these proceedings describe new detection systems that have become operational since the last conference. Other papers discuss plans for new instrumentation to be used in studies of relativistic heavy ions — a new frontier of heavy ion science. The 27 contributed papers that were presented at the conference appear in a separate publication (ORNL publication No. CONF-841005) that was distributed to all the conference participants.

The meeting was overshadowed by the untimely death of Dr. J. L. C. Ford, whose contributions to the field of nuclear instrumentation are numerous and who also conceived and initiated this conference. The first few papers are devoted to our friend and collaborator Jim and to his recent work. The participation of many scientists active in the field of instrumentation and the excellent presentations delivered by the speakers provided for a very stimulating atmosphere.

I want to thank all the members of the organizing committee, Drs. J. Blankenship, R. Robinson, E. Spejewski, F. Plasil, I. Y. Lee, D. Hensley, and J. Beene. Assistance in planning the program provided by the program committee is greatly appreciated. The aid and advice given by the ORNL conference office in planning and organizing the meeting are also greatfully acknowledged. Financial support by the Physics Division at ORNL and sponsors from the instrumentation industry provided the necessary "bread and butter" to make the conference a pleasant affair. Special recognition goes to J. Blankenship, who organized the instrumentation exhibits associated with the conference.

Thanks are due to Mrs. Christine Wallace and Mrs. Kay Thacker, who were helpful in many ways, and especially to Ms. Regina Webber, the conference secretary.

Reference

1. W. VON OERTZEN, Detectors in Heavy Ion Physics, *Lecture Notes in Physics*, 178, (Berlin, 1982).

Dan Shapira

PROGRAM ADVISORY COMMITTEE

H. C. Britt
Los Alamos National Laboratory and
U.S. Department of Energy

T. M. Cormier
University of Rochester

W. K. Dawson
TRIUMF, University of British Columbia

C. K. Gelbke
Michigan State University

H. H. Gutbrod
GSI, Darmstadt

D. L. Hendrie
Lawrence Berkeley Laboratory

W. Henning
Argonne National Laboratory

B. Herskind
Niels Bohr Institute, Copenhagen

T. W. Ludlam
Brookhaven National Laboratory

W. von Oertzen
Hahn Meitner Institut, Berlin

J. Peter
GANIL, Caen

W. J. Willis
CERN, Geneva

THE ORNL GAS-JET TARGET

S.T. THORNTON and R.L. PARKS
Institute of Nuclear and Particle Physics
University of Virginia
Charlottesville, VA 22901, U.S.A.

D. SHAPIRA, J.L.C. FORD, JR., B. SHIVAKUMAR,[†] and
R. NOVOTNY[+]
Oak Ridge National Laboratory[*]
Oak Ridge, TN 37831, U.S.A.

A supersonic gas jet target built for studying heavy-ion induced nuclear reactions with high energy resolution is described. The device is installed as part of a magnetic spectrograph and has produced image sizes in the focal plane as small as 0.4 mm using 100 MeV ^{16}O beams collimated to an area of about $1mm^2$. Target densities of 30 $\mu g/cm^2$ (^{40}Ar) and 15 $\mu g/cm^2$ (^{14}N) have easily been obtained. The target has already been used for two nuclear reaction experiments.

1. INTRODUCTION

Several years ago we decided to build a supersonic gas jet target because of the many difficulties encountered with solid targets. Our goal was to construct a simple, easy-to-use target that would be mobile and could be used at several beam stations at the HHIRF. The present gas target has already been used in an Enge split-pole magnetic spectrograph and a standard 48 cm ORTEC scattering chamber, and could also be used with the Elbek magnetic spectrograph and the time-of-flight spectrometer.

The advantages of a gas jet target are

1) Pure, clean targets
2) Uniform target thickness
3) Nondestructible targets-intense heavy ion beam currents can be used
4) Localized target area
5) Absence of entrance and exit foils
6) Variable target thickness

In this paper the HHIRF supersonic gas jet target facility will be described. Test results will be shown, and two recent experiments will be discussed.

2. GAS JET DESCRIPTION

The elevation and plan schematic views of the gas jet are shown in Figs. 1 and 2. Many of the mechanical details have been omitted for clarity. The gas jet is formed with a convergent-divergent Laval nozzle[1] which allows for rapid supersonic expansion of gas at high pressures into the target region. The gas flows upward, and near the nozzle an intake tube pumps more than 90% of the gas flowing out of the Laval nozzle (stage 1).

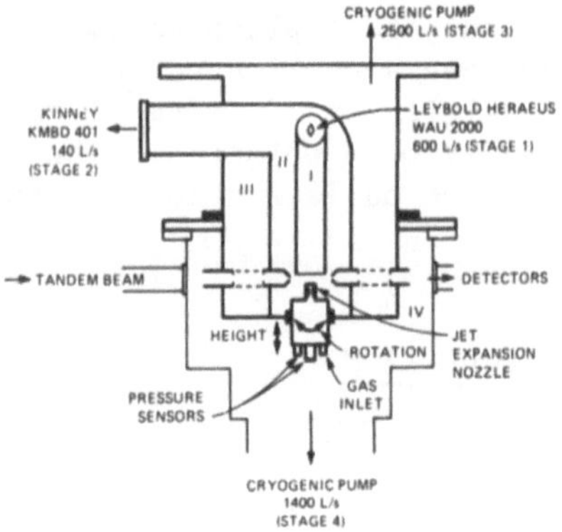

Fig. 1. Schematic view of gas jet device. The central part (heavier line) may be removed as one unit and can fit in a different chamber. The four different stages of the vacuum system and their pumps are indicated. The gas flows upward from the jet.

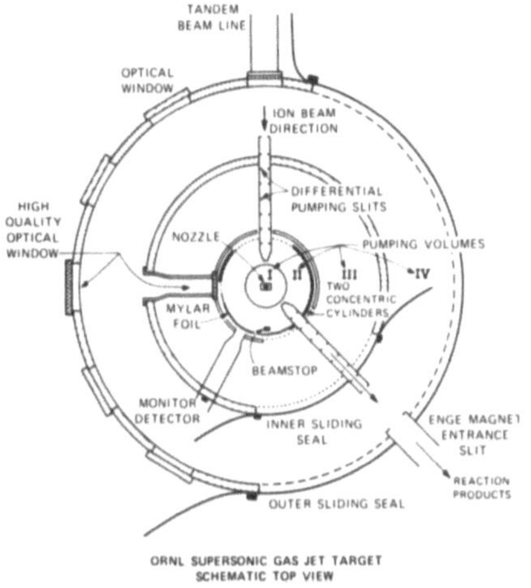

Fig. 2. Horizontal cutaway view of gas jet. Note that volumes II, III and IV are connected only through differential pumping slits which are needed for beam entrance and reaction product exit.

Most of the remaining gas (10%) is pumped from stage 2, and differential pumping apertures separate stages 2 from 3 and also 3 from 4. The entire gas jet assembly is compact and can be inserted from above into a scattering chamber by use of a suitable adaption flange.

The crucial ingredient of any gas jet target is the vacuum pumping system. The various pumps and pumping speeds are indicated in Fig. 1. Most of the gas is removed by the mechanical pumps of stages 1 and 2, cryogenic pumps can be used for stages 3 and 4. The pressure drops by about a factor of 1000 from stage 2 to 3 and a factor of 100 from stage 3 to 4. Since the great majority of the gas is collected by mechanical pumps, provision has been made to collect, clean, and reuse the output gas in a recycling mode. This feature has not yet been used.

The gas handling system used to control the gas flow through the jet is shown schematically in Fig. 3. The gas pressure P_0 is the pressure at the nozzle entrance and for any fixed nozzle geometry is the only variable that controls the target thickness. For normal operation P_0 is controlled

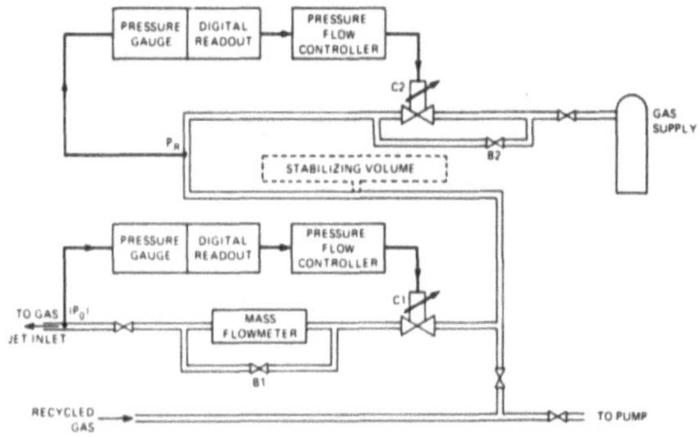

Fig. 3. Schematic drawing of gas handling system.

by valve C1 (bypass valve B2 is left open). When recycling the gas, both C1 and C2 are used. Valve C2 then adds any additional gas needed to maintain the pressure P_o. The MKS Baratron system allows us to set the desired pressure, P_o, and wait a short time until the automatic system has reached stability.

The Laval nozzle used in the present gas jet is shown in Fig. 4 along with several specifications. The predicted shape of the gas jet is indicated. The formulae on which this shape was calculated can be found in the literature.[1,2] We have used the volume next to the nozzle exit as our target region. Conditions at this first high density knot depend to first order only on P_o (and temperature) and the nozzle shape. Our nozzle has been designed to produce a thin sheet of dense gas, with the sheet orthogonal to the beam direction.

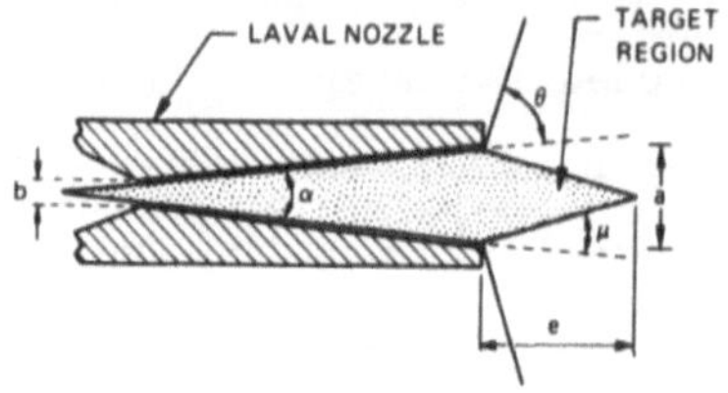

SUPERSONIC GAS JET FLOW PARAMETERS

GAS	MACH NO. M	θ (deg)	μ (deg)	TARGET HEIGHT e (mm)	MASS FLOW RATE M (g/s)	TARGET THICKNESS t ($\mu g/cm^2$)
N_2	3.17	77.43	18.36	2.25	0.214	14.19
Ar	3.8	41.76	15.26	2.83	0.262	21.27

Fig. 4. The shape of the gas jet and several jet parameters. The data indicate the gaseous region.

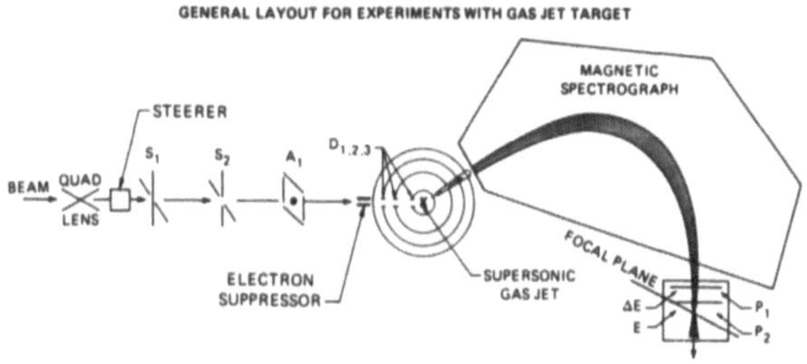

Fig. 5. The general layout of the gas jet beam line and data taking system. S_1 and S_2 are two four-jaw slits with current read out. A_1 is a circular aperture and $D_{1,2,3}$ are the differential pumping slits

This emulates a solid target. Since the exact position of the high density knot with respect to the beam is difficult to establish mechanically, we allowed for externally controlled vertical positioning of the nozzle with respect to the beam.

Since the size of the gas jet is small, the position of the beam must be reproducible and small. The differential pumping apertures allow a target area less than 2 x 2 mm^2. The beam itself is stopped inside the gas jet assembly. The whole gas jet assembly serves as a Faraday cup and is isolated from ground. The tight beam collimation ensures that it passes through the differential pumping apertures unscathed. Figure 5 shows the schematic experimental layout. The collimating slits S_1, S_2, and A_1 are located 1.7, 0.9, and 0.7m upstream from the target, respectively. The nuclear reaction products exit through a port mounted on a sliding seal capable of moving between lab angles of 0° to 60°. A solid state detector mounted at 45° and separated from the gas region by a thin mylar foil serves as a monitor.

3. OPERATION AND TESTING OF GAS JET

Extensive tests have been made with the gas jet to insure that absolute cross section measurements can be made. These tests prompted several changes, such as the tight collimation of the beam and the isolation of the entire gas jet to produce a "Faraday cup." Comparison of theoretical calculations for the target area densities with experimental measurements are shown in Fig. 6. The experimental measurements were made by varying the gas pressure P_o and by measuring the energy loss of elastically scattered particles with a magnetic spectrograph. These measurements were also compared with measured cross sections at low bombarding energies (Rutherford scattering). The energy loss measurements and Rutherford scattering show good agreement with the calculated target density. Normally a small, but predetermined amount of xenon gas is mixed with the target gas to allow an absolute normalization using Rutherford scattering from xenon. This has proven to be a successful technique.[3]

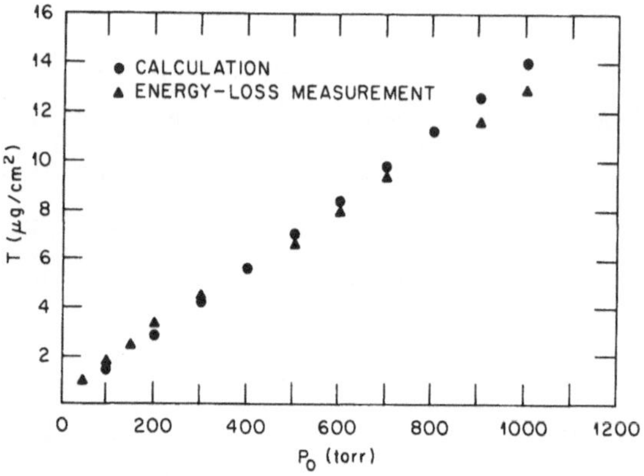

Fig. 6. The variation of target thicknesses as a function of inlet pressure. Shown are target thicknesses determined experimentally (triangles) by measuring the energy loss of 12 MeV ^{16}O in the target gas and also calculated values.

The shape of the gas jet itself has been

investigated by moving the jet nozzle up and down with respect to the beam and by rotating the nozzle at an angle with respect to the beam. Those results confirm that the shape of the gas jet was as predicted.

4. GAS JET USE

Besides initial tests using the University of Virginia CN Van de Graaff accelerator and the ORNL EN tandem Van de Graaff accelerator with an ORTEC 48 cm scattering chamber, the gas jet has only been used with the HHIRF split-pole magnetic spectrograph as shown in Fig. 5. The focal plane detector[4] is capable of measuring the total energy, energy loss and two position signals along the focal plane region for reaction products.

The gas jet facility has been used already for two experiments. The first[5] was a study of peripheral reactions induced by 100 MeV ^{16}O on ^{40}Ar. The target was natural argon gas with areal density of 15 μg/cm^2. The solid angle was 2.4 msr with incident beam intensities of 10 particle nA. The detector system allowed us to separate out-going reaction products with different A and Z. Figure 7 shows a one dimensional plot of the ^{17}O spectra along the focal plane with a resolution of about 100 keV. This resolution corresponds to a position resolution along the focal plane of about 1.5 mm. Since the object size expected was about 1.2 mm (the width of the nozzle's exit), we expect an image size along the focal plane of about 0.4 mm, because the magnetic spectrograph has a horizontal demagnification of three. The focal plane detector was clearly limiting the resolution. We remeasured the resolution using a solid

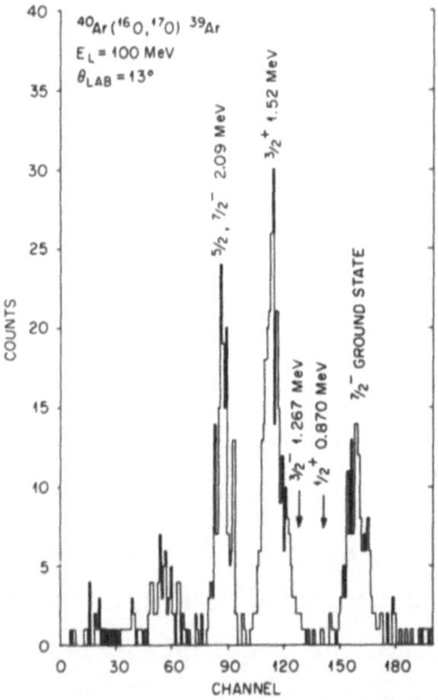

Fig. 7. A projected momentum spectrum of outgoing ^{17}O products. The excited states of ^{39}Ar are shown and an overall energy resolution of ∿120 keV is obtained.

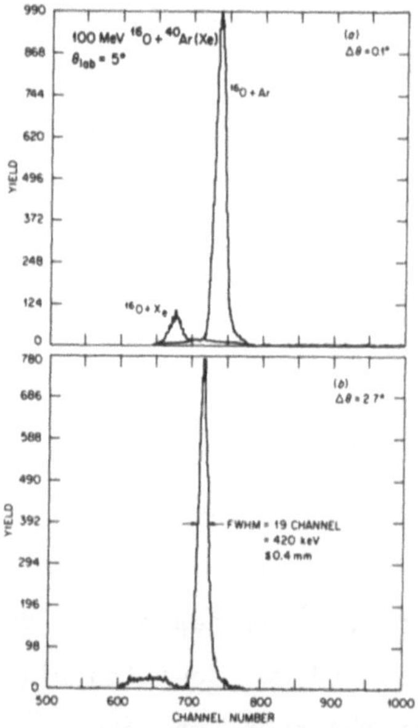

Fig. 8. High resolution position spectra for the elastic scattering of ^{16}O on a mixture of ^{40}Ar and Xe.

state position sensitive detector capable of ~0.4 mm resolution. These data are shown in Fig. 8 for $\Delta\Theta = 0.1°$ and $\Delta\Theta = 2.7°$. The resolution for $^{16}O + {}^{40}Ar$ elastic scattering is now 42 keV, corresponding to a position resolution of ~0.4 mm. These data demonstrate the great usefulness of such a gas target when coupled with a high-resolution magnetic spectrograph. Momentum resolution better than 5 parts in 10^4 for heavy ion reaction studies is possible.

The second experiment[6] performed with the gas jet target was a study of recoiling target-like reaction products from the bombardment of a ^{14}N target with ^{28}Si beams. Such measurements probe scattering and reaction products emitted in the backward hemisphere (c.m. frame) from the bombardment of ^{28}Si by ^{14}N. We have previously shown[7] that in the collisions between such light nuclei a long-lived rotating dinuclear system is formed. Extensive experimental studies focused on $^{28}Si + {}^{12}C$.[7,8] The $^{28}Si + {}^{14}N$ entrance channel is of particular importance because there are many channels which become energetically open at low bombarding energy. This multitude of open channels affords us the possibility of studying mass and charge flow in such reactions.[9]

Natural nitrogen gas (99% ^{14}N) was bombarded by 150 MeV ^{28}Si. The target density was 15 μg/cm^2. The reaction products are dominated by ^{16}O, ^{14}N, and ^{12}C, with the most dominant being the $^{12}C + {}^{30}O$ channel, and not $^{28}Si + {}^{14}N$. This result shows that the orbiting process lasts long enough for energy, mass and charge flow to reach equilibrium.

There seems to be a wide variety of experiments including studies of continuum and high resolution spectra of products from heavy ion induced reactions for which a gas jet target is very useful. The present gas jet facility is easy to operate and maintain. Its mobility allows it to be a versatile device. J.L.C. Ford, Jr. who passed away before the target was fully utilized has contributed much of his time and effort to build this facility. We shall miss his inspiration and guidance.

REFERENCES

1. J.A.D. ACKROYD, SRC Daresbury Report DL/NSF/R12 (1975) unpublished.
2. T.C. ADAMSON, JR. and J.A. NICHOLLS, J. Aerospace Sci. 26, 16 (1959). S. CRIST, P.M. SHERMAN, and D.R. GLASS, AIAA J., 4, 68 (1966).
3. D. SHAPIRA, R.A. DAYRAS, J.L.C. FORD, JR., J. GOMEZ DEL CAMPO, A.H. SNELL, P.H. STELSON, and R.G. STOKSTAD, Nucl. Instr. and Meth., 163, 325 (1979). D. SHAPIRA, et al., Phys. Rev. C, 28, 1148 (1983).
4. D. SHAPIRA, G. BOMAR, J.L.C. FORD, J. GOMEZ DEL CAMPO, and L.C. DENNIS, Nucl. Instr. and Meth., 169, 77 (1980).
5. B. SHIVAKUMAR, K.A. ERB, J.L.C. FORD, JR., J. GOMEZ DEL CAMPO, D. SHAPIRA, F. PAUGHEON, and S.T. THORNTON, Bull. Am. Phys. Soc., 28, 645 (1983).
6. B. SHIVAKUMAR, D. SHAPIRA, J. GOMEZ DEL CAMPO, J.L.C. FORD, JR., P.H. STELSON, B.A. HARMON, and S.T. THORNTON, Bull. Am. Phys. Soc., 29, 626 (1984).
7. D. SHAPIRA, et al., Phys. Lett., 114B, 111 (1982).
8. D. SHAPIRA, D. SCHULL, J.L.C. FORD, JR., B. SHIVAKUMAR, R.L. PARKS, R.A. CECIL, and S.T. THORNTON, Phys. Rev. Lett., 53, 1634 (1984).
9. D. SHAPIRA, P.H. STELSON, J. GOMEZ DEL CAMPO, B. SHIVAKUMAR, B.A. HARMON, R.A. PARKS, and S.T. THORNTON, Proceedings of the Eighth International Conference of the Application of Accelerators in Research and Industry, Nov. 12-14, 1984, to be published in Nucl. Instr. and Meth.

†Permanent address: A.W. Wright Nuclear Structure Laboratory, Yale University, New Haven, Connecticut 06520. Supported in part by contract No. DE-AC02-76ER0374 with the U.S. Department of Energy.

+Present address: II Physics Institute, University of Giessen, D-6300 Giessen, West Germany.

*Operated by Martin Marietta Energy Systems, Inc. under contract DE-AC05-84OR21400 with the U.S. Department of Energy.

A NEW FOCAL PLANE DETECTOR SYSTEM FOR THE BROAD RANGE SPECTROMETER

T. P. SJOREEN
Oak Ridge National Laboratory*
Oak Ridge, Tennessee 37831

A focal plane detector system consisting of a vertical drift chamber, parallel plate avalanche counters, and an ionization chamber with segmented anodes has been installed in the Broad Range Spectrometer at the Holifield Facility at Oak Ridge. The system, which has been designed for use with light-heavy ions with energies ranging from 10 to 25 MeV/amu, has a position resolution of ~ 0.1 mm, a scattering angle resolution of ~ 3 mrad, and a mass resolution of ~ 1/60.

1. INTRODUCTION

It is appropriate that this paper presented at the memorial session for J. L. C. Ford, Jr. is about a focal plane detector system. Jim Ford had been involved in research using magnetic spectrometers for nearly two decades and was widely recognized for his contributions to focal plane instrumentation. The focal plane detector system that Jim had been working on was for the Broad Range Spectrometer (BRS) at Oak Ridge. The BRS is a spectrometer which had been used

*Operated by Martin Marietta Energy Systems, Inc. under contract DE-AC05-84OR21400 with the U.S. Department of Energy.

extensively during the late 1960's and early 1970's for light-ion spectroscopy. However, with the installation of the new 25 MV tandem, it became apparent that a new focal plane detector system would be needed for heavy-ion spectroscopy.

During 1980 Jim Ford and M. V. Hynes had the idea of using the vertical drift chamber (VDC) as a heavy-ion focal plane detector. Because the VDC was originally designed for nuclear spectroscopy with high energy electrons,[1] it was necessary to build a prototype counter for use with heavy ions. After the counter had been successfully tested, Jim decided to proceed with the installation of a complete focal plane detector system, which could determine the mass, atomic number, charge, and energies of the reaction products. For a discussion of particle identification with magnetic spectrometers see Refs. 2-3. The new detector system which was completed before Jim's death has worked well and Jim's efforts certainly contributed to this success.

2. DETECTOR SYSTEM REQUIREMENTS

There were several requirements which the new detector system had to meet if it was to be useful with the high-energy, high-resolution beams of HHIRF. 1) The position resolution had to better than .25 mm. 2) The detector system had to be able to measure the angle of the particle trajectory. 3) The active length of the detector along the focal plane had to be long enough to measure a 10% range in particle momentum. 4) The system had to be able to function with 10-25 MeV/amu heavy ions. 5) It had to have sufficient mass resolution for studies with projectiles up to mass 40.

The 0.25 mm position resolution was necessary for two reasons. First, the dispersion of the BRS is only about 2 cm/(dp/p)%. Secondly, the Holifield Facility is capable of producing high-resolution beams with $dE/E \approx 1/3500$ and energies up to 25 MeV/amu. Thus energy shifts could be as large as several hundred keV/mm. Under such conditions a detector which would contribute more than a few tenths of a mm to the overall linewidth would be useless for high-resolution spectroscopy.

The system had to be able to measure particle trajectory angles so that ray tracing could be performed. This was necessary because the entrance angle aperture of the BRS spans nearly 4° in scattering angle which corresponds to nearly 14° at the focal plane. Thus ray tracing would supply several benefits. 1) Cross sections could be binned in angle. 2) Corrections could be made for magnet aberrations. 3) Time-of-flight (TOF), hence mass resolution, could be improved with path length calculations. 4) Corrections to the energy loss dE/dX could be made to improve both mass and charge resolution.

3. THE DETECTOR SYSTEM

The detector system which was designed to meet the above requirements is shown schematically in Figure 1. The BRS consists of an Elbek dipole[4] later modified by the addition of a vertically focusing quadrupole,[5] that increased the maximum solid angle to 5 msr. The detector spans about 35 cm along the focal plane with a vertical acceptance of 5 cm. This acceptance is compatible with operating the BRS in either the single focusing mode (quad off, vertical image 5 cm) or the double focusing mode (quad on, vertical image about 1 cm). Particles intersect the focal plane at angles

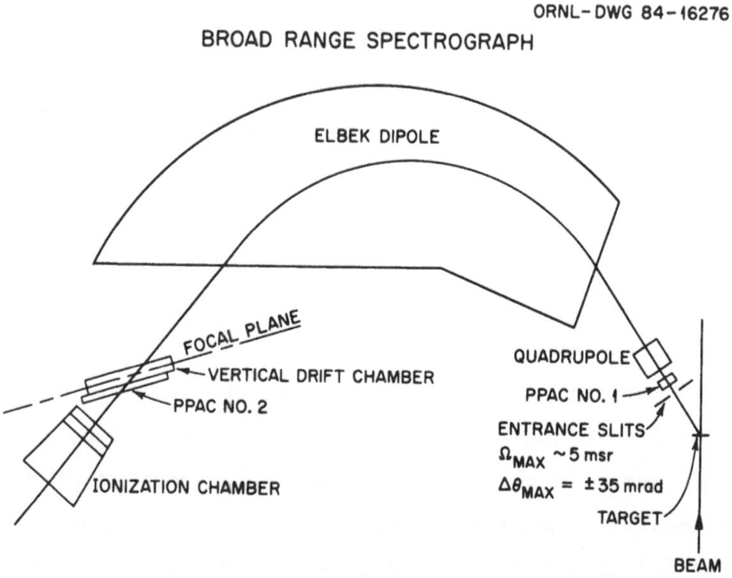

FIGURE 1 Sketch of the BRS and focal plane detectors.

between 30° and 44°.

The VDC is the focal plane counter of this system for measuring the magnetic rigidity B-rho and angle of the reaction products. Behind the VDC is a large parallel plate avalanche counter (PPAC) which provides the start signal for the TOF and the fiducial start for the VDC. Behind these detectors, oriented so that the central ray is normal to the entrance window, is the ionization chamber. This detector measures total energy (E), energy loss (dE/dX), and vertical position. The fourth detector is a small PPAC located just after the entrance aperture and before the quadrupole. It provides the stop signal for TOF.

Details of the operating principle, construction, and performance of the heavy-ion VDC have been described in detail in Refs. 6-7. Here the operating principle and performance are briefly discussed.

The VDC is a multi-wire proportional counter. As schematically shown in Figure 2, the counter consists of an anode wire plane symmetrically spaced between two high-voltage, aluminized-mylar, cathode planes. The wire plane consists of active sense wires, each separated by two guard wires. The larger diameter, lower-gain guard wires effectively serve to divide the electrostatic volume of the counter into a series of drift cells, each centered around a sense wire. The counter is positioned so that the wire plane is perpendicular to the bending plane of the spectrometer. Thus, particles pass through the first gas-pressure bearing window (not shown), pass through the first high-voltage cathode plane, traverse 4 or 5 adjacent drift cells, and exit through the other cathode plane and pressure bearing window. The free electrons then drift to the anodes where signals are produced. Because the electrostatic field of the drift cells are set so that the drift velocity is uniform, the drift times are proportional to the distance between the trajectory and anode. Thus position and angle can easily be determined if the drift times are measured.

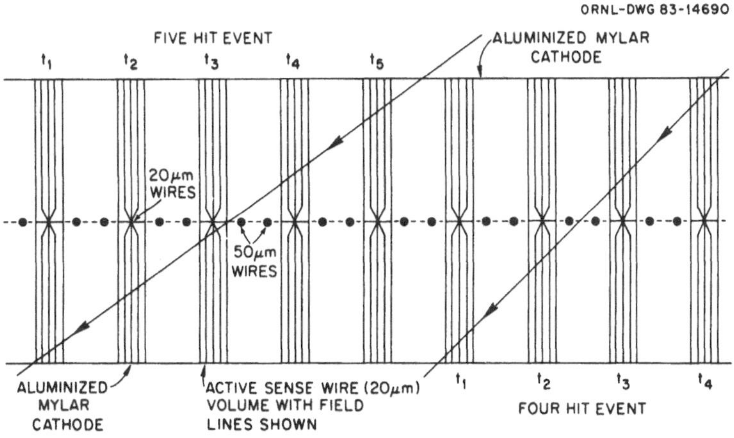

FIGURE 2 Schematic cross sectional view of the vertical drift chamber perpendicular to the anode wires.

The operating characteristics of the VDC were measured with 129-MeV ^{12}C, 176-MeV ^{14}N, and 140-MeV ^{16}O ion beams from the Oak Ridge Isochronous Cyclotron. Measurements were made with isobutane gas at 50, 100, and 200 Torr. The intrinsic cell precision, which is the accuracy to which a trajectory position can be measured in a drift cell, was determined to be about 75 μm. Based on this value we predicted that for trajectories which traverse 4 adjacent drift cells, the position resolution of the VDC should be 0.12 mm FWHM and the angle resolution 6.6 mrad FWHM.

An example of the excellent position resolution of the VDC is shown in Figure 3, which is the distribution of elastic events obtained by bombarding a 100 $\mu g/cm^2$ ^{197}Au target with 176-MeV ^{14}N. The width of this peak is 69 keV FWHM (0.55 mm) for a $dE/E \approx 1/2500$. After considering target and beam contributions to the linewidth, the VDC's contribution to the linewidth is estimated to be ~ 0.1 mm, a value that agrees well with the prediction based upon the 75 μm cell precision estimate.

The angle measuring capability of the VDC is shown in Figure 4, where the distribution of counts obtained by elastically scattering ^{14}N from ^{208}Pb is displayed as a function of the spectrometer entrance angle. To estimate the angular resolution, a mask consisting of four equally spaced wire strips (each 1 mm wide) was positioned between the target and the entrance aperture. The relative positions and widths of the wire mask in the angle spectrum are indicated by the cross hatched areas in Figure 4. The scattering angle subtended by these wires was 3 mrad. From the excellent correlation between the position of the wires and the attenuation of counts in the angular distribution, the scattering angle resolution is estimated to be about 3 mrad.

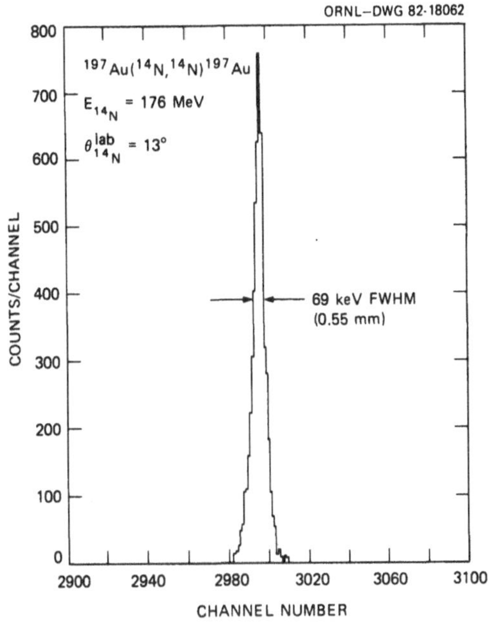

FIGURE 3 Elastic peak obtained with the VDC by scattering 176-MeV ^{14}N from 100 μg/cm^2 ^{197}Au.

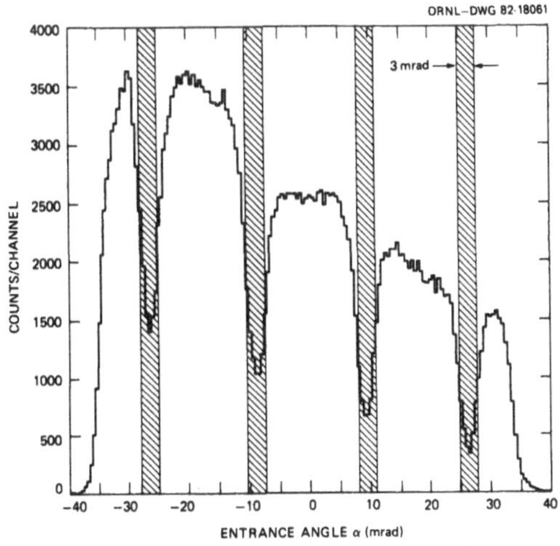

FIGURE 4 Angular distribution of elastically scattered particles as a function of the spectrometer entrance angle. See text for details.

Because there is an angle magnification of about 3.5 between the object (target) and image (focal plane) in the BRS, the 3 mrad value corresponds to about 10.5 mrad at the focal plane. This 10.5 mrad value is about 50% larger than that predicted with the cell precision estimate. This difference is probably due to the finite size of the beam spot.

The ionization chamber for the BRS was designed to be as small as possible, with the conditions that it be able to stop 25 MeV/amu $^{12}C^{6+}$ ions and detect all particles that pass through the VDC and PPAC. Thus the active gas volume of the ion chamber is 40 cm deep, 26 cm wide at the front, 36 cm wide at the back, and 5.4 cm high between the cathode and Frisch grid. The anode-cathode spacing is 6 cm. In addition the anode is segmented into three parts, the first two each 5 cm deep and the last 30 cm, so that energy loss (dE/dX) and total energy (E) can be measured. The chamber has been tested to two atmospheres of gas pressure. Of course, the gas pressure, plate biases, and window thicknesses will depend on the ion species and its energy. For example, with 350 MeV ^{18}O ions, the ion chamber required 900 Torr of isobutane gas, a 50 μm thick Kapton window, and +5600 V and -4000 V biases on the anode and cathode, respectively. Energy loss and total energy resolutions of 2.5% and 1.3%, respectively, have been obtained. A photograph of the ion chamber is shown in Figure 5.

The start detector is a 42 cm long by 5 cm high PPAC which mounts directly on the back of the VDC housing. The detector consists of 3 μm mylar pressure bearing windows and 3 μm aluminized mylar electrodes. The electrode spacing is 3.2 mm. Comparison of the count rates in this detector with those in the ionization chamber indicate a detector efficiency of essentially 100% for 400 MeV ^{16}O ions when

FIGURE 5 Photograph of the ionization chamber.

using 7 Torr of isobutane and a bias of 600 V.

The stop detector, also a PPAC, is 2.54 cm by 2.54 cm. This detector consists of 1.5 μm thick mylar windows and 1.5 μm aluminized mylar electrodes. The electrode spacing is 1.6 mm. It operates with 7 Torr of isobutane at a bias of about 500 V. The total thickness is about 600 $\mu g/cm^2$. This thickness did not affect the resolution of particles detected at the focal plane during tests with 400 MeV ^{16}O.

The timing resolution of the two PPACs is about 600 ps FWHM; this corresponds to a TOF resolution of about 1% for 25 MeV/amu particles.

Figure 6 is a photograph of the three focal plane detectors, i.e. the VDC, the large PPAC, and the ionization chamber. The electronics for the VDC are described in Ref. 7. The electronics for the PPAC and ionization chamber are conventional fast and slow NIM modules. The gas handling system is outlined in Ref. 7 also.

4. SOME RESULTS

After some initial tests with 400-MeV ^{16}O ions, the new detector system was used in two experiments.[8,9] Although

FIGURE 6 Photograph of the VDC, start PPAC, and the ionization chamber.

the analysis of the data is far from complete, the initial indication is that the detector worked quite well, although there were some problems. The next several figures give some indication of the capability of this system.

Figure 7 shows data from the study[8] of spin orbit effects in the $^{28}Si(^{11}B,^{10}B)^{29}Si$ reaction at a 93-MeV bombarding energy. Part (a) shows the correlation between the total energy E and the energy loss in the second anode dE2 for the reaction products in the ion chamber. This is the typical E-dE plot which separates the reaction products into their respective atomic numbers. As shown, the B, Be, and Li products are well separated. Part (b) shows the correlation between focal plane position and E for those boron reaction products which fall inside the contour drawn in part (a). This correlation separates the isotopes for a given element. Part (c) shows those ^{10}B reaction products, that fall inside the contour of part (b), as a function of focal plane position binned into 5 angles. Each angle bin has an acceptance of 0.8 degrees. These spectra were

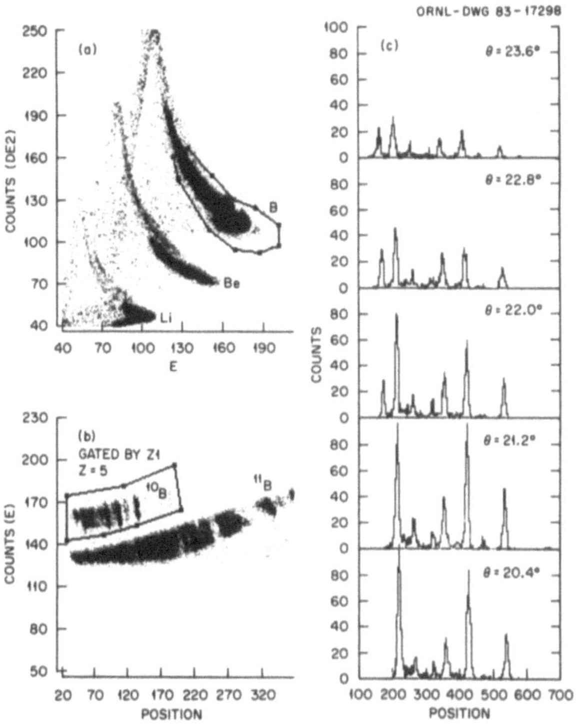

FIGURE 7 Spectra obtained by bombarding ^{28}Si with 93-MeV ^{11}B. See text for a description.

obtained with one angle setting of the spectrometer. The 150-keV FWHM linewidths are due almost entirely to target thickness and beam energy spread.

Figure 8 shows several spectra from the charge exchange study[9] with 350-MeV ^{18}O ions. These spectra, which were taken from a "tune up" run with a Au target, demonstrate the TOF capability. Part (a) is E-dE2 correlation similar to that shown in Fig. 7a. Part (b) is the raw TOF spectrum, with a time resolution of ~ 800 ps. Part (c) shows mass spectra for each of three elements (oxygen, nitrogen, and carbon) using the element identification contours

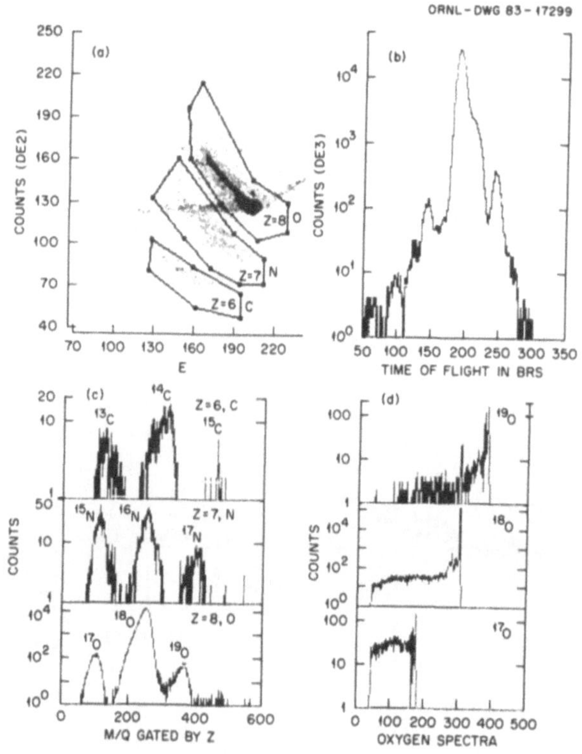

FIGURE 8 Spectra obtained by bombarding ^{197}Au with 350 MeV ^{18}O. These spectra demonstrate time of flight capability. See text for details.

of part (a). The mass distributions were calculated with TOF, E, and angle information. The mass resolution is about 1/60. Part (d) shows focal plane position spectra obtained by gating each of the oxygen isotopes in part (c).

Figures 9 and 10 are more detailed energy spectra obtained during the same experiment with 350-MeV ^{18}O ions on a ^{28}Si target. Figure 9 is a (^{18}O,^{18}O') inelastic spectrum at 6.9 deg (lab), while Figure 10 is an (^{18}O,^{19}F) spectrum at 2.1 deg (lab). The energy resolution of these spectra is about 250-keV FWHM.

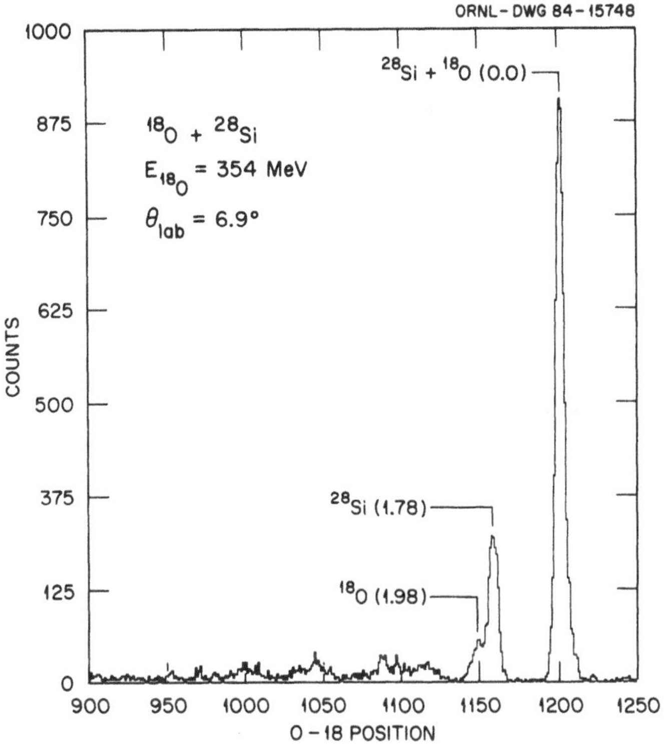

FIGURE 9 Energy spectrum for the $^{28}Si(^{18}O,^{18}O')^{28}Si$ reaction at 350-MeV bombarding energy.

As indicated in the spectra discussed above the detector system has worked well with lighter mass ions $A \leqslant 20$ for energies between 10 and 25 MEV/amu. With a mass resolution of 1/60, we anticipate no significant problems with studies up to mass 40.

Recently, the detector system has been used in an experiment with 913 MeV ^{58}Ni on a ^{208}Pb target. It appears that the detector system can be used under these conditions, if care is taken in choosing the operating parameters of the system. In addition, it will be necessary to carefully correct the data to optimize the charge and mass resolution.

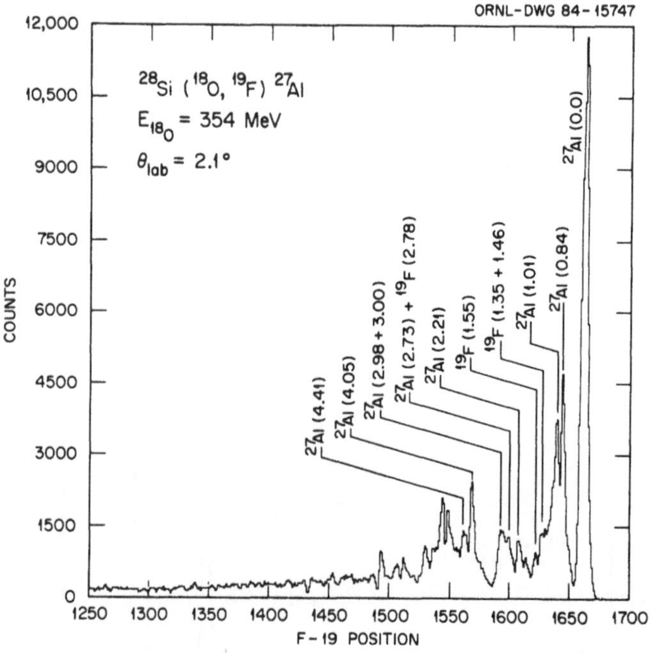

FIGURE 10 Energy spectrum for the $^{28}Si(^{18}O,^{19}F)^{27}Al$ reaction at 350-MeV bombarding energy.

Up to now, for projectiles with A ⩽ 20, mass identification has been straightforward.

5. CONCLUSIONS

The new detector system for the BRS has met or exceeded our original expectations. The excellent position resolution of 0.1 mm makes it possible to do high resolution spectroscopy with a low dispersion spectrometer such as the BRS. With a scattering angle resolution of ~ 3 mrad (~ 10.5 mrad at focal plane), cross sections can be binned in angle. In addition, corrections for magnet aberrations, TOF pathlengths, and energy loss can be made to increase mass resolution and the usable solid angle of the

BRS. The detector system can easily handle projectile energies between 10 and 25 MeV/amu as well as masses up to A=40 or possibly A=60. With a 10% range in momentum for a given field setting, large ranges of excitation energy can be studied. For example, with 400-MeV ^{16}O on ^{208}Pb, the range of excitation is nearly 100 Mev. In addition multiple charge states and reactions can be detected simultaneously. The versatility of the new system assures much use in the foreseeable future.

ACKNOWLEDGEMENTS

The successful commissioning of the BRS focal plane detector is the result of the efforts of many people. They are R. L. Auble, F. E. Bertrand, J. L. Blankenship, E. E. Gross, D. C. Hensley, and R. O. Sayer of ORNL. D. Schull while on leave from GSI, Darmstadt, and M. V. Hynes provided valuable assistance during the initial tests of the VDC. I would like to thank B. L. Burks for providing Figures 9 and 10.

REFERENCES

1. W. BERTOZZI, M. V. HYNES, C. P. SARGENT, C. CRESWELL, P. C. DUNN, A. HIRSCH, M. LEITCH, B. NORUM, F. N. RAD, and T. SASANUMA, Nucl. Instr. and Meth., 141, 457 (1977).
2. FRED S. GOULDING and BERNARD G. HARVEY, Identification of Nuclear Particles, Ann. Rev. Nucl. Sci., edited by Emilio Segre (Annual Reviews Inc., Palo Alto, CA 94306, 1975), pp. 167-240.
3. J. L. C. FORD, JR., Nucl. Instr. and Meth., 162, 277 (1979).
4. J. BORGGREEN, B. ELBEK, and L. PERCH NIELSON, Nucl. Instr. and Meth., 24, 1 (1963).
5. E. E. GROSS, Nucl. Instr. and Meth., 121, 297 (1974).
6. M. V. HYNES, J. L. C. FORD, JR., T. P. SJOREEN, J. L. BLANKENSHIP and F. E. BERTRAND, Nucl. Instr. and Meth., 224, 89 (1984).

7. T. P. SJOREEN, J. L. C. FORD, JR., J. L. BLANKENSHIP, R. L. AUBLE, F. E. BERTRAND, E. E. GROSS, D. C. HENSLEY, D. SCHULL, and M. V. HYNES, Nucl. Instr. and Meth., 224, 421 (1984).
8. C. F. MAGUIRE, ZHAO KUI, K. TEH, J. L. C. FORD, JR., F. E. BERTRAND, E. E. GROSS, R. O. SAYER, D. SHAPIRA and T. P. SJOREEN, to be published.
9. D. J. HOREN, R. L. AUBLE, F. E. BERTRAND, B. L. BURKS, J. L. BLANKENSHIP, J. L. C. FORD, JR., E. E. GROSS, D. C. HENSLEY, R. O. SAYER, D. SHAPIRA, and T. P. SJOREEN, to be published.

GAS FILLED IONIZATION CHAMBERS

H. SANN
Gesellschaft für Schwerionenforschung
Darmstadt, Germany

ABSTRACT: For several reasons Gas Ionization Chambers have proven their enormous versatility in experiments where heavy ions have to be detected. The detector design can easily be adapted to the experimental problem. It is possible to build detectors with very large sensitive areas. Extremely thin energy loss detectors can be built, as well as detectors which record all the kinematical quantities (E,dE,t,x,y) of a heavy reaction product. Using sampling ADC systems gas detectors can be provided which are able to measure several particles simultaneously. The first results of the MUltiple Sampling Ionization Chamber (MUSIC), a new large area (1x2m^2) detector measuring the charge of relativistic heavy ions, are presented.

1) INTRODUCTION

Three main developments have taken place in the last 10 years leading to the renaissance of gas filled Ionization Chambers: the availability of heavy ion beams, the development of extremely low noise preamplifiers (noise equivalent

≤ 600 electrons) and the availability of very thin polyesther foils.

The main advantages of gas detectors compared to solid state detectors are as follows: no radiation damage in the counting medium; the possibility to built them position sensitive; almost complete independence as far as size and geometrical form of the detector are concerned.

As long as alpha particles are the heaviest reaction products and gas counters have to be used for their detection a counter with gas amplification is needed. The utilization of gas Ionization Chambers begins to be useful whenever the energy deposited in the detector is larger than ~1 MeV. So they are useful whenever the detected particles become heavier than alpha particles. It is unnecessary to preamplify the larger signals generated by heavy ions by means of a gas amplification process, because the primary electron statistics is no longer limiting the resolution. Therefore, the uncertainties produced by the gas amplification process (variation of wire diameter, avalanche statistics etc.) can be avoided.

The resolution of an Ionization Chamber mainly depends on the collection efficiency of the primary electrons and on the quality of the following low noise charge sensitive preamplifier. The only purpose of the electric field in the chamber consists of collecting the electrons as completely and as fast as possible. This is achieved if the reduced electric field E/p in the sensitive volume is chosen, in such a way that the velocity of the electrons reaches a maximum, but is still low enough that charge multiplication is avoided. The position and the height of the maximum is strongly dependent on the gas type and its purity. The low lying molecular rotational and vibrational states determine the electron temperature in the gas and, thus, the electron drift velocity under the influence of an external electric field. That is the reason why methane and tetrafluormethane have the highest drift velocities among all known gases, the maximum being reached at a reduced field strength of E/p~1.0 [V/cm/Torr]. The drift velocity dependence is shown in Fig.1. for the three gases which are best suited for gas Ionization Chambers. For a mixture of 90% argon and 10% methane the maximum of the drift velocity of v_d= 5cm/µsec is already observed at E/p=0.3. This high drift velocity at a very low field strength allows to operate the detector at substantially lower voltages. This fact and the very low price make this gas mixture best suited for very big detectors. Pure methane, a very fast gas with v_d= 10 cm/µsec, is cheap and can be obtained with very high purities, the reason, why it is commonly used. For all applications, where high pressures are necessary, it is not recommendable for its high explosibility. The ideal gas for Ionization Chambers is tetrafluoromethane [1]). It is very fast and has a very high stopping power (~4 times higher than methane), but it is at least 10 times more expensive. In addition it is rather difficult obtaining the right purity specifications.

The collection of the primary electrons generated by the detected particle in the active volume of the detector is most effective if the homogenous electric field in the active volume is adjusted in order to reach the maximum of the drift velocity. To meet this requirement the cathode voltage is determined by the geometry, the gas and the pressure. The gas pressure is determined by the energy of the particles which shall be stopped in the chamber. If the chamber is operated this way, the diffusion of the electrons is minimized. A typical value of the diffusion of the primary electrons is $\sigma \leq 0.2$ mm per 1 cm of drift path [3]). The most important by-product of the small electron diffusion is that the entire information stored in the spatial distribution of the primary electrons is conserved during the electron drift process. This fact can be used to deduce the spatial position of the primary track by measuring the drift times to destinated grids or electrodes, or to determine the range and the charge of the particle by collecting sections of the track on different anodes. If the detector dimensions become so large (meters), that the diffusion of the primary electrons limits the result, it is possible to put a magnetic field parallel to the electric field lines. Thus the diffusion can be reduced. Even if the drift distance exceeds some meters the entire information of the primary track can be conserved. This is done in the Time Projection Chambers, and can in the same way be done in an Ionization Chamber.

There are two effects working strongly against a fast and efficient electron collection. The first are inhomogenities in the collecting electric field. By taking extreme care in the design, it should be possible to avoid all uncontrolled charges which lead to a deformation of the field in the chamber. Problematic to this effect is the window area. There the enormous fluxes of x-rays and electrons, generated by the passage of a heavy ion beam through the target, are of great importance. A high voltage (15 kV or more) at the target normally reduces these problems, because the effect of the electrons is diminished. The second point leading to a reduced electron collection efficiency is the capture of primary electrons at electronegative gas impurities, like watervapor and oxygen. If such gas impurities are well below 1 ppm, even electron drift paths of meters are possible. At this purity level the electron losses are in the range of a percent over a drift distance of a meter.

If the detector geometry does not allow to generate a homogeneous field, e.g. if large solid angles have to be covered (4π detectors), the main problem arises from the position dependent electron collection times. This leads to very complicated signal forms. In order to solve such problems, the readout system has to be adapted. For such detectors the use of sampling ADC-systems, which record the complete time evolution of the signal, is a necessity. A first solution of such an Ionization Chamber is presented in the last chapter.

2) THE TWIN IONIZATION CHAMBER [3][4].

At energies near the Coulomb barrier heavy ion reactions have predominantly a binary character. The strong kinematical energy variation of the reaction products dictates position-sensitive detectors. In addition these processes are totally determined by the complete measurement of one of the two reaction products. A reconstruction of the collision is possible, if the total energy E, the positions x,y, the charge and the mass of the fragment are determined. The general layout of a position sensitive Ionization Chamber is shown in Fig.2. The counter is built symmetrically to the cathode. Both sensitive volumes are terminated by a Frisch grid, which has the purpose to shield the following electrodes from the effect of the positive ions. It is followed by a θ-grid and finally the dE,E-anodes. The sensitive area of the detector shown is 20 by 40 cm. The depth of the detector is determined by the dynamical range of the typical experimental problem and can reach up to 150 cm in our case.

A particle entering the sensitive volume of the detector is stopped in the counter gas and creates a column of electron-ion pairs. Under the influence of the electric field the charges drift across the active volume inducing a prompt signal on the cathode. As soon as the moving electron cloud passes the Frisch grid, a signal is influenced on the θ-grid and on the anode. The signal on the θ-grid reaches its maximum, as the cloud goes through the grid surface, and it disappears when all the electrons are collected on the anodes. Fig.2. shows the form and time dependence of the signals seen at the different electrodes.

The position information contained in the signal of the θ-grid is read out through an LC-delay line. The perpendicular space coordinate is obtained from the time delay between the cathode and the anode signals. The x,y position resolution is ≤2 mm for particle energies ≥30 MeV. The arrival time of the particle can be derived from the central cathode, and a resolution of ~1 nsec can be achieved for particle energies above 100 MeV. This is not sufficent for a good mass determination but very useful in coincidence experiments. In a kinematic coincidence measurement for binary reaction products this resolution allows a mass determination with a resolution of ~4 %. In all γ-particle coincidence experiments this timing resolution is normally sufficient.

The charge is determined by the dE,E measurement. The resolution strongly depends on the particle energy. For elastically scattered uranium ions of 4 MeV/u, a charge resolution of 0.8 charge units can be obtained, which proves the good internal resolution of these gas detectors. In a normal experiment however this is an artificial number. The charge resolution is strongly dependent on the particle velocity. Furthermore the energy loss shows a strong mass dependence, which cannot be corrected for unless the mass is known. In Fig.3 a,b our results for lighter ions are demonstrated. In the energy range above 3 MeV/u the resolution is sufficient to separate even the isotopes for these light heavy ions.

The total energy resolution of an Ionization Chamber is typically around 1 %, and is mainly determined by the energy straggling in the entrance window of the detector. Thus, the resolution improves with the particle energy and becomes worse with increasing charge. In ref.[6]) a nice systematical investigation of the limits of the resolutions is given. The authors give a best value of $\delta E/E$ = o.38 % for ^{32}S ions with an energy of 165 MeV. Their results are shown in Fig.4,5.

3) NEW METHODS FOR Θ AND CHARGE DETERMINATION [6]).

It is very difficult to determine the Θ-scattering angle and the charge of a particle having a short range in the detector or even stopping above the first dE-anode. In ref.[6]) methods are described which in some cases may allow to overcome this problem.

For the determination of the scattering angle a subdivision of the dE-anode in the form of a saw tooth is proposed. If two adjacent segments of this position sensitive anode (PSA) are added the energy loss signal is obtained. The result of a division of the two signals determines the particle position. The intrinsic resolution of the PSA is o.7 mm, that is a lot better than normally needed in a typical experiment. This method is helpful for particles with smaller ranges in the chamber. For longer ranges the doubling of the number of anode sections may be a disadvantage.

In order to determine the charge of a particle with the dE,E-method the particles must at least reach the second anode section. For those particles which do not meet this requirement, a new technique is used, where the nuclear charge is determined by a Position Sensitive Frisch grid (PSFG). This interesting method is based on a Frisch grid whose front part has been modified to be range sensitive. The PSFG consists of 13 segments. Each segment contains a certain number of equally spaced wires. The number of wires varies with the distance along the grid as is schematically shown in Fig.6. The graded density segments are alternatingly connected to a right and a left preamplifier. The ratio R of these preamplifier signals contains the information of the particle range. For further details see [6]). The scatter plots shown in Fig.7 compare the two alternative charge determination methods, and demonstrate that the events which could not be identified by the dE-E method, because the particles stopped above the first dE-anode can be identified with the PSFG method as long as their energy is greater than 1 MeV/u. If equal energies per nucleon are compared the resolution is better with the dE-E method.

4) LOGARTHMIC HYBRID DETECTORS.

During the last years the developments in heavy ion accelerators opened the completely new field of high energy heavy ion physics. If the projectile energy is well above 10 MeV/u the reaction mechanism is no longer dominated by binary processes. Above these energies multiparticle breakup processes constitute the dominant part of the reaction cross section. This changes the experimental situation completely. The particle multiplicities and the dynamic range in charge and energy increase strongly. In order to solve this experimental problem hybrid detectors with a 'logarithmic' density increase within the detector are needed. These counters should have a low threshold for heavy ions but should allow at the same time the detection of protons of some 100 MeV.

A large variety of combinations of silicon, plastic and NaI detectors is used in this field. In this context I want to present a telescope with an extreme density variation [7]). The telescope design was adopted to our experiment with 84 MeV/u ^{18}O ions at the CERN SC, for a precise particle identification and an energy measurement between o.5 and 75 MeV/u. The hybrid detector consists of a very thin gas Ionization Chamber, the thickness is equivalent to only 3 µm Si, followed by a stack of solid state detectors. The electric field of the Chamber is directed parallel to the particle track, in order to collect the electrons at the central anode. The stack consists of a 50 µm Silicon dE-detector, a 300 µm Silicon dE-detector, a 4 mm Si-Li detector and finally a 1 cm BGO-scintillator. The telescope is schematically shown in Fig.8. The signal of the BGO crystal is recorded with a photodiode [8]).

Fig.9. shows the charge spectrum detected at 40° in the reaction ^{18}O + Au at 84 MeV/u by this telescope. The upper spectrum shows the charge spectrum where the 50 µm Silicon detector is used as a dE-counter. In the lower part the Ionization Chamber serves as a dE-counter. Charge identification is possible up to Z = 16; the typical resolutions are o.2 charge units for Helium and o.4 for Carbon.

The signal rise time of 200 nsec in the dE-Ionization Chamber limits the count rates in the telescope to ~ 50 kHz. The short radiation length and the fact that the crystal is not hygroscopic makes the use of a BGO crystal very advantageous, although the light signal is a factor 10 smaller than in a NaI crystal. The photodiode used to record the BGO-signal helps to keep the size of the telescope small. Its total length runs up to only 10 cm. In addition the photodiode offers a good longtime stability and a linearity over some orders of magnitude; it can furthermore be operated in strong magnetic fields and serves as a veto counter for all particles not stopped in the telescope. The main disadvantage is that the active area is limited to the size of the silicon dE-detectors. If telescopes with large active areas are needed, a combination of a gas dE-Ionization Chamber and a phoswich detector consisting of a thin fast plastic dE-section followed by a slow plastic E-section seems to be the best solution (see: R.G.Stokstad, this conference). The main problem in dealing with such a telescope may be the energy calibration.

5) THE MULTIPLE SAMPLING IONIZATION CHAMBER (MUSIC) [9].

In connection with the HISS spectrometer at the BEVALAC in Berkeley (see: D.Greiner, this conference) a detector was needed in order to measure the charge of projectile fragments in relativistic heavy ion reactions. In this experiment the momenta of the particles are defined by ray-tracing the fragments in a huge magnetic dipole field. The particle identification is accomplished through the measurement of the rigidity R (= momentum/charge), the charge Z and the velocity β. A plastic scintillator wall defines the time-of-flight of the particles. The energy loss signal in this wall is sufficient to determine the element number for all fragments lighter than Carbon. Concerning the heavier fragments the charge can no longer be determined by a single energy loss measurement, for in that case the energy loss distributions for adjacent charges are no longer separated. For this charge range a new type of large-area detector is needed which is able to measure the energy loss of each detected fragment many times in order to allow a reconstructing the parent distribution for each particle. The mean value, or the truncated mean value if the parent distribution is asymmetric, of this set of measurements determines the fragment charge. Monte-Carlo calculations indicated the necessity of measuring 50 energy loss samples in order to reconstruct the parent distribution with a precision good enough that the resulting charge resolution is well below one charge unit even for uranium. As all the heavy projectile fragments are emitted into a small angle range around the beam axis, the detector must allow a simultaneous detection of many (up to 20) heavy reaction products. These requirements made the developement of the MUltiple Sampling Ionization Chamber (MUSIC) necessary.

In a typical experiment at HISS the momenta of the fragments reach values up to 2 GeV/c/u. The detector size and position is then determined by the 6 Tesla-meter bending power of the HISS-magnet. In order to separate two adjacent charges with the same energy by more than 2 cm the detector must be positioned more than 6 m behind the target. The detector, shown in Fig.10, needs an active area of $1x2m^2$ in order to accept all charged particles heavier than carbon produced in a typical projectile fragmentation reaction. Because of this size and the large energy loss of heavy ions, a gas ionization Chamber is well suited to meet this problem.

The detector consists of a stainless steel barrel with a diameter of 3.2m and a length of 2m. The vessel is evacuated before filling in order to reach the required gas purity. The entrance and exit window consist of two mylar-copper-mylar foils, which are separated by a 1.5cm gas gap. The gas purity in the detector is maintained by purging a clean gas through this gap. The sensitive volume has a size of $1x2x1.7m^3$. The field cage consists of a Mylar foil onto which horizontal copper stripes are evaporated. They are connected by a 20 MΩ/cm resistor chain. So the negative cathode voltage (<80kV) is divided down to the grounded Frisch grid. The setup is shown schematically in Fig.11.

The anode structure is mounted 2cm below the Frisch grid and operates at a positive voltage. The anode configuration consists of two symmetric parts with 64 dE divisions each. To be prepared for the big dynamic range of the signals expected in relativistic heavy ion collisions the preamplifiers are installed outside the tank. In this way the number and size of the samples can be adapted to the experimental needs. In the low charge range ($6<Z<15$), a small number of bigger samples is preferable because the charge resolution is noise-limited for these fragments, whereas the resolution of the heavy fragments improves with the square root of the number of samples.

In order to record multiple tracks (<20) in the detector, the time dependence of the amplitudes of all anode signals has to be measured over a time of 20 μsec after a trigger has occured. During that period the amplitudes of the anode signals are strobed every 100 nsec for 2 nsec into an analog storage device (CCD). If a trigger has occurred these amplitudes are digitized and written on tape. This readout system derived from the LBL-TPC detector [11]). Fig.12 shows the reconstructed amplifier signals produced by two heavy Lanthanum fragments which passed the detector simultaneously. The particles where moving through the detector at a height of 50cm above the anode plane and were seperated by ~6cm.

The first experiment was done with a La beam. In Fig.13 the energy loss spectrum of La projectile fragments detected in a single cell of 2cm width is shown. The distribution is very symmetric and does not show the high energy tail of a Landau distribution, predicted for the energy loss of heavy ions in the passage of a thin absorber. The energy deposited in the detector is discribed by a more symmetric distribution, because all delta rays above a certain energy ε_d are not stopped in the detector cell, and are not contributing to the measured signal [12]). The effect of the escaping knock-on electrons is to make the distribution more Gaussian in shape, because the statistical fluctuations are due to the high energy δ-rays.

The fact that the parent distribution measured with a single cell detector is nearly symmetric, does not mean however, that a single detector with the full detector length gives the same results. The maximum energy ε_d of the δ-electrons contributing to the signal is dependent on the detector thickness. Increasing the detector length results in an increased asymmetry of the parent distribution which leads to a reduction of the resolution. Only in the case of Gaussian parent distribitions sampling is not necessary. In our case, however, the form changes with detector size. Under these circumstances the resolution is expexted to depend inversely proportional to the square root of the number of samples measured.

The resolution for the signal deposited in the detector is ~4.5 %. This value agrees with the results given in ref.[13]) and is shown in Fig.14. The charge resolution with a single 2cm cell detector is ~2.3 %. In Fig.15 the charge spectrum measured with MUSIC for La projectile fragmentation is presented. In the analysis 60 cells of the detector are used in order to determine the most probable value for the energy loss of the fragment. A truncation is changing the result only below the 10% level. The charge resolution is $dZ < 0.3$ charge units or 0.53 %. This proves that the charge resolution improves inversely proportional to the square root of the number of samples taken. From the results of the single cell measurement a charge resolution of 0.3% could be expected. This discrepancy is not yet understood completely.

6) CONCLUSION.

I hope to have demonstrated that Gas filled Ionization Chambers are an adequate tool whenever heavy ions have to be detected. Some of the reasons are: the detectors have a very linear response to the energy deposited; a very small mass dependence of the measurements, because plasma recombination effects are negligible due to the fact that the charge densities are 1000 times lower than in solid state detectors; charge carrier losses are mainly due to gas impurities, and thus to a large extend controllable; the detectors can be built in very large sizes and it is easy to adjust the detector to the dynamic range of the experimental problem by choosing the detector length; the effective thickness can be regulated by the gas pressure and type.

The most important impact on the application of these detectors in the future is the developement of sampling ADC systems. The systems under discussion , either the CCD analog storage method or the use of ADC flash encoders, are interesting in this respect. A higher digitizing accuracy of 11 bits is necessary in order to be adapted to the big dynamic range to be expected in heavy ion collisions. This developement will broaden the possibilities of the gas filled Ionization Chambers.

In connection with relativistic heavy ion experiments these detectors have the useful feature that they are not sensitive to minimum ionizing singly charged particles. To my opinion it is extremely useful that detectors are available which are not sensitive to every particle produced in uranium uranium collisions at 1, 10 or 100 GeV/u.

For the higher beam energies the hybrid detector [7]) presented in chapter 4 is a very promising development because these detecors have a very high sensitivity and can be built in very small dimensions. A system consisting of an Ionization Chamber dE counter combined with a fast and a slow plastic scintillator phoswich detector seems to be the ideal element for a 4π Gas Ball detector.

Especially the Bragg spectrometers will benefit from the further development of sampling ADC systems. They will allow detailed analysis of the complete Bragg curve. By analyzing the details of the energy loss near the maximum range of the particles, there may be a slight chance in using such a device to determine the element number of the detected particle also if the energy is below o.5 MeV/u, an hitherto unsolved problem.

References:

1) P.Kienle private communications.
Ch.Schiessl, W.Wagner, K.Hartel, P.Kienle, H.J.Körner, W.Mayer K.E.Rehm, Nucl.Instr.and Meth. 192 (1982) 291.

2) L.G.Christophorou, D.L.McCorkle, D.V.Maxey and J.G.Carter, Nucl.Instr.and Meth. 163 (1979) 141.

3) D.C.Cheng, W.A.Kozanecki, R.L.Piccioni, C.RUBBIA, L.R.Sulak,H.J.Weedon, and J.Wittacker, Nucl.Instr.and Meth. 117 (1974) 157.

4) H.Sann, H.Damjantschitsch, D.Hebbard, J.Junge, D.Pelte, B.Povh, D.Schwalm and D.B.Tran Thoai, Nucl.Instr.and Meth. 124 (1975) 509.

5) U.Lynen, H.Stelzer, A.Gobbi, H.Sann and A.Olmi, Nucl.Instr.and Meth. 162 (1979) 657.

6) J.Pochondzalla, R.Butsch, B.Heck, G.Hlawatsch, A.Miczaika, H.J.Rabe and G.Rosner, Nucl.Instr.and Meth. (1984) in press.
B.Kolb et al. Nucl.Instr.and Meth. 188 (1981) 555.

7) R.Trockel, Diplomarbeit, Universität Münster.(1984).

8) R.Glasow et al. sub. to Nucl. Inst.and Meth.(1984).

9) H.Sann, MUSIC at HISS, XXI International Winter Meeting on Nuclear Physics, Bormio 1983, p.597.
H.Sann, Nuclear Science Division Annual Report, LBL-15207, UC-34(1983), p.254.
B.Christie et al. Nucl.Instr.and Meth. to be published.

10) S.Eberwein private communications.

11) C.Jared LBL-reports, LBL-13183,4 (1981).

12) G.D.Badhwar, Nucl.Instr.and Meth. 109 (1973) 119.

13) K.Nagata, J.Kikuchi, T.Doke, and C.R.Gruhn, Report of Sci. and Eng. Res. Lab., Waseda University, Dec 1980.

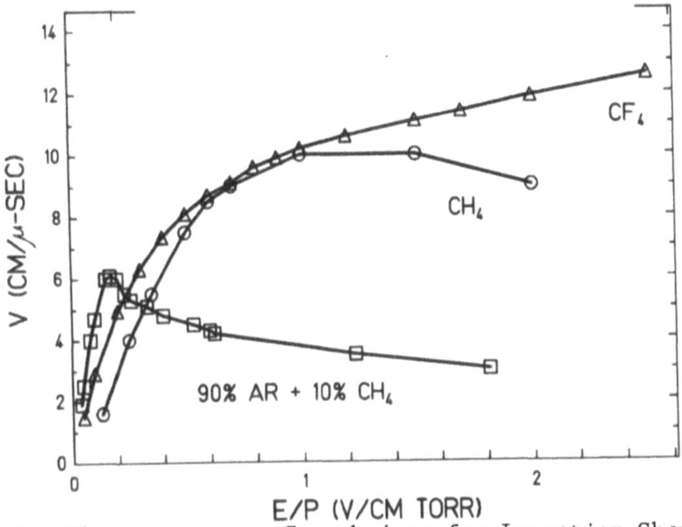

Fig.1): The most common Gas choices for Ionzation Chambers:

$90\%Ar+10\% CH_4$: slow, cheap.

CH_4 : fast, cheap, explosive.

CF_4 : fast, expensive, high stopping power [1][2].

Fig.2): The Figure shows a cross section through a Twin Ionization Chamber for Heavy Ions. It is built symmetrically around the central Cathode. The two active volumes are followed by the Frisch-Grids, the θ-Grids and the segmented dE,E-Anodes.

Below the time sequence and the form of the signals derived from the different electrodes is shown. The Cathode signal appears promptly with the particle entering the detector. The θ-grid signal starts after the time t_ϕ, needed by the electrons to drift from their origin to the Frisch-grid. The θ-Grid signal is delayed by an LC-delay line by the time t_θ. On the Anode the electrons are collected delivering the Energy and the Energy loss signals. The Times t_ϕ and t_θ are used to determine the particle positions. If a current sensitive preamplifier is used at the cathode, the time of flight can be derived.

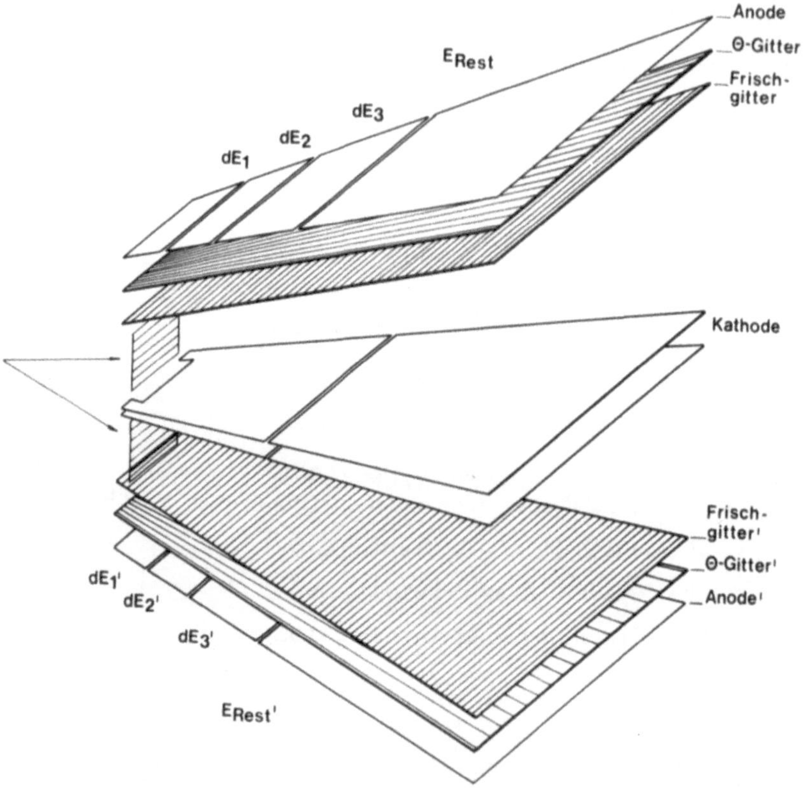
Anode
Θ-Gitter
Frisch-
gitter
E Rest
dE3
dE2
dE1
Kathode
Frisch-
gitter'
Θ-Gitter'
Anode'
dE1'
dE2'
dE3'
E Rest'

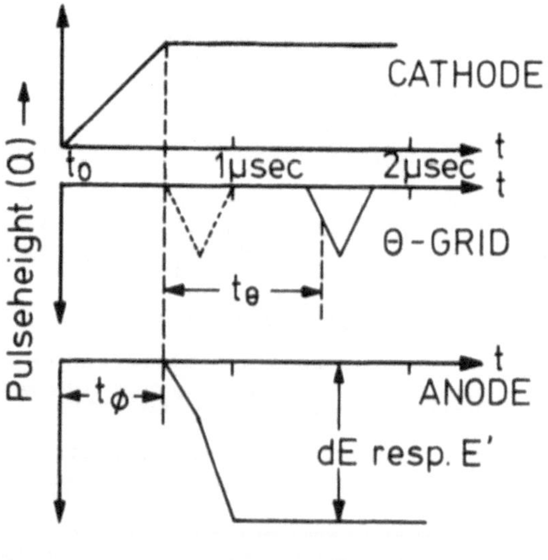
Pulseheight (Q) →
CATHODE
t
t0
1μsec
2μsec
t
Θ-GRID
tΘ
t
ANODE
tφ
dE resp. E'

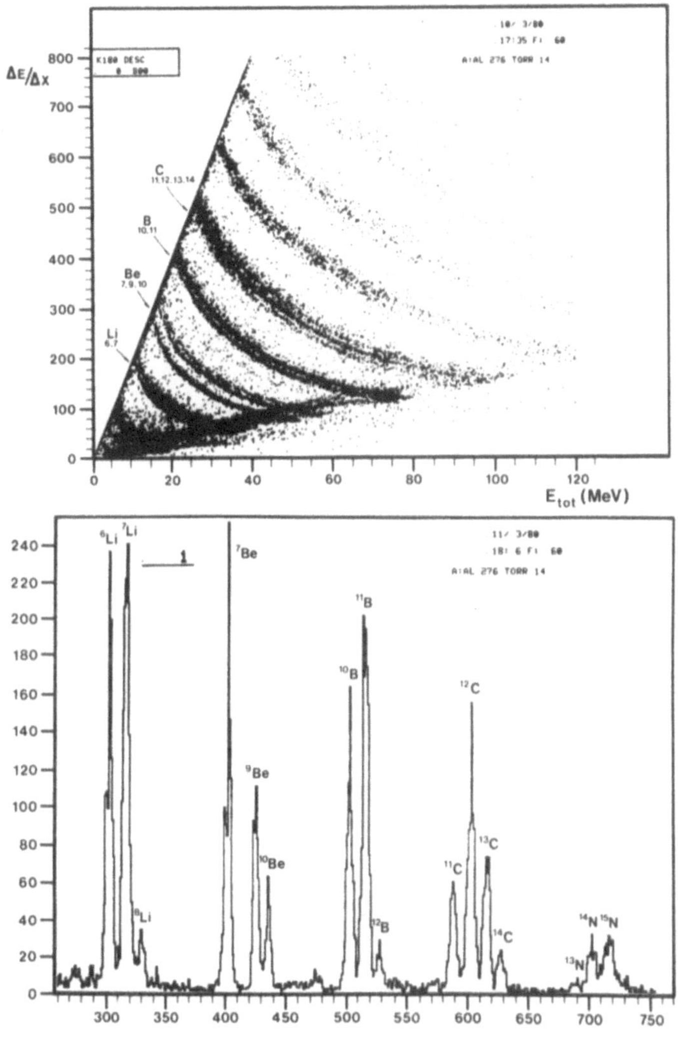

Fig.3): The two dimensional dE,E plot shows the typical result for the Charge resolution in an Ionization Chamber. The reaction is 84 MeV/u ^{12}C + Au. If the particle energy is above 3 MeV/u the resolution is good enough to separate the isotopes for these light ions. This is well demonstrated in the lower part of the picture, where the projection on the Charge axis is shown [3)][4)].

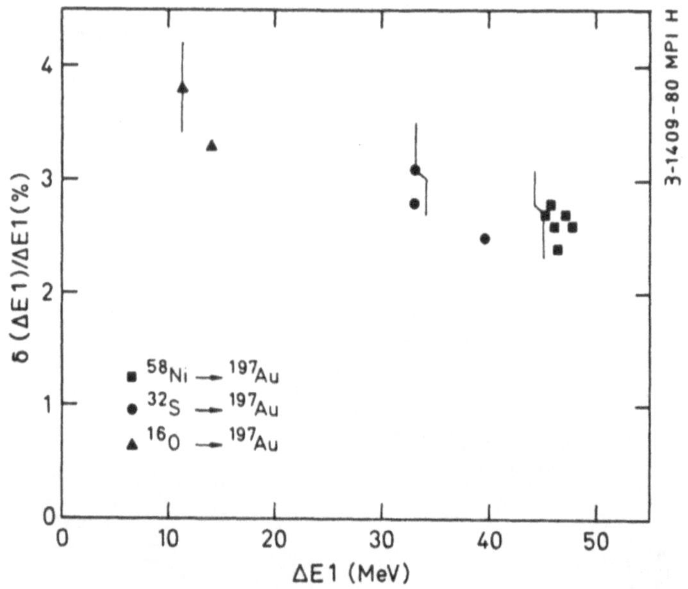

Fig.4): Relative energy loss resolution δ(ΔE1)/ΔE1 as a function of energy loss ΔE1 across the first ΔE anode for various ions observed in the Ionization Chamber is shown [6]).

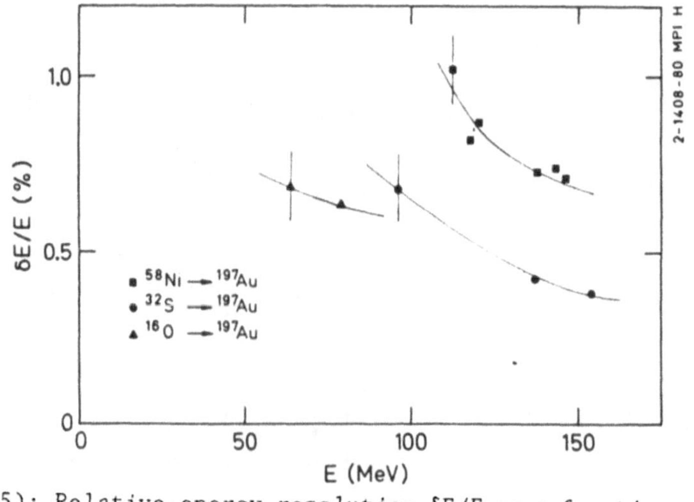

Fig.5): Relative energy resolution δE/E as a function of the total energy E for various elastically scattered particles scattered on a thin gold target [6]). The lines are drawn to guide the eye.

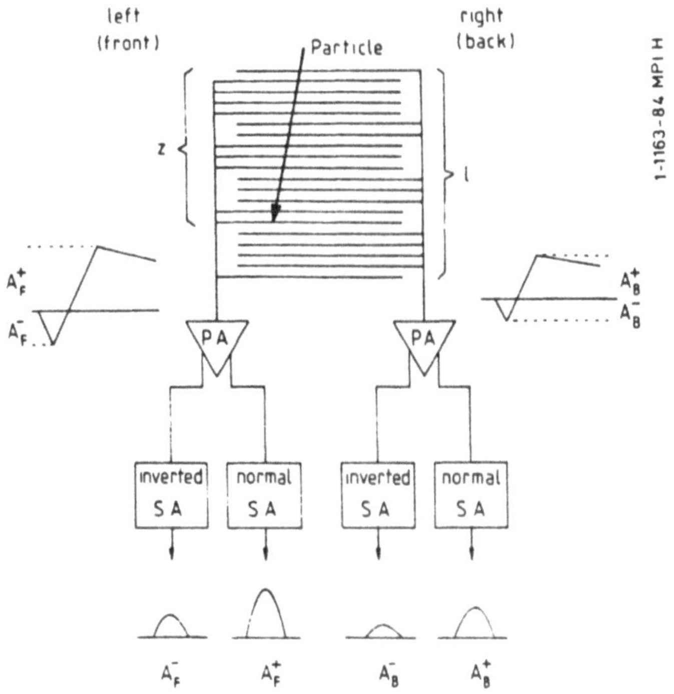

Fig.6): Shematic design of the graded density Position Sensitive Frisch Grid (PSFG). The frist 10 cm of the grid consists of segments which have a length which changes as is shown. The segments are connected alternatively to a right and a left amplifier. The ratio R of these preamplifier signals contains the information of the particle range. For details see [6]).

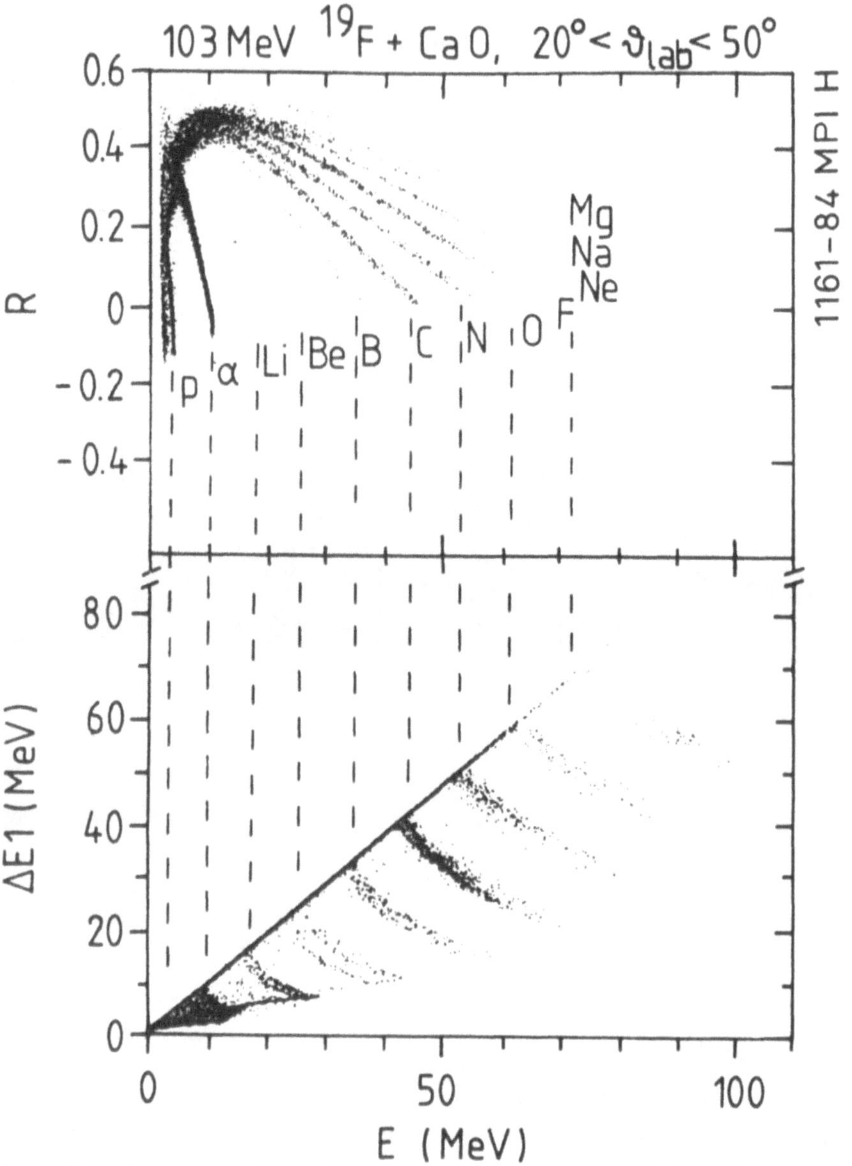

Fig.7): Scatter plots for the two alternative charge determination methods: dE-E (bottom) and range related R-E (top). The plot shows the reaction products from 103 MeV ^{19}F + CaO in the angular range from 30° to 50°. The upper part of the figure displays the events which could not be identified by the dE-E method.

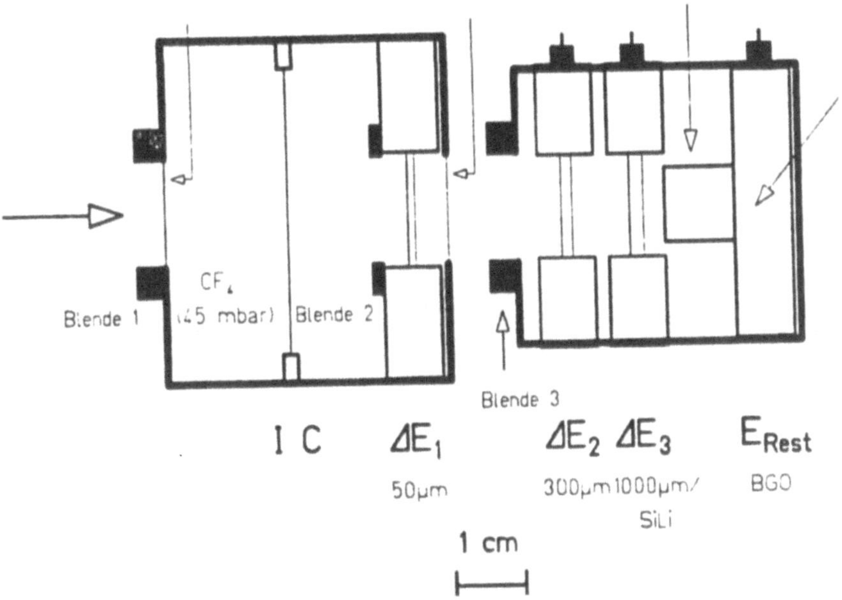

Fig.8): The 'logarithmic hybrid' detector consists of a Gas Ionization Chamber with a central Anode, followed by a stack of solid state detectors. This consists of two Silicon dE-counters of 50 μm and 300 μm thickness, followed by a Si-Li dE-detector of 4 mm thickness, and finally a 1 cm BGO-scintillator with measure the rest energy. The BGO crystal signals are recorded with a photodiode.

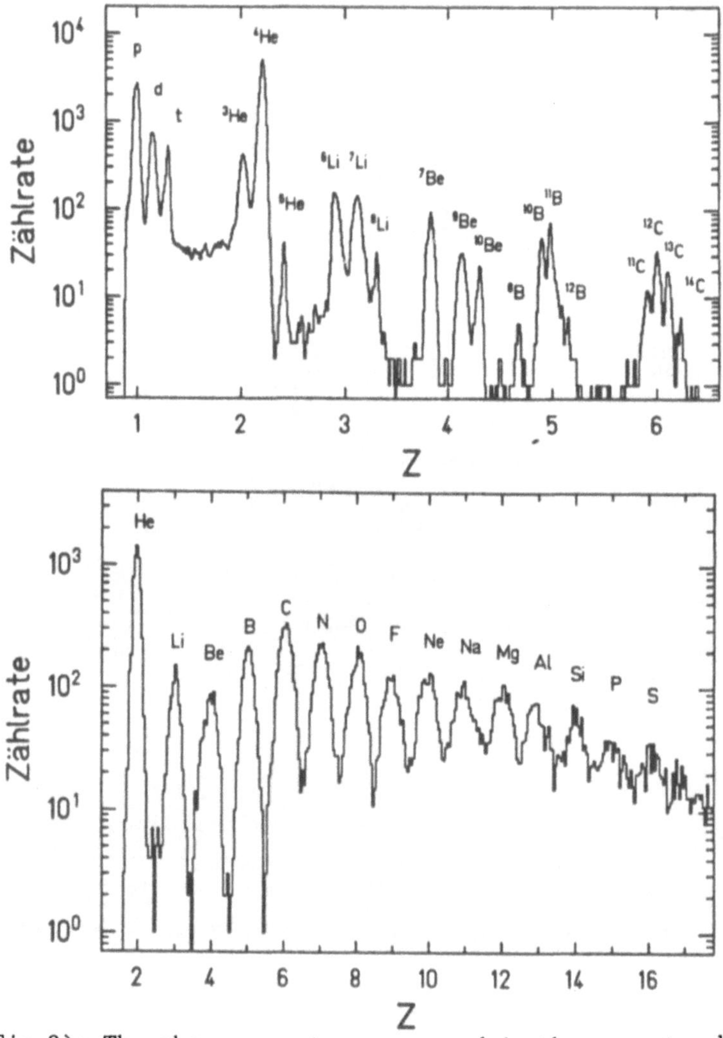

Fig.9): The charge spectrum measured in the reaction ^{18}O +Au at 84 MeV/u, measured with the hybrid detector, is shown. The upper spectrum shows the charge spectrum where the 50 µm Silicon detector is used as a dE-counter. The lower part shows the same spectrum when the Ionization Chamber is used as a dE-counter. The energy spectra can be measured down to 0.5 MeV/u. Charge identification is possible up to Z = 16; the typical resolutions are .2 charge units for Helium and .4 for Carbon.

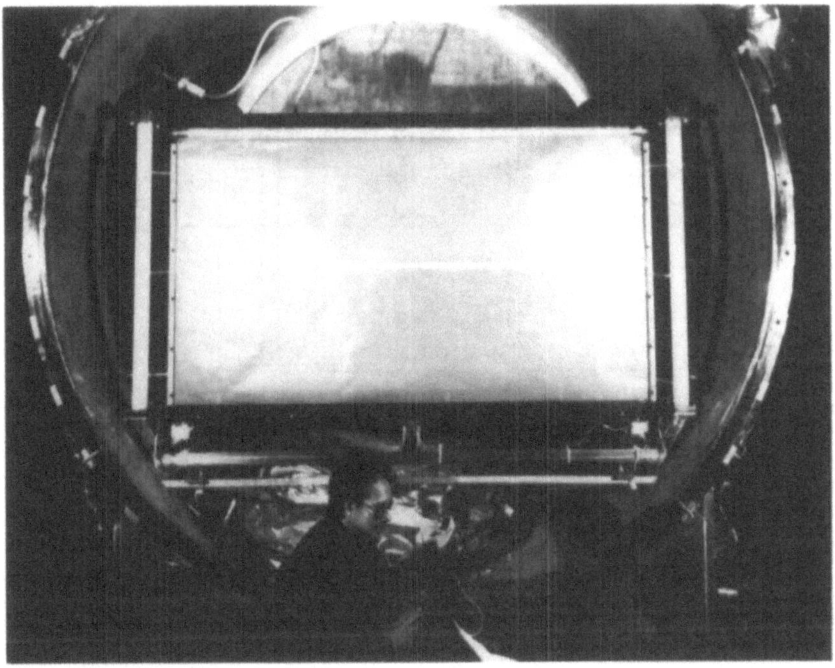

Fig.10): The Figure shows the MUSIC detector after removing the endcap and the entrance windows. The 1x2m field cage, the high voltage feedthrough and the anode structure can be seen. The amplifiers are mounted outside the gas tank. This allows to adapt the size and number of the anode segments according to the charge range of the particles expected in the different experiments. The active volume of the detector is $1x2x1.7m^3$.

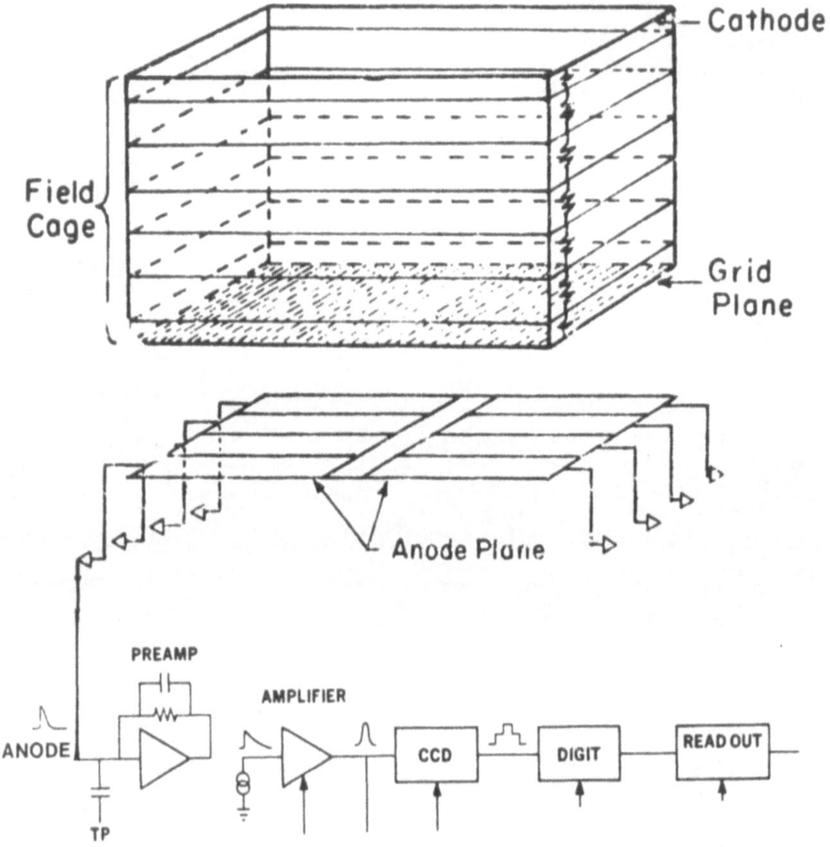

Fig.11): The relativistic particles cross the MUSIC detector without being stopped. They loose energy at a constant rate which is ~Z^2. The generated primary electrons move under the influence of the electric field in the cage in the direction of the Frisch grid and are collected on the anode plane. Each anode section is connected to a charge sensitive preamplifier followed by a shaping amplifier. The amplitude of each signal is then strobed every 100 nsec for 2 nsec into an analog storage device (CCD). After a trigger has occurred, these amplitudes are fed into an ADC and written via CAMAC on tape.

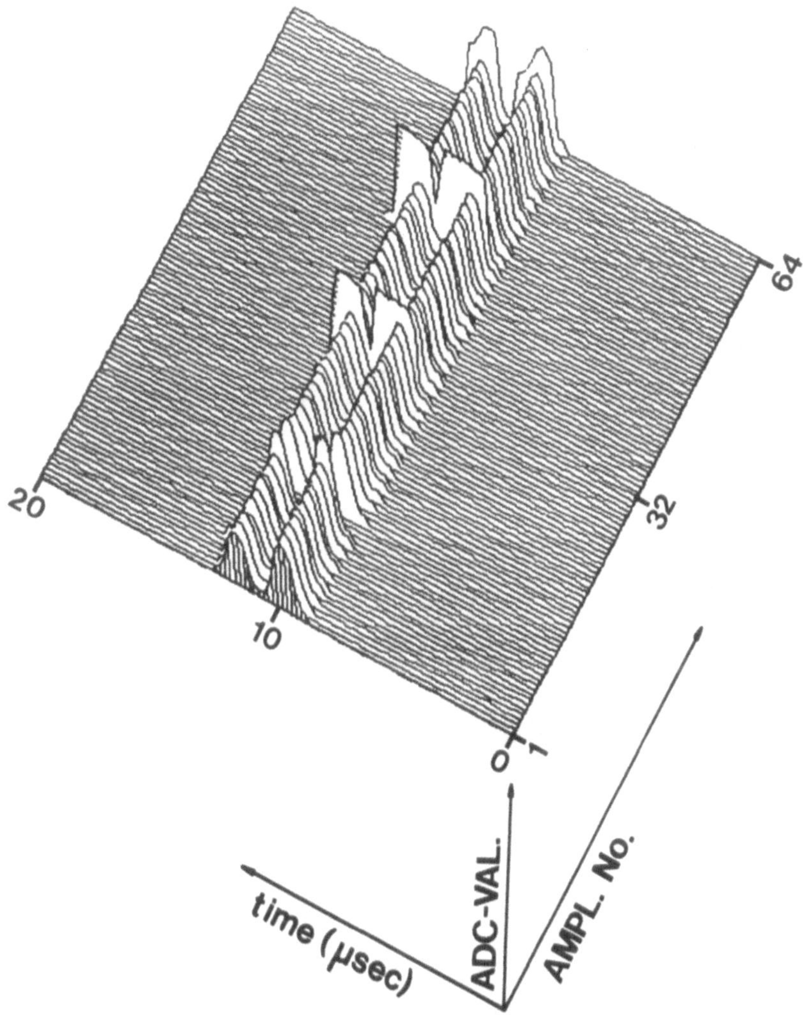

Fig.12): The MUSIC response to a fissioning La projectile. The detector was triggered by an event where two heavy fragments crossed the detector. The time evolution of the signals of the 64 anode amplifiers are shown. All amplifiers show very small signals, until after 10 μsec corresponding to a drift distance of 50cm the electrons generated by the first fragment arrive at the anodes. After 1.2 μsec (~6cm) appears the signal of a second fragment on all amplifiers.

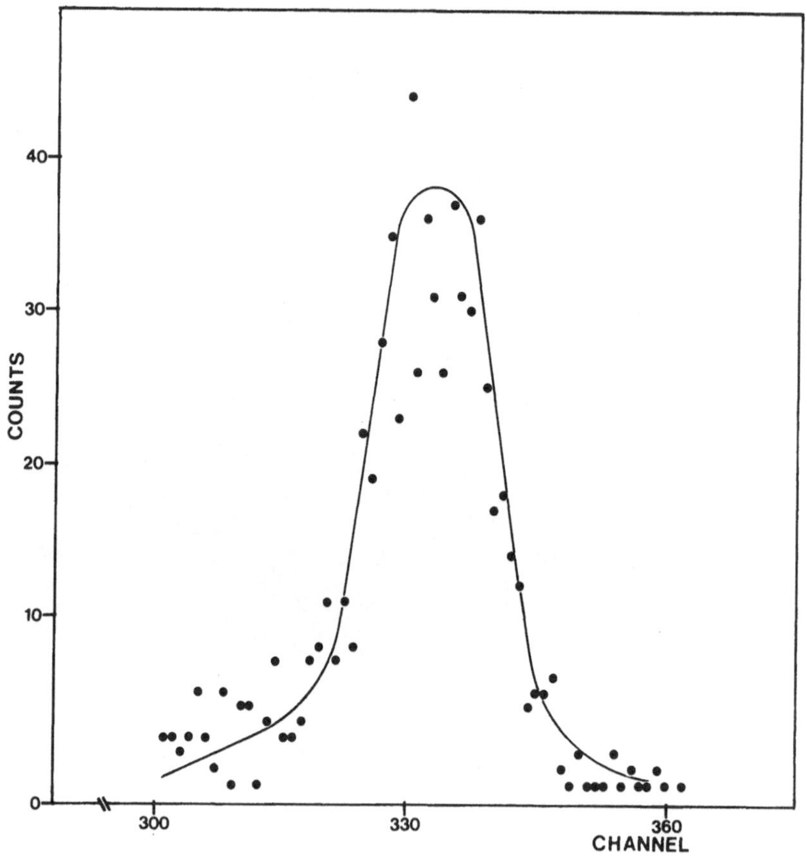

Fig.13): The parent distribution for La measured at a single Ionization Chamber cell of 2 cm width. The mean energy deposited is ~2 MeV. The energy deposited in the detector is discribed by a symmetric distribution, because all delta rays above a certain energy ε_d are not stopped in the detector cell, and are not contributing to the measured signal. The width of the measured distribution is ~4.5 %. This is in agreement with the measurements shown in [12]). A single, 2 cm cell detector would give then a charge resolution of of ~2.25 %.

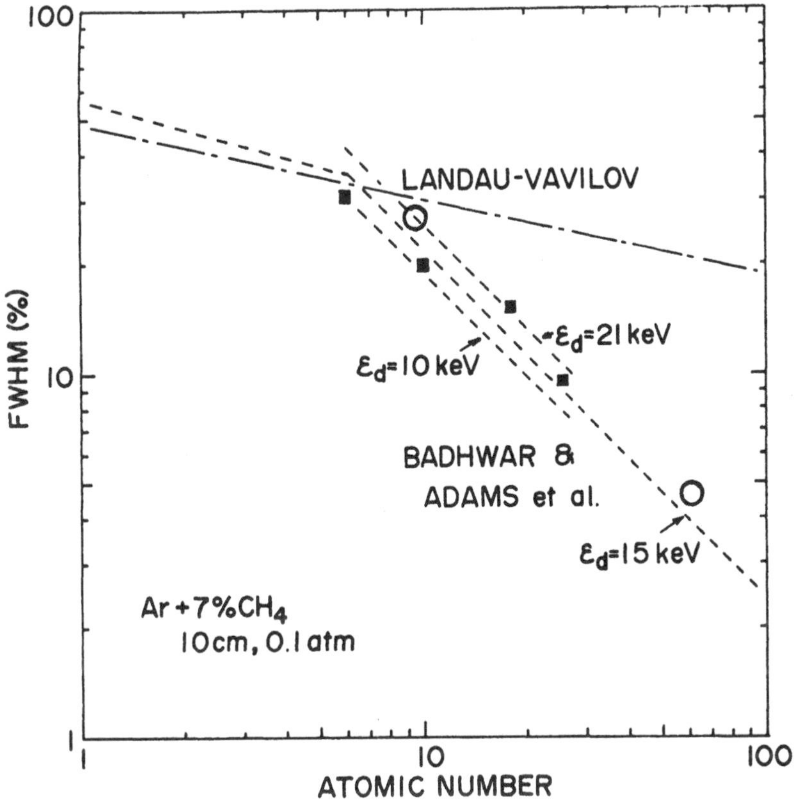

Fig.14): The deposited energy loss distributions for different projectiles are measured with a 10cm single cell Ionization Chamber in [13]). From these distributions the fwhm's are extracted and plotted vs. the atomic number of the particle. The squares show the experimental results measured in that paper. The chain line shows the theoretical fwhm's of the Landau-Vavilov distribution. The broken lines show the theoretical fwhm's calculated by Badhwar [12]). The expectations for different electron escape values ε_d are indicated. The results obtained with the MUSIC chamber are indicated by circles.

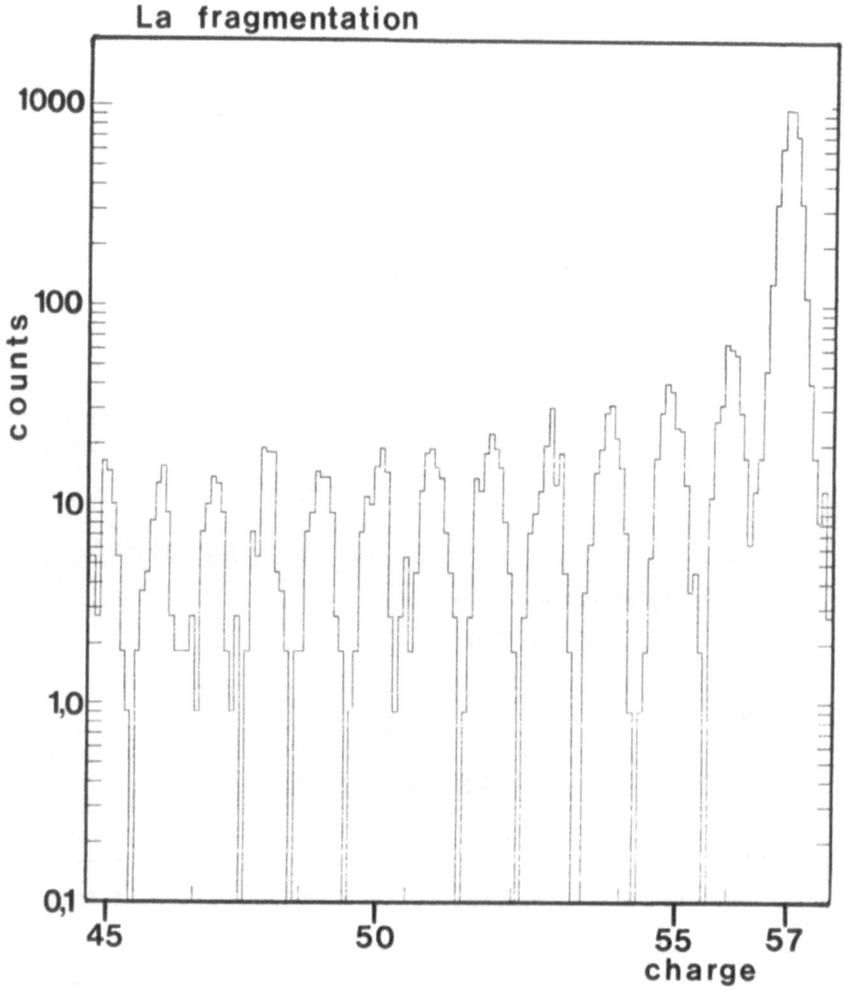

Fig.15): The charge spectrum measured with MUSIC for La projectile fragmentation. In the analysis 60 cells of the detector are used in order to determine the most probable value for the energy loss of the fragment. The charge resolution is dZ < 0.3 charge units or 0.53 %. The charge resolution improves roughly inversely proportional to the square root of the number of samples taken.

A Bragg Curve Spectrometer for a Multi Detector Array

J.E. Yurkon, G.D. Westfall, J. van der Plicht,
Z.M. Koenig, B.V. Jacak, R. Fox, G.M. Crawley, M.R. Maier,
B.E. Hasselquist
National Superconducting Cyclotron Laboratory
Michigan State University
East Lansing, MI 48824 USA

R.S. Tickle
Department of Physics
University of Michigan
Ann Arbor, Michigan 48109 USA

D. Horn
Chalk River Nuclear Laboratory
Chalk River, Ontario, K0J1J0, Canada

A Bragg Curve Spectrometer has been developed for detection of heavy ions with energies of a few MeV/nucleon. The detector is a hexagonal pyramid which makes it suitable for close packing into a 4π sphere. A field shaping grid gives the Bragg Curve Spectrometer a radial field. The charge is read off a Frisch gridded anode. The detector has been successfully tested so far for particles with $1 \leq Z \leq 18$.

1. Introduction

In recent years experiments have established the importance of measuring light particles ($A \leq 4$), medium mass nuclear fragments ($A \leq 40$), fission fragments, and projectile-like fragments resulting from the collisions of high energy nuclei[1]. In addition it has been shown that experiments measuring one particle from each reaction suffer from integration over the unseen particles.

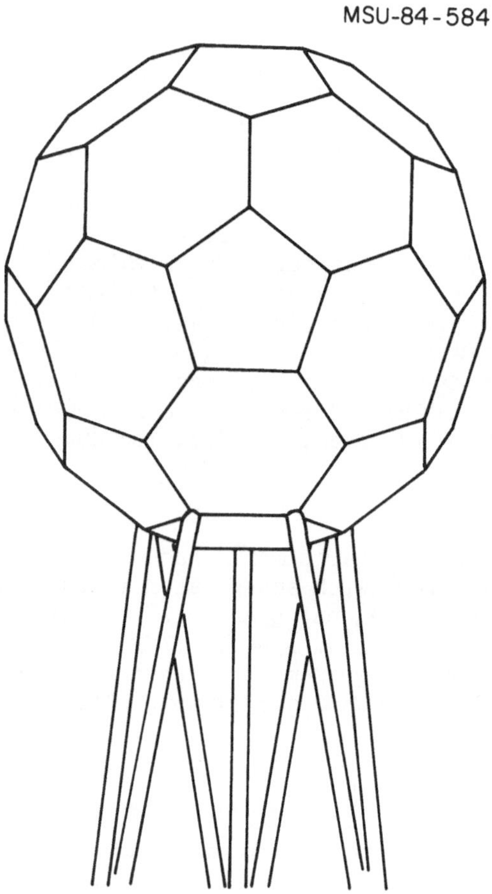

Figure 1 The NSCL multi-detector array, consisting of 20 hexagons and 12 pentagons.

Moreover, at phase II NSCL energies (up to 200 MeV/nucleon), it is important to study not only light particles and projectile-like fragments, but intermediate rapidity medium mass fragments and fission fragments. Therefore we are constructing a device that is capable of simultaneously detecting most of the particles emerging from nuclear collisions occurring at energies available (up to 200 MeV/nucleon) from the phase II NSCL facility. The basic component of this device is the "logarithmic" detector. Its response is designed to increase logarithmically in stopping power with the range of penetration of the observed fragment. Thus fission fragments will be detected in thin front elements (low pressure MWPC's), and highly energetic light particles will be detected using thick range energy scintillator telescopes at the back.

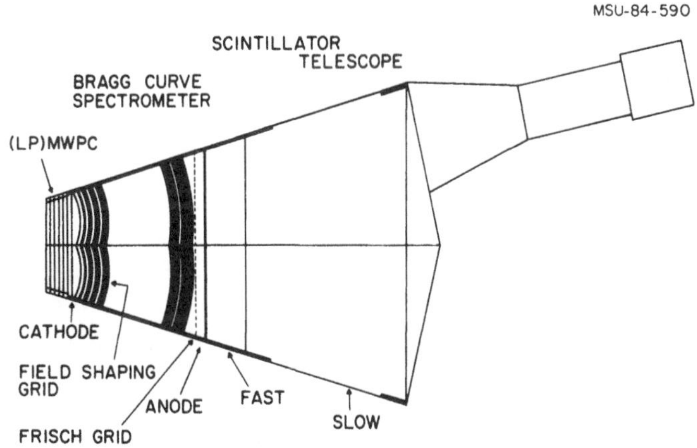

Figure 2 Schematic representation of one of the 30 subarrays that will make up the multi-detector array.

The intermediate energy, medium mass nuclear fragments will be detected with Bragg Curve Spectrometers (BCS's), placed between the MWPC - and scintillator detectors as the second layer of the array.

The requirement of covering a large fraction of 4π with MWPC's, BCS's, and scintillator telescopes constrains the shape of the detectors. The shape chosen is based on a thirty two face truncated icosahedron containing 20 regular hexagonal faces and 12 regular pentagonal faces - a "soccer ball" geometry. An example of this geometrical object is shown in Figure 1 which shows the proposed multi detector array now under construction at NSCL. This array of counters will have sufficient granularity to handle light particle multiplicities up to 50 with reasonable efficiency. The geometry of one subset of counters presented here is a truncated hexagonal pyramid and is shown schematically in Figure 2. We will report here on the performance of the hexagonally shaped Bragg Curve Spectrometer.

2. Construction

We have constructed and successfully tested one prototype detector element (i.e. one MWPC and BCS, followed by an array of 6 scintillator telescopes) as is shown in Figure 2. The distance from the target to the center of the front of the MWPC is 15 cm. The total thickness of the MWPC is 1.6 cm and the total length of the BCS is 15 cm. The total thickness of the prototype scintillator telescope is, at present, 15 cm. The solid angle of this subarray is about 0.3 sr.

The idea of a Bragg Curve Spectrometer was originally suggested by Gruhn et. al. who also constructed the first operational BCS[2]. Since that time several groups have constructed BCS's along similar lines[3-5]. The BCS takes advantage of the fact that the maximum specific ionization of a stopping ion is proportional to the atomic number of the particle. By measuring the maximum of the ionization one obtains the Z and by measuring the integral of the ionization one obtains the energy of the particle. In principle, since one has obtained multiple ΔE measurements, also the A of the particle should be accessible. Also, the range of the particles that are stopped in the detector is obviously determined.

The main structure of the BCS is a hexagonal pyramid made from 6.35 mm G10 fiberglass epoxy laminate. The entrance window of the BCS has a minor diameter of 10 cm. The pressure window of the BCS is made of 6 μm thick aluminized mylar supported by a wire grid with 1 cm spacing. This window acts as the cathode.

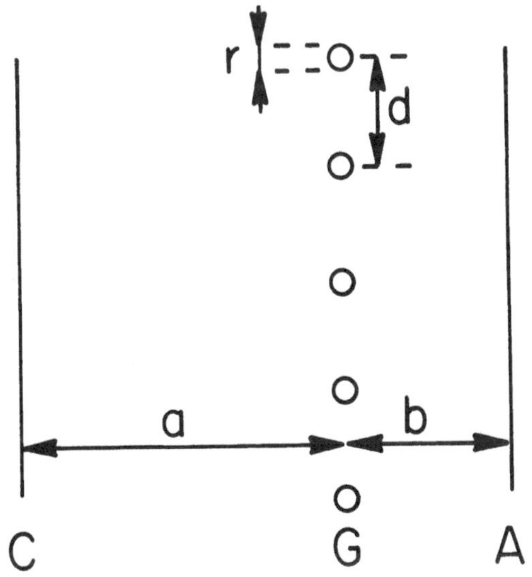

Figure 3 Schematic representation for ion chambers.

The distance between the cathode and the Frisch grid is 14 cm. The Frisch grid is made of 12.5 μm gold plated tungsten wire with a 0.5 mm spacing. The anode is a similar wire grid located 1 cm behind the Frisch grid. The rear pressure seal is formed by the scintillator telescopes. The BCS usually operates at a pressure of 500 torr $ArCH_4$ (90/10%) gas. The Frisch grid/anode geometry determines the shielding inefficiency, which for ion chambers in general is defined as:

$$\sigma = \frac{d}{2\pi b} \ln \frac{d}{2\pi r}$$

where b is the grid/anode distance, and d and r are the spacing and the radius of the grid wires. For our detector (see Figure 3), this works out to $\sigma = 2\%$.

In order to drift the electrons parallel to the trajectory of the particles, a field shaping grid was installed inside the BCS to approximate a radial field. This is illustrated schematically in Figure 2. The calculations were done using the POISSON program in cylindrical geometry.

3. Operation

The charge created by a stopping charged heavy ion is drifted to the Frisch grid shielded anode. One measures the charge collected at the anode as a function of time, and thus obtains the complete energy loss distribution of the stopping particle.

In Figure 4, calculated Bragg curves (i.e. a plot of energy loss versus penetration depth) are shown for ^{6}Li, ^{9}Be, ^{11}B and ^{12}C ions with energies of 2.22, 2.6, 2.87 and 3.4 MeV/A respectively. The top part of Figure 4 shows the theoretical Bragg curves for a 15 cm deep volume of P10 gas at a pressure of 500 torr. In the plots of the bottom part of Figure 4, the Frisch grid and electronics time constant of RC = 0.1 μsec are included.

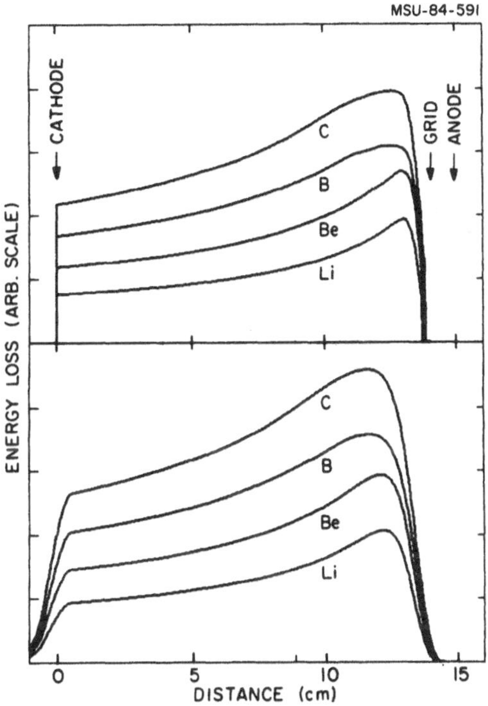

Figure 4 Calculated energy loss as a function of penetration depth in the Bragg Curve spectrometer for four particle species.

The positions of the electrodes in the detector are also indicated. Figure 5 shows the drift velocity of electrons for P10 gas (Ar/CH_4 90/10%) as a function of the reduced electric field (E/p). Also indicated is the regime where the BCS typically operates. It can be seen that the drift velocity can change by as much as 50% over the detector depth.

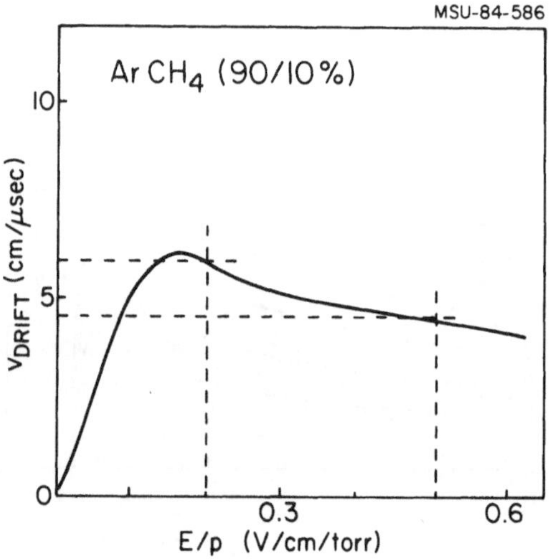

Figure 5 Drift velocity dependence on reduced field for Ar/CH_4(90/10%).

This is obviously influencing the curves as shown in Figure 4, and therefore also the behavior of the detector (as will be discussed in chapter IV).

The BCS is readout as is shown schematically in Figure 6. After integration by a charge sensitive preamplifier, the anode signal of the BCS is split into two amplifiers with different time constants τ. One amplifier is set at τ_1 = 0.25 μsec, and therefore basically measures the height of the peak of the signal which is proportional to the charge Z of the particle. The second amplifier is set at τ_2 = 6 μsec, and therefore measures the area of the total signal which is the energy E of the particle.

Alternatively, the BCS can be read out by a flash encoding ADC as is indicated in Figure 6 by dashed lines. The flash encoder integrates and digitizes the charge in 75 nsec bins and thus provides a "snapshot" of the particle going through the detector.

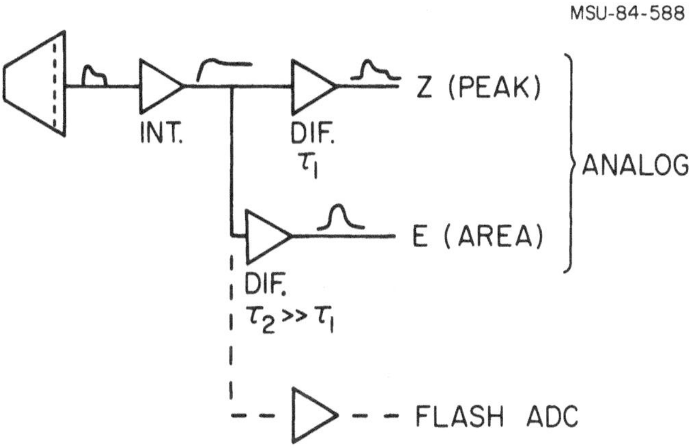

Figure 6 Schematic diagram illustrating the readout of the Bragg Curve Spectrometer.

Also here, the peak of the distribution is a measure of Z and the integral of the curve is the energy E.

4. Results

The Bragg Curve Spectrometer, as part of the prototype array as is shown in Figure 2 has been tested extensively, both with radioactive sources (notably ^{228}Th) and with heavy ion beams. Experiments were performed with beams of 360 MeV ^{12}C and 490 MeV ^{14}N at the National Superconducting

Cyclotron Laboratory of Michigan State University, with 230 MeV ^{35}Cl at the Tandem/Linac accelerator of Argonne National Laboratory, and with 4 GeV ^{40}Ar at the Bevalac of Lawrence Berkeley Laboratory.

As mentioned in chapter III, the Bragg Curve Spectrometer can be read out using two different techniques.

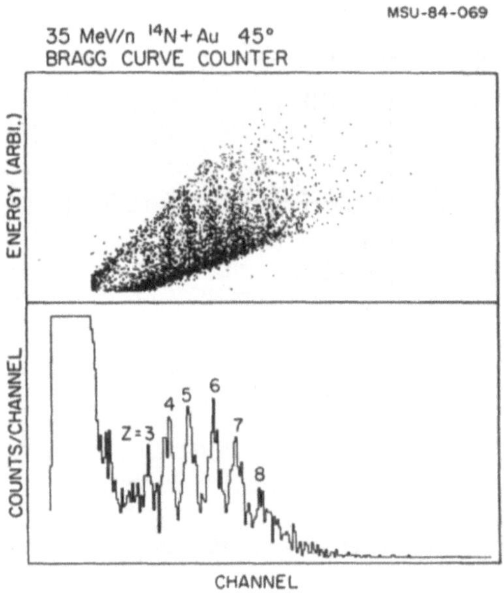

Figure 7 Top: scatter plot of Z (horizontal) versus E (vertical) for ^{14}N + ^{197}Au at 35 MeV/A.
Bottom: projected Z spectrum for the same reaction.

Firstly, from the measured Bragg curve the peak amplitude and the total energy can be deduced using amplifiers with short and long shaping times, respectively. Secondly, the particle can be followed through its Bragg curve as a function of time using a flash ADC.

Figure 7 shows an example of the results of the first (analog) method. Heavy fragments resulting from the reaction Au(^{14}N,X) were detected.

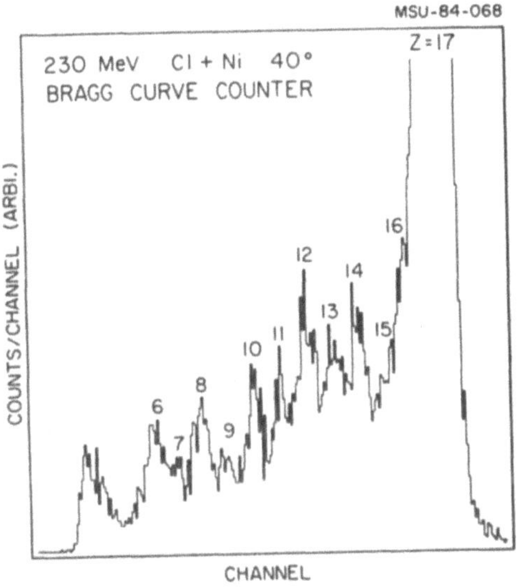

Figure 8 Charge spectrum, measured for the reaction ^{35}Cl + ^{58}Ni at 230 MeV/A.

The beam energy was 35 MeV/A, delivered by the superconducting cyclotron of NSCL/MSU. The detector was placed at an angle of 45°. The top part of Figure 7 shows a two - dimensional plot of the charge Z (horizontal axis) versus energy E (vertical axis). The integration times of the amplifiers were 0.25 μsec for Z and 6 μsec for the E. The individual charges of the particles stopped in the counter are clearly isolated. They show up as peaks in the one-dimensional Z spectrum on a background of particles that barely get into the detector, and particles that punch through. This is illustrated in the bottom part of Figure 7.

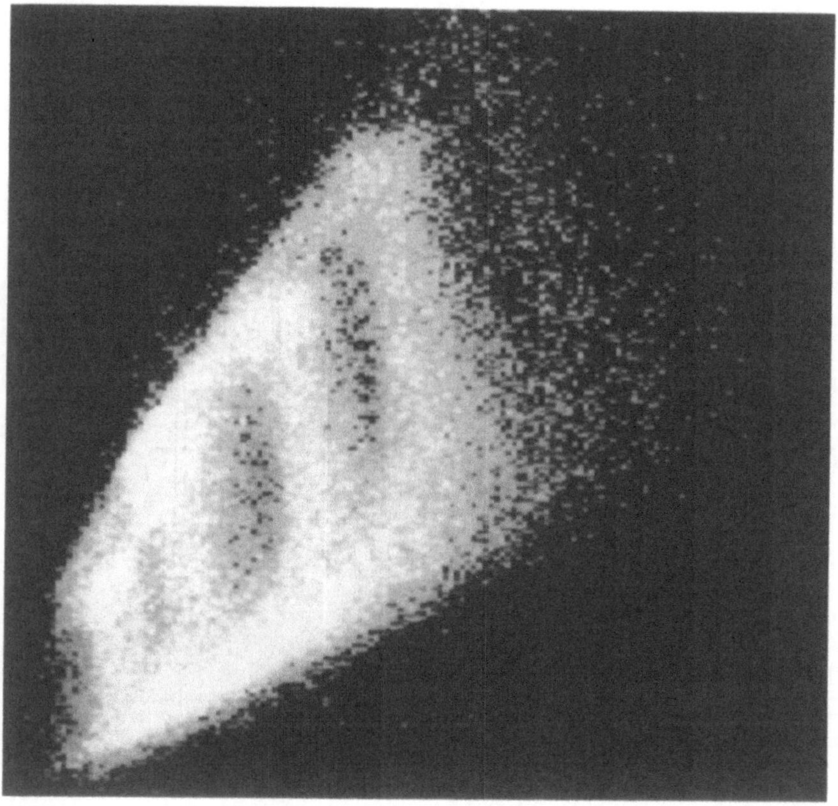

Figure 9 Contour plot of E (vertical) versus Z (horizontal) of the BCS as measured for the reaction ^{12}C + ^{12}C at 30 MeV/A.

In Figure 8, results from the reaction ^{35}Cl + ^{58}Ni are presented. The beam of 230 MeV ^{35}Cl was provided by the tandem/linac of Argonne National Laboratory. In this case, charges up to Z=17 were resolved.

A two-dimensional picture of charge Z (vertical) versus energy E (horizontal) as obtained from the reaction ^{12}C + ^{12}C is shown in Figure 9. The beam of 30 MeV/nucleon ^{12}C was produced by the K500 superconducting cyclotron of NSCL/MSU. The resolution of this plot is better than the one shown in Figure 7. It appears that the horizontal Z lines are sightly curved.

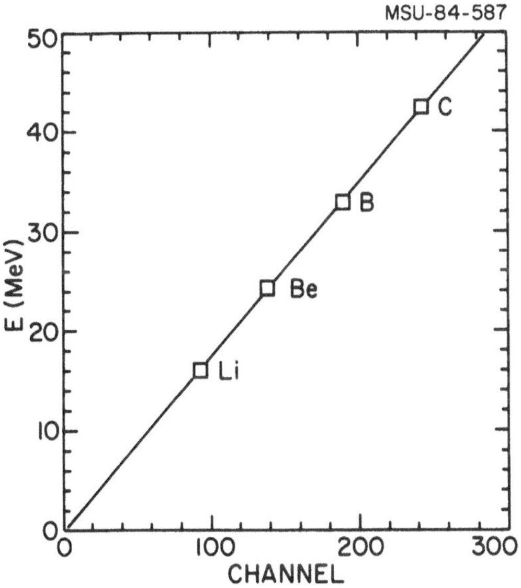

Figure 10 Calibration curve for ^{6}Li, ^{9}Be, ^{11}B and ^{12}C fragments from the reaction ^{12}C + ^{12}C at 30 MeV/A.

This is due to a non-perfect radial electric field in the pyramidal Bragg Curve Spectrometer. Also, the individual Z lines curve back near the punch through. This can be explained by observing the theoretical Bragg curves in Figure 4. As the particle starts to punch through the peak energy loss information is not immediately cut off while the total integrated charge is being reduced. Then, as the Bragg peak starts to punch through one obtains a peak signal that is still somewhat proportional to the charge of the particle and inversely proportional to the energy.

Despite these effects, the individual charges remain resolved as can be seen in Figure 9. The energy calibration for the 30 MeV/A ^{12}C + ^{12}C experiment, as obtained from the punch through line, is shown in Figure 10.

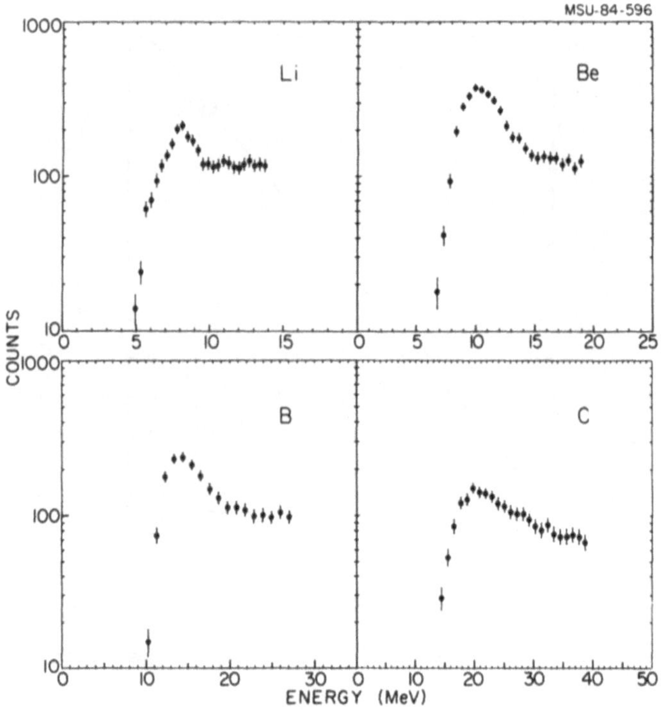

Figure 11 BCS energy spectra for ^{6}Li, ^{7}Be, ^{11}B and ^{12}C fragments from the reaction ^{12}C + ^{12}C at 30 MeV/A.

The BCS energy spectra for these particles (^{6}Li, ^{9}Be, ^{11}B, ^{12}C) are shown in figure 11. The dynamic range for the fragments is about 0.8 - 2.5 MeV/A.

The current signal of the Bragg Curve Spectrometer for α particles from a ^{228}Th source is shown in Figure 12. The signal was amplified by a timing filter amplifier.

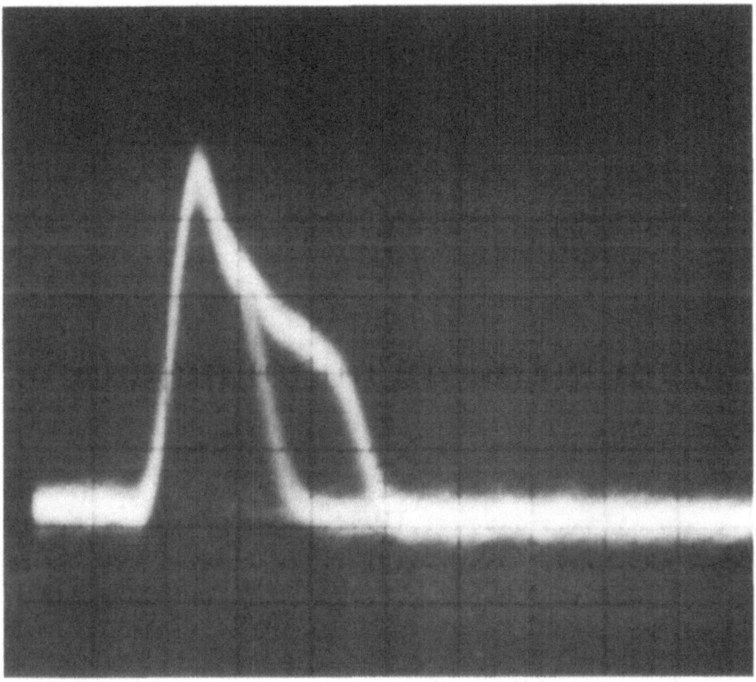

Figure 12 Bragg curves for α particles from a ^{228}Th source, as obsvered on an oscilloscope (amplitude 50 mV/div, timebase 1 μsec/div).

The Bragg curves for the two α - lines (8.78 and 6.09 MeV) are clearly visible. The second method of reading out the BCS utilizes this current signal. This signal is digitized by a flash encoder that integrates the charge in 30 - 100 nsec timebins. Again, the peak of the distribution can be related directly to the Z of the passing ion, and the integral of the curve is the total energy. A typical Bragg curve, measured with the flash ADC (set at 75 nsec) for the reaction Au(^{14}N,X) at 35 MeV/A, is shown in Figure 13.

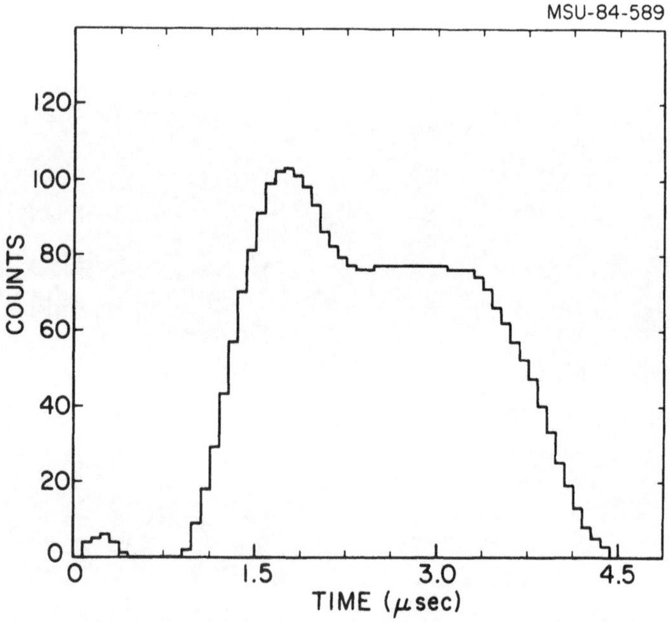

Figure 13 "Snapshot" of the BCS, i.e. the digitized current output as measured with the flash encoder for a ^{6}Li fragment.

5. Conclusions

A logarithmic detector for nuclear experiments with up to about 200 MeV/nucleon beam energies has been built. It consists of a stack of a low-pressure MWPC for slow and heavy fragments, a Bragg Curve Spectrometer for intermediate energy fragments and scintillator telescopes for light particles.

This detector array is a hexagonal shaped cone and is a prototype for a multiparticle array with a solid angle close to 4π.

Results of test experiments with the hexagonal Bragg Curve Spectrometer are reported in this paper. The detector worked satisfactorily, and unit charge resolution has so far been observed for $1<Z<18$.

References

1. B.V. Jacak, G.D. Westfall, C.K. Gelbke, L.H. Harwood, W.G. Lynch, D.K. Scott, H. Stöcker, M.B. Tsang, and T.J.M. Symons, Phys. Rev. Lett. 51, (1983) 1846.
2. C.R. Gruhn, M. Bimini, R. Legrain, R. Loveman, W. Pang, M. Roach, D.K. Scott, A. Shotter, T.J. Symons, J. Wouters, M. Zisman, R. deVries, Y.C. Peng and W. Sondheim, Nucl. Instr. Meth. 196 (1982) 33.
3. Ch. Schiessl, W. Wagner, K. Hartel, P. Kienle, H.J. Korner, W. Mayer and K.E. Rehm, Nucl. Instr. Meth. 192 (1982) 291.
4. J.M. Asselineau, J. Duchon, M.L. Haridon, P. Mosrin, R. Regimbart and B. Tamain, Nucl. Instr. Meth. 204 (1982) 109.
5. R.J. McDonald, L.G. Sobotka, Z.Q. Yao, G.J. Wozniak and G. Guarino, Nucl. Instr. Meth. 219 (1984) 508.

LOW MASS DETECTORS

A. BRESKIN
Department of Nuclear Physics
The Weizmann Institute of Science
76100 Rehovot, Israel

This article reviews the properties of several low-pressure gaseous detectors, multiwire chambers, multistep counters and the TREC (Tracking Range and Energy Chamber) developed during the last few years. They cover most of the particular requirements of low energy heavy ion physics, being capable of detecting and identifying particles over a broad range of specific ionizations - from heavily ionizing ions down to single electrons. Present detection devices are described and potential applications in the fields of minimum ionizing particles and energetic heavy ions, based on single electron counting, are discussed.

1. INTRODUCTION

After about two decades of intensive development, Nuclear Radiation Detectors based on ionization and charge collection or amplification in gaseous media have become frontier tools not only in particle physics but also in the field of nuclear physics and particularly in heavy ion research.

With modern accelerators and present day physics detector requirements have become more severe which is stimulating the ongoing development of new techniques. On the other hand, new developments in the field continuously stimulate new ideas and new applications in Physics. Various techniques are often combined for the efficient detection and identification of reaction products, capable of measuring with great accuracy parameters such as velocity, scattering angle, specific energy loss or total kinetic energy of one or several simultaneous particles. In the new field of relativistic heavy ion collisions, very similar detection problems to those encountered in particle physics arise, implying the use of the same techniques[1]. At lower energies, besides some dedicated heavy ion instruments such as large position-sensitive ionization chambers[2] or Bragg Curve Spectrometers[3], many devices are based on techniques commonly employed in high energy particle physics. Heavy particles are often identified by time-of-flight (TOF) techniques using localization and time measuring devices such as drift chambers[4], parallel plate avalanche counters[5,6], multiwire proportional chambers (MWPC)[7] and recently multistep chambers (MSC)[8]. These detectors have lost in size and pressure on their way to heavy ion applications, where low mass detectors are required.

The low pressure operation mechanism of detectors like MWPCs and MSCs is characterized by a rapid avalanche growth and high gain, resulting in excellent timing and imaging characteristics and in several other remarkable properties. They are applied over a broad spectrum of ionizing particles ranging from heavily ionizing ions down to single electrons. Due to their high sensitivity to single electrons they have regained interest in high energy physics where they are presently applied to UV photon detection in calorimetry and in Čerenkov ring imaging. Several applications to energetic heavy ions and minimum ionizing particles are foreseen. The properties and applications of low pressure MWPCs and MSCs will be reviewed in the present article.

I will present some recent results from the TREC, a low-pressure Tracking Range and Energy Chamber, designed for high efficiency heavy ion identification. It is a 4π instrument capable of three-dimensional track localization, and of measuring the range energy and sampled specific ionizations along the path of particles, arising from a central source or target, fully stopped in its sensitive volume.

2. LOW-PRESSURE MULTIWIRE PROPORTIONAL CHAMBERS (LPMWPC)

LPMWPCs are widely employed as accurate timing and localization devices in heavy ion research, atomic physics and recently for photo-electron imaging in high energy physics calorimetry. The rapid and particular avalanche formation mechanism that occurs in MWPCs operating at gas pressures as low as a few Torr is the source of the several attractive properties of these detectors. I will summarize the most important of them, referring the reader to previous works in this field, and present some recent applications.

2.1. Mechanism of Operation and Properties

Charges released by an ionizing particle in the sensitive volume, confined between the cathode planes and the anode wires, start an avalanche process already at their formation place due to the high reduced electric field strength, E/p, which is of the order of several hundreds of V/cm.Torr. A second amplification step occurs when the preamplified electron swarm reaches the anode wires. This combined mechanism[9)] enables very high gains to be reached in comparison to parallel plate avalanche chambers operating at similar conditions[10)]. Amplification factors of $10^5 - 10^6$ are reached at pressures as low as a few Torr of isobutane. The avalanche has an exponential growth with a dominant contribution from charges released near to the cathodes. The time resolution depends on the fluctuation in the position of these charges with respect to the cathodes. The time resolution depends also on the geometrical parameters and the detector size. Best results were achieved with detectors having 10μm diameter anode wires, 1 mm apart, and an anode-cathode spacing of about 3 mm. Resolutions obtained with a few cm^2 area counters, used as START detectors for Time-of-Flight experiments, are of the order of 100 ps (fwhm), as shown in fig. 1. Typical resolutions of larger, position-sensitive detectors having active areas of 80 cm^2 are of the order of

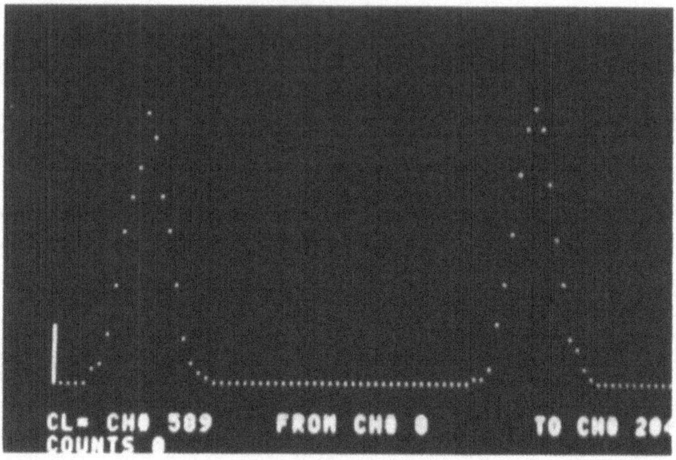

Fig. 1. Time resolution of a 4 cm^2 MWPC, measured with 160 MeV ^{80}Br ions, against a surface barrier detector p=2 Torr of isobutane, anode potential 500 V. The two peaks are 1 ns apart. The intrinsic (corrected) time resolution is better than 100 ps (fwhm).

130 ps (fwhm). During the amplification stage, the growing electron cloud expands and reaches a lateral size of about 1-2 mm (fwhm), depending on the gas pressure, as shown in fig. 2[11]. Therefore, contrary to normal-pressure operation, the avalanche is not confined around a single wire but is spread over several of them. This allows a good position interpolation between neighbouring wires using anode or cathode induced pulses. The distribution of the cathode induced charges is quite narrow in comparison with normal pressure operation and reaches values of 2.4-1.6 mm (fwhm) for respective pressures of 1-20 Torr as shown in fig. 3. We have found that a very accurate position determination can be done when measuring charges induced on cathode wires or strips, connected to commercial tapped delay lines. A resolution of about 100 μm (fwhm) was obtained with fission fragments and a 8x10 cm^2 detector, as shown in fig. 4[11]. The resolution is of the same order in the two dimensions

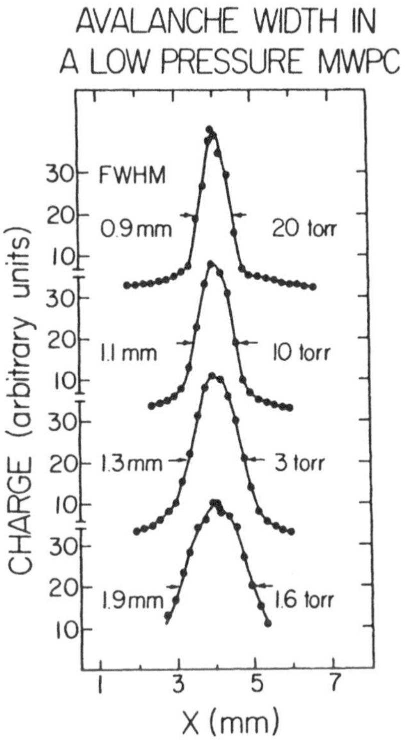

Fig. 2.

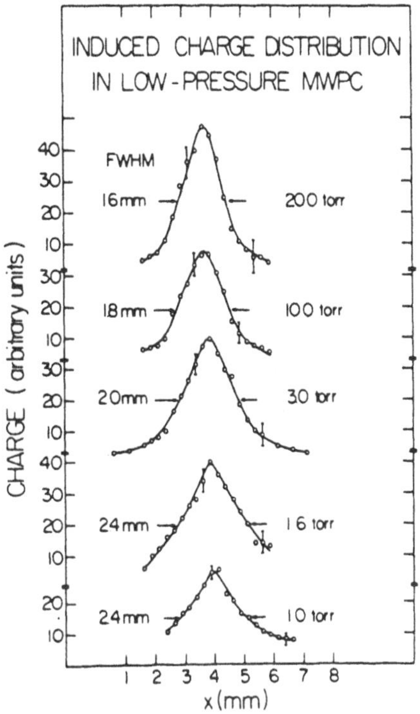

Fig. 3.

Fig. 2. Avalanche width in a MWPC operated with isobutane at a pressure range of 1.6-20 Torr, measured with a collimated α-source. Anode wires are of 10 μm dimater, 1 mm apart; the anode-cathode gap is of 3.2 mm.

Fig. 3. The width of the cathode-induced charge, measured at the conditions quoted in fig. 2.

parallel and orthogonal to the anode wires, which demonstrates the efficient interpolation between the anode wires. The overall resolution and linearity are visible in fig. 5 showing the image of a mask placed in front of the detector. The image contains the whole raw data and shows no parasitic background effects.

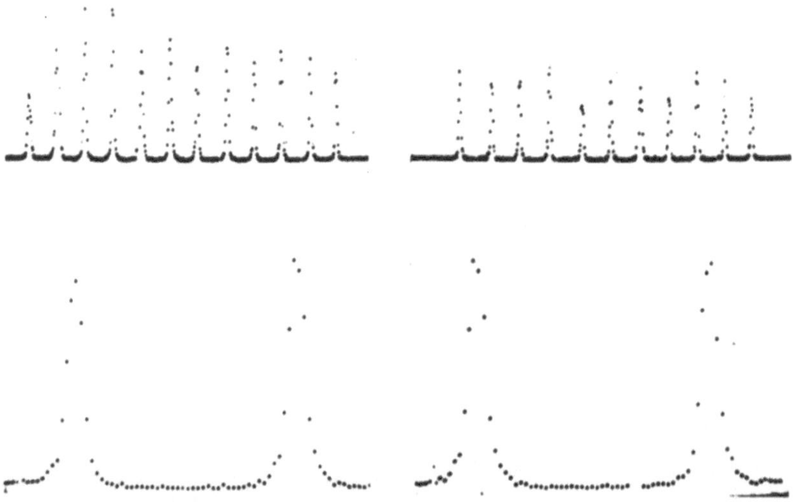

Fig. 4. Position resolution of an 80x100 mm ^{2} MWPC measured with a ^{252}Cf source and a set of 150 μm wide slits, 2.1 mm apart. p=3 Torr, anode potential 560 V. On the left, the resolution in a direction perpendicular to the anode wires (interpolation between wires); on the right, the resolution along the wires. The fwhm of the distribution is of the order of 180 μm in both coordinates. The intrinsic resolution is of about 100 μm (fwhm). Data accumulated over 12 h.

The relatively narrow distribution of the cathode induced charge offers the possibility of detecting several simultaneous particles with a good double track resolution. This can be achieved by correlating direct and induced signals recorded from anode wires and cathode strips running in various directions. Other read-out methods can be used in such cases. A digital read-out of each wire or strip will provide coordinates correlation via avalanche cluster size but with rather moderate accuracy. An analog method consisting of the direct computation of the centre of gravity of direct and induced charges[12] will provide a better coordinate correlation and an excellent position resolution but at a rather high cost. The individual anode wire signals or signals from groups of wires can be used for fast timing purposes. These detectors are very slightly sensitive to background radiation, especialy that of electrons or x-rays emerging from the target, and which often limit the detector gain. They can therefore operate efficiently as START detectors at the vicinity of the target.

Detectors of this type can operate at very high counting rates. Due to high values of the reduced electric field, positive ions produced during the avalanche process are collected within about 1 μs, more than an order of magnitude faster than at normal gas pressure. This fast collection time considerably reduces space charge effects, typical to avalanche formation, thus allowing a high rate operation to the order of 10^5c/s·mm^2 10).

Amplification factors to the order of 10^7 could be reached with single photoelectrons photo-produced on the detector cathodes[7)]. This property had triggered the application of low-pressure MWPCs to UV photon detection using "liquid" photocathodes and photosensitive gas media[13)].

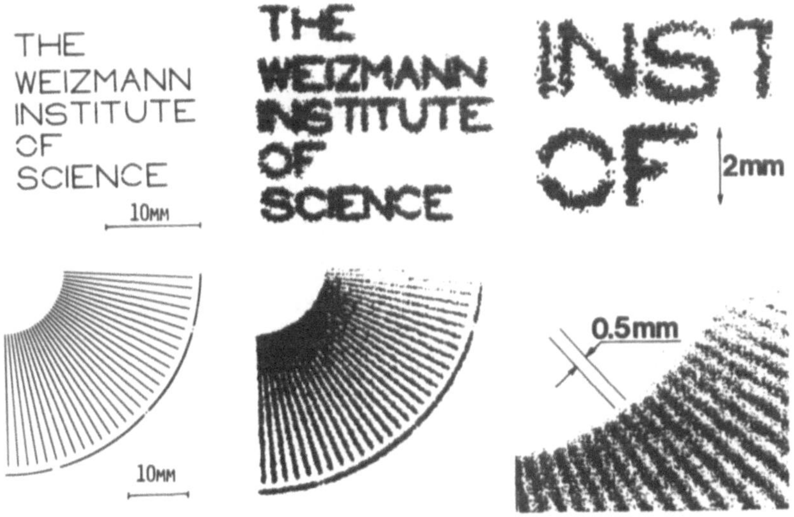

Fig. 5. A bidimensional image form the MWPC irradiated through two masks, shown on the left hand side of the figure. The line widths are of the order of 150-200 μm. 100 MeV ^{80}Br ions, p=3 Torr, HV=520 V.

2.2. Applications

A large variety of detectors having sensitive areas ranging from 2 to 1000 cm^2 and total thicknesses of 100-150 $\mu g/cm^2$ were built in the past few years at the Weizmann Institute and are successfully employed as timing or timing and imaging devices in various experimental set-ups. Main application fields are in heavy ion reactions, proton and heavy ion induced fission studies, Coulomb explosion of molecular ions and dating experiments using accelerator mass spectroscopy. One particular application, namely the search for exotic heavy relic particles from the Big Bang by observing anomalous backscattering of heavy ions (uranium or lead) from appropriate target samples[14)], has motivated the design of a large area detection system based on this technique. The system is composed of 18 position-sensitive and timing detectors covering the backward hemisphere of the target with a solid angle close to πStr (see fig.6). Six START detectors, 8x30 cm^2 MWPCs, placed in a semicircle at 20 cm from the target and twelve STOP detectors, 26x39 cm^2 MWPCs, placed in two rows at 75 cm from the target provide the velocity, specific energy loss and the

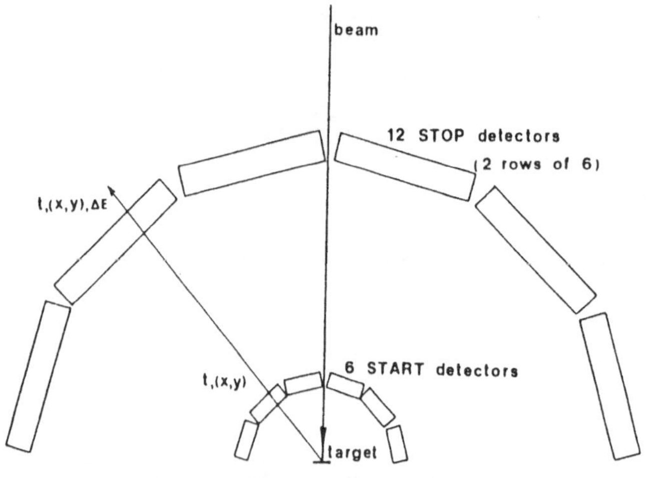

Fig. 6. A schematic view of the detection system designed for the search for heavy relic particles from the Big Bang.

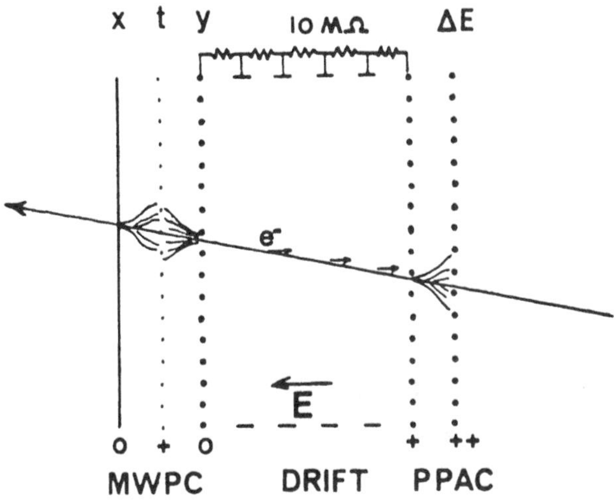

Fig. 7. A schematic view of the STOP detector. The MWPC provides time and position information. Electrons released across the drift volume are collected and multiplied in the PPAC to provide ΔE information. The two detector parts are independent.

scattering angle of the reaction products. The anode planes, made of 10 μm diameter wires, 1 mm apart, are subdivided into 4 (START detector) or 8 (STOP detector) timing sections to reduce capacitance effects. Cathode planes are made of 50 μm diameter wires 1 mm apart. Groups of 2-5 wires, depending on the detector coordinate, are connected to commercial tapped delay-lines to provide position information. The polypropylene window thickness in both detectors is 50-70 $\mu g/cm^2$; it is supported by 100 μm stainless steel wires 30 mm apart. The STOP detector, shown in fig. 7 has an additional and independent conversion drift space, 20 mm long, coupled to a PPAC stage for the measurement of specific energy loss. Full size prototypes of both detectors (see fig.8) have been constructed and tested with 120 MeV ^{58}Ni ions. They operated at a pressure of 2.5 Torr of isobutane with an anode potential of 480 V. The time resolution measured with independent timing sections is of about 250 ps (fwhm)

a)

b)

Fig. 8. Photographs of the two types of detectors used in the system shown in fig. 6. The START detector (8a) is of 8x30 cm^2. The STOP detector (8b) has a sensitive area of 26x39 cm^2. The window supporting wires and the integrated delay-lines are visible in the figures.

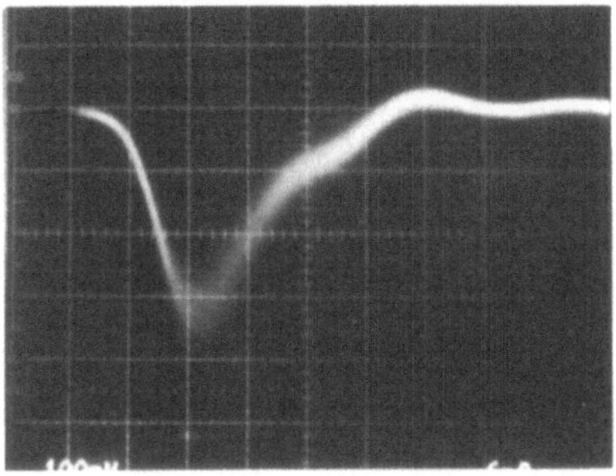

Fig. 9. Timing signals after a fast current amplifier (gain 200) from one cell (5x26 cm^2) of the STOP detector. The time scale is 5 ns/div. 120 MeV ^{58}Ni ions, 2.5 Torr, HV=450 V.

for both detectors. It deteriorates to about 350 ps (fwhm) when interconnecting two adjacent timing sections. Fig. 9 shows timing pulses from a single section of the STOP detector (26x5 cm^2). The position resolution is about 1 mm (fwhm) with 3 cathode wires connected to each delay-line top, it deteriorates to about 1.5 mm (fwhm) for 5 wire groups. Fig. 10 shows the energy loss distribution from the full detector area; a local resolution of about 10% (fwhm) is obtained with an operation potential of 500 V.

A totally different application of low pressure MWPCs is in the field of photoelectron detection. The imaging of UV photons is of great importance mainly in the domains of high energy physics and astronomy[15]. Several devices based on the detection of secondary electrons photo-produced on photosensitive surfaces[13,16] and gas media[17,18] were proposed. In the recent few years there has been a continuous effort in the development of a BaF_2 Scintillator coupled to a photosensitive wire chamber for eventual applications in high energy physics calorimetry and γ-ray detection[19,20]. The basic idea is of detecting the fast and short-wavelength component, peaking at 225 mm, of the high density BaF_2 Scintillator with a low-pressure MWPC[19].

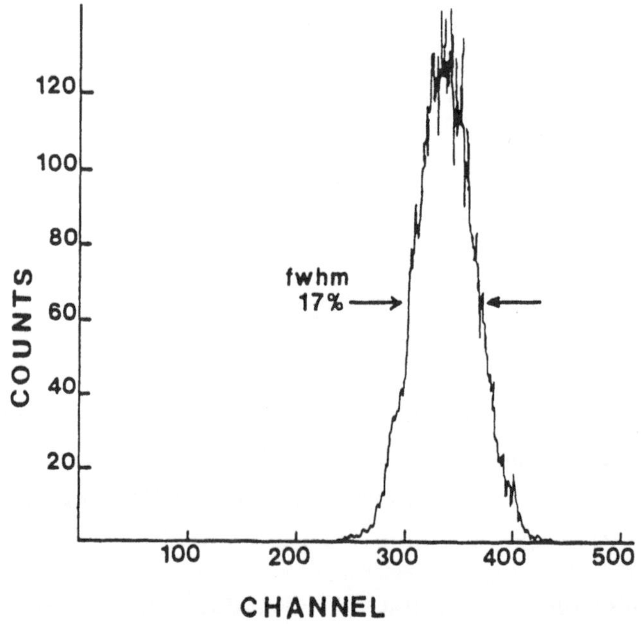

Fig. 10. Energy loss distribution of 120 MeV ^{58}Ni ions scattered from a 100 $\mu g/cm^2$ Ni target. The distribution is from the whole surface of the STOP detector. Drift field: 20 V/cm·Torr, HV of the PPAC=500 V.

The gas mixture contains TMAE (Tetrakis(dimethylamine) ethylene) as photosensitive material, having a low photo-ionization potential of 5.4 eV, and a low vapor pressure of 0.35 Torr at 20°C. UV photons from the scintillator can photo-ionize TMAE, used in a form of gas, liquid condensed on a cold cathode surface or an adhered surface. There are several advantages to detecting the photoelectrons with LPMWPC: an enhanced extraction efficiency from the photocathode surface, fast timing, high rate capability and reduced sensitivity to beam particles. A Test Calorimeter Tower has recently been built consisting of 14 BaF_2 crystals coupled to 15x15 cm^2 imaging MWPCs operated at 3 Torr of isobutane and TMAE vapor having a partial pressure of about 0.3 Torr. The module was tested with high energy electrons yielding interesting results as far as concerns energy resolution and the measure of the lateral shower development[20].

3. LOW-PRESSURE MULTISTEP CHAMBERS (LPMSC)

Multistep avalanche chambers[21)] are based on the preamplification of an initial charge and the transfer of the primary avalanche to a second stage for further amplification and position or time measurement. Such detectors have been proved to be successful detection devices at normal gas pressures for minimum ionizing particles at high counting rates due to the gating possibility on selected events[22)]. The high attainable gain led to their application to UV photon detection, using photosensitive gases, in Čerenkov ring imaging[18)]. An efficient transfer of the preamplified charge to a second amplifying stage can only occur if the initial avalanche has a substantial lateral size. At normal gas pressure some binary gas mixtures allow an extended photon-mediated avalanche formation. At low pressures, electron diffusion leads to wide avalanches[11)]. An efficient multistep amplification process can therefore occur even in pure organic gases as in isobutane at pressures of 2-20 Torr, with an increase in total gain of one to two orders of magnitude as compared to a single step detector[23,8)]. In addition to the high attainable gain, reaching values larger than 10^8, LPMSCs present several other interesting properties: excellent timing and imaging resolutions, wide dynamic range, high rate operation, gating possibility on events of interest, high efficiency for detecting low ionization particles and single electrons[24,25)], and capability of operation with photosensitive gases[26)]. These properties lead to numerous applications in detection of heavily ionizing particles, low ionization heavy and light particles and single photoelectrons. Various devices based on the efficient single electron detection are foreseen.

3.1. Application to Very Low Energy Heavy Ions

Heavy nuclei of very low kinetic energy are often produced in heavy ion reactions. For example, in fusion of light projectiles with heavy target nuclei at energies close to the Coulomb barrier, the evaporation residues are produced with a kinetic energy of less than 10 MeV. Conventional identification methods that consist of measuring ΔE-E, or time-of-flight and E cannot be applied for the study of the fusion process at this low energy range. Furthermore, in such cases the evaporation residues are concentrated in a very narrow cone around zero degree and are superimposed on a huge elastic scattering yield. We have found that a LPMSC is particularly suited for this application[8]. It can operate at a very low pressure, 1-2 Torr, and therefore with the thinnest foils available (40 $\mu g/cm^2$ of polypropylene in our case). The high gain enables efficient measuring of the time and impact position, without suffering from the high background of more than 10^6 elastically scattered ions per second.

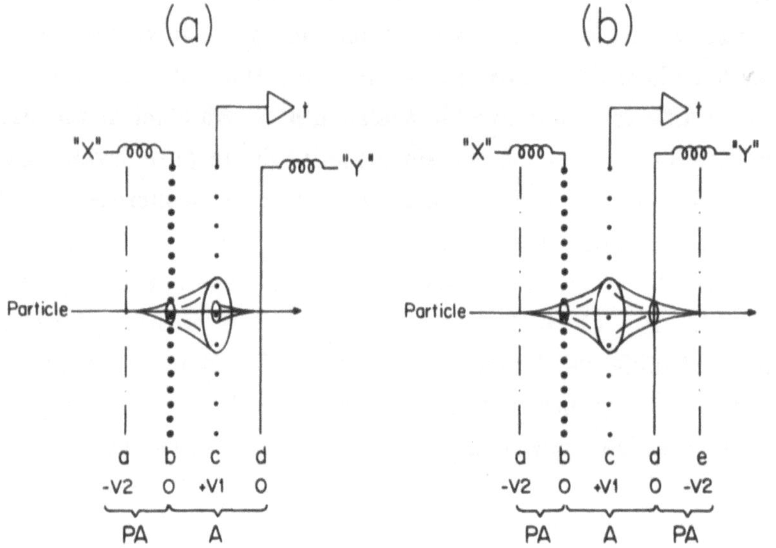

Fig. 11. A schematic view of the MSCs. a) A MWPC amplifier (A) with a single preamplification (PA) step. The primary charges deposited in the PA volume are preamplified, transferred to the MWPC and are further amplified. The contribution of the primary charges deposited in the MWPC is negligible. b) A MWPC with two PA steps. The total amplified charge is twice as high as in the first scheme.

The detector built for the present application having a sensitive area of 8x10 cm^2 is shown in fig. 11a. A preamplification stage (PA) operating in a parallel plate avalanche mode is directly coupled to a MWPC amplifying element. In the version shown in fig. 11b two preamplifying elements are used on both sides of the MWPC. Both versions perform equally well except for the doubling of the gain in in the second structure. It has been found that a transfer region between the elements, usually existing in normal pressure detectors of this type for the absorption of parasitic photons and the reduction of the number of back-drifting ions, can be omitted due to the efficient photon quenching by isobutane. The suppression of the transfer region enables better timing properties. It will be shown, however, in the next section that the transfer region is necessary when operating at gains higher than 10^8 for single electron detection. The cathode of the PA element, made of a stainless steel mesh, is held at a negative potential. The anode, which is also the front cathode of the MWPC as well as the rear cathode, are made of 50 μm diameter wires, 1 mm apart, held at ground potential. The anode of the MWPC is made of 10 μm diameter wires, 1 mm apart, held at a positive potential. Distances between adjacent electrodes are of 3.2 mm. Particles enter through a front window made of 40 $\mu g/cm^2$ stretched polypropylene supported by 100 μm stainless steel wires 20 mm apart. The cathode wires running at directions parallel or orthogonal to the anode wires are connected to tapped delay-lines to provide bidimensional localization. A ΔE element similar to that described in section 2.2 follows the timing and localization part. It was not used for this particular application.

Laboratory and beam tests made with heavily ionizing particles, have shown time and position resolutions of the same order as that obtained with MWPCs operated in similar conditions. Fig. 12 shows a time spectrum of the LPMSC, measured against a solid state detector, with a 120 MeV ^{58}Ni beam. A resolution of 150 ps (fwhm) is obtained with the MWPC stage operating without any preamplification (12a). The full multistep operation yields a resolution of 180 ps (fwhm)

a)
b)

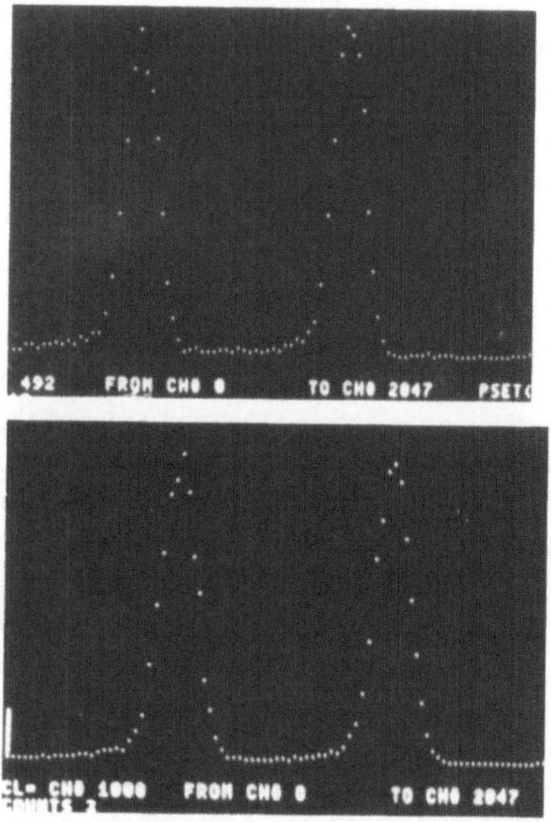

Fig. 12. Time resolution measured against a surface barrier detector. p=2 Torr, 120 MeV ^{58}Ni ions (1 ns between peaks). a) A single step MWPC, 150 ps (fwhm). b) The full MSC, 180 ps (fwhm) (not corrected for the contribution of the surface barrier detector).

(12b). In both cases the contribution of the solid state detector was not corrected for. Fig. 13 shows a slit image measured with the LPMSC operated with fission fragments. The intrinsic resolution is of the order of 100 μm (fwhm). The detector efficiency was determined using ^{160}Gd ions with kinetic energies in the range of 0.8-10 MeV. The experimental method is presented elsewhere[8]. Fig. 14 shows the measured efficiency as function of the energy. It is seen that the

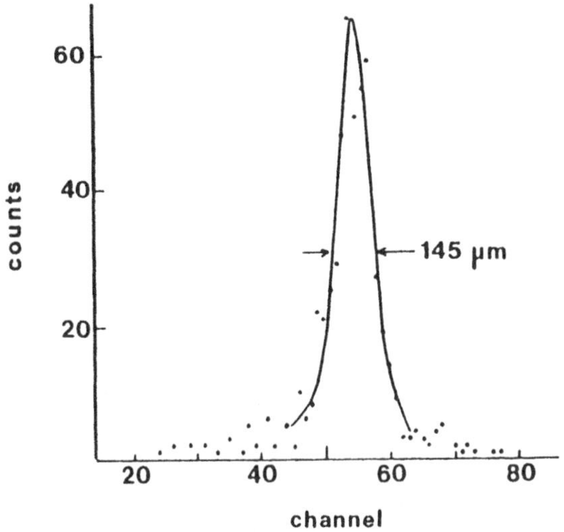

Fig. 13. Position resolution of the MSC of fig. 11a measured at 2 Torr with fission fragments and a set of fine slits in front of the detector. The slit's image on the first grid (electrode a) is 90 μm wide. The intrinsic resolution is of about 100 μm (fwhm). Data accumulated over 72 hours.

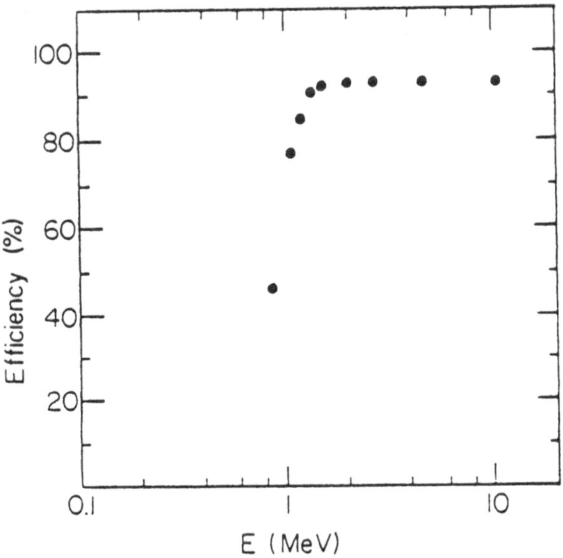

Fig. 14. Efficiency of the MSC for ^{160}Gd ions as function of their kinetic energy E.

detector is fully efficient for ^{160}Gd ions with kinetic energy down to 1.3 MeV. The limit of 93% results from the transparency of the first cathode grid. The efficiency drops to 50% at 0.85 MeV which is compatible with the calculated range of the ions in the entrance foil.

The detector was used for the study of sub-barrier fusion of oxygen ions on various tin isotopes at kinetic energies close to the Coulomb barrier[27]. The experimental set-up is shown in fig. 15. The LPMSC is centered at 0°, 30 cm from the target, in the plane perpendicular to the beam direction. A small tantalum beam stopper covering all angles $\theta \leq 1.2°$ is located in front of the detector. The MSC covers an angular range $\theta = 1.2°$-$9°$. The evaporation residues are identified only by their TOF measured against a pulsed beam of the 14UD Pelletron accelerator. An example of TOF plot is shown in fig. 16. The fusion peak has a fwhm of 6 ns, essentially due to the energy spread of the evaporation residues. The average rate on the

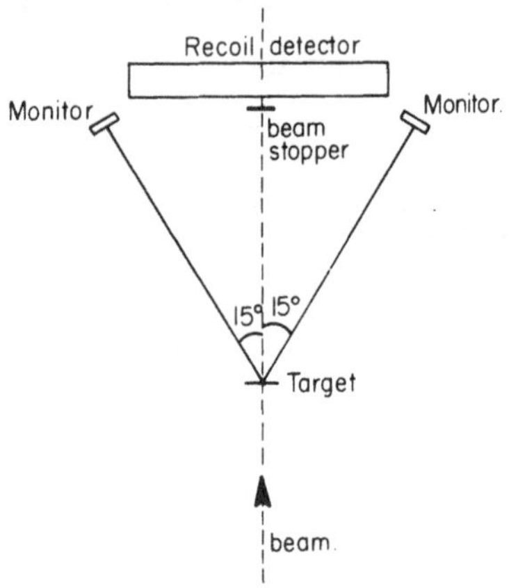

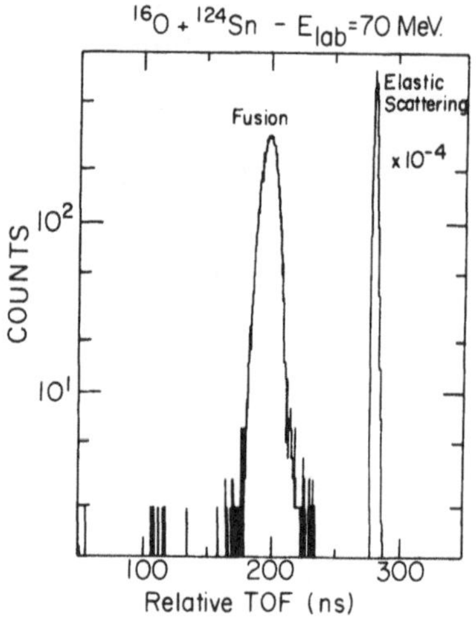

Fig. 15. Experimental set-up for the sub-barrier fusion measurements of oxygen ions on tin isotopes.

Fig. 16. An example of a time of flight (TOF) spectrum for ^{16}O + ^{124}Sn at incident energy $E_{lab} = 70$ MeV.

detector was of the order of $10^5 - 10^6$ particles per second, most of it suppressed electronically using a delayed coincidence between the pulsed beam signal and that of the detector. Though the specific energy loss of the elastically scattered light ions and that of the residues is of the same order of magnitude, the high rate did not cause any particular operation problems.

3.2 Detection of Low Energy Light Ions

Being sensitive to low ionization radiation LPMSCs may be successfully applied to the detection of low energy light ions. A particular application is in the field of Coulomb explosion of molecular ions[28]. There is a heavy demand for devices enabling the accurate time and position determination of several simultaneous fragments, in most cases protons and heavy nuclei emerging from the explosion of molecular ions passing through thin absorbers. They have nearly the same velocity but very different ionization densities. Several detectors based on drift chamber and parallel plate avalanche chamber techniques were successfully used during the past few years at our laboratory for the study of OH^+ and H_3^+ ions[29]. MWPCs are presently being used for the study of various few-atom molecules.

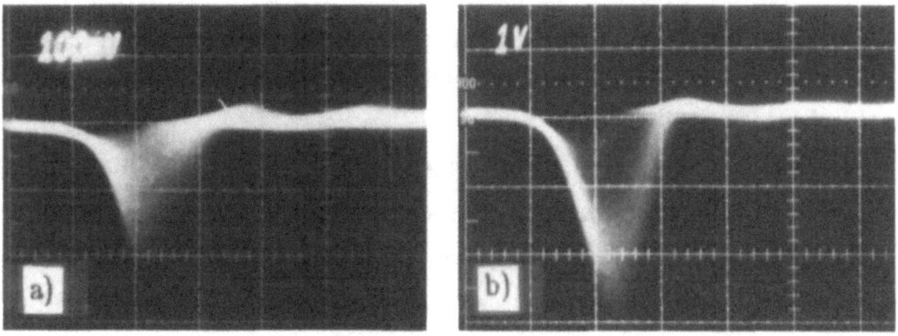

Fig. 17. Time signals from the MSC of fig. 11b after a current amplifier (gain 200) for 0.7 MeV protons (a) and 2 MeV carbon ions (b). p=3.2 Torr, $V_2 = -350V$, $V_1 = +340V$. 5 ns/div.

The advantage of using LPMSCs in this domain is the higher gain, enabling the simultaneous efficient detection of protons in the energy range of a few hundreds of KeV and a few MeV ions like carbon or oxygen, at the same operation conditions. Fig. 17 shows time pulses of 0.7 MeV protons and 2 MeV carbon ions measured at a pressure of 3 Torr of isobutane. A rather poor time resolution of the order of 1 ns (fwhm) was measured, and is entirely due to the inadequate response of the solid state detector used as a time reference. Another important advantage of a multistep detection structure is the possibility of employing several electrodes for the unambiguous position and time read-out of multiple events. A large detector of the type presented in fig. 11b is presently under construction at Argonne National Laboratory for the study of molecular ions. Cathode-induced charges are measured with a delay-line technique to provide position information while independent anode wires provide the time response. The system can efficiently detect up to 10 simultaneous particles[30].

3.3. Single Photoelectron Imaging

The importance of UV photon imaging was already discussed in section 2.2. An efficient detector based on the extraction of photoelectrons and their detection requires high quantum efficiency for the photoelectric effect, good photoelectron extraction efficiency, high amplification capability and often good localization properties. LPMSCs, having gains of the order of 10^8, present several advantages in this field of application.

The properties of LPMSCs were studied with photoelectrons emitted from a copper photocathode, directly coupled to detector, illuminated by an UV photon source[24]. A schematic view of the experimental set-up is shown in fig. 18. Photoelectrons emitted from a copper photocathode, coupled to the LPMSC, are directly multiplied in the PA region and the electron swarm is further transmitted to the MWPC amplifying element. Fig. 19 shows the total amplification curves obtained in this structure with photoelectrons. In the same figure are also shown results from a different amplifying structure -

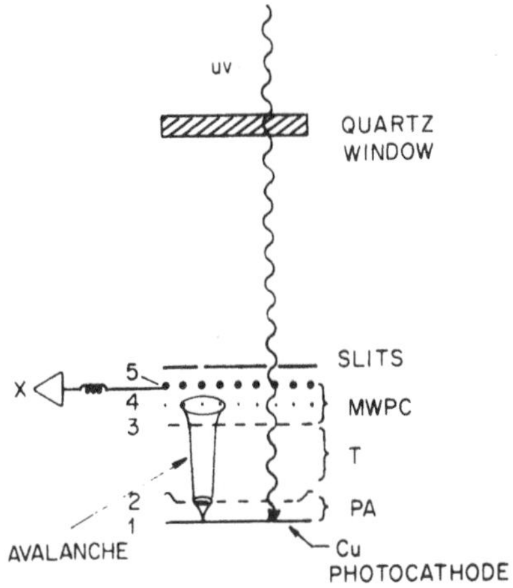

Fig. 18. **Schematic diagram of the MSC coupled to a copper photocathode. The two amplification stages are separated by an intermediate transfer region.**

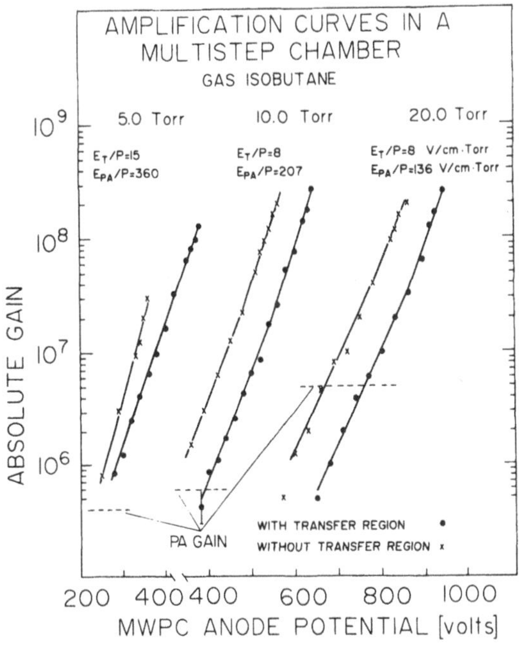

Fig. 19. **Amplification curves measured with single photoelectrons in a MSC operated with isobutane.**

without a transfer gap. It yields slightly lower gains, particularly at the low pressure range, due to secondary effects. The imaging capability was tested with a set of two parallel slits placed in front of the detector, illuminated with a point UV source. The image of the slits was measured using cathode-induced charges and a delay-line readout method. The measured position distribution is shown in fig. 20. The fwhm is of 260 μm corresponding to an intrinsic resolution of about 200 μm (fwhm), after deducing the contribution of the slit image. The resolution in both dimensions is of the same order. Fig. 21 shows a bidimensional image of a mask placed in front of the detector, demonstrating the linearity and the resolution. Due to the high gain the imaging was performed with about 99% of the electrons detected by the MWPC. Fig. 22 shows the pulse-height distributions of the total number of the detected electrons (a) and that of electrons

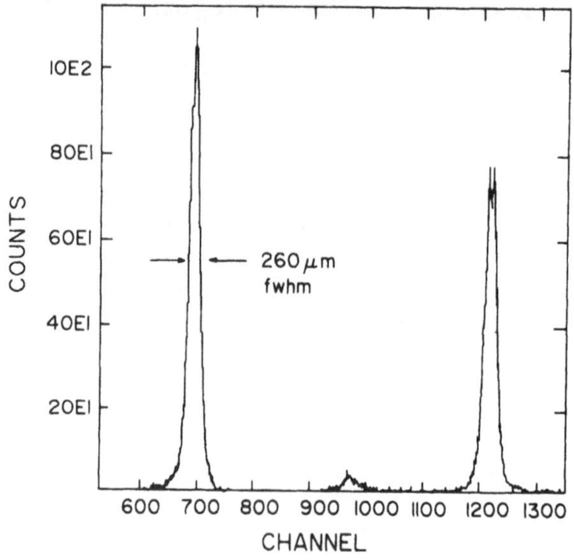

Fig. 20. Position distribution of photoelectrons traversing two slits 100μm wide, in front of the detector of fig. 18. 20 Torr of isobutane $V_1 = -1160$ V, $V_2 = -300$ V, $V_4 = +700$ V. The measured distribution has a width of 260 μm, i.e. intrinsic resolution of 210 μm (fwhm).

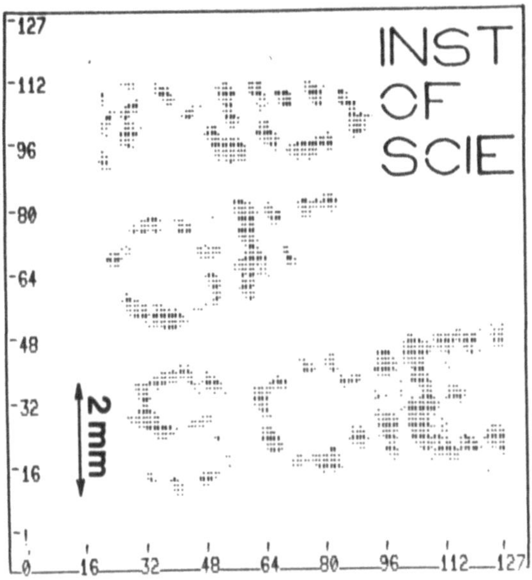

Fig. 21. A bidimensional image of a mask placed in front of an 80 cm^2 LPMSC irradiated with a UV source. The structure of this detector is similar to the one described in fig. 18 with the exception of the copper cathode being replaced by an 86% transparent wire mesh. The letters on the mask are 2 mm high and their width is 200 μm. (The width of the image on the cathode is of 250 μm). The granularity in the image of the letters is due to photoelectrons produced on the mesh.

for which the position was recorded (b). Only about 1% of the events is missing. As the preamplification and charge transfer mechanism was proved to be fully efficient at low gas pressures we may assume that practically all photoelectrons extracted from the cathode were measured.

It has been recently established that LPMSCs can operate efficiently with a photosensitive mixture of low-pressure isobutane and a few percent of TMAE[26)] (the vapor pressure of TMAE is about 0.3-0.7 Torr at temperatures of 20°-30°C). The maximum attainable

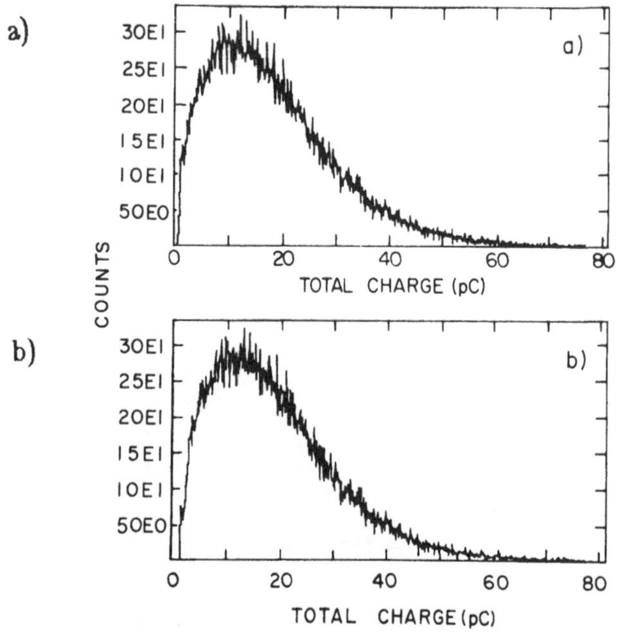

Fig. 22. Single photoelectron pulse-height distribution. at a pressure of 10 Torr, in a MSC with a transfer gap. The operating potentials are: $V_1 = -715$ V, $V_2 = -80$ V, $V_4 = +525$ V. The charge at the peak of the saturated distribution is of 12 pC, corresponding to a total gain of about $7 \cdot 10^7$. a) A distribution of all the electrons. b) A distribution of the electrons for which the position was recorded. Only about 1% of the total number of events is missing.

gain is about four times smaller than that obtained in pure isobutane at a pressure of 5 Torr. A stable operation was obtained in a pressure range of 5-20 Torr, using photoelectrons produced in the gas by an external UV source. Due to the relatively long absorption length of TMAE (about 20 mm at 20°C) the photoelectrons are produced in a "conversion region" coupled to the preamplification stage of the LPMSC. Only electrons photo-produced in this region are subject to the full double amplification process. Parasitic secondary avalanches arising mostly from photons produced in the head of the avalanche, could be totally suppressed at pressures above 10 Torr, by the incorporation of a transfer region of 30 mm length between the amplifying elements. The localization capability of such a photosensitive detector is expected to be similar to that measured in pure isobutane. A preliminary resolution of about 1 mm (fwhm) was obtained at a pressure of 10 Torr with a "wedge and strip" read-out method[31)] that may be interesting for recording multiple events. This type of photosensitive detector can be efficiently applied to Čerenkov ring imaging. It will have some considerable advantages over normal pressure MSCs presently being used for this purpose[32)]: low sensitivity to parasitic beam particles crossing the chamber, higher gains and thus higher detection efficiency and cheaper electronics, high localization accuracy, fast timing which allows the measurements of photons absorbed at various depths and thus a high multiplicity read-out[33)]. Further intensive investigations are presently being carried out at the Weizmann Institute and at FNAL in view of this application. LPMSCs may also improve the characteristics of BaF_2 Scintillators presently coupled to low-pressure photosensitive MWPCs[20)] for application in high energy physics calorimetry and γ-ray detection.

3.4. Further Applications with Single Electron Detectors

The high efficiency of detecting single electrons opens up the way to several potential applications. One may imagine a variety of devices based on the detection of primary or secondary electrons released by particles traversing gaseous or solid media. A few years ago a method was proposed for identifying minimum ionizing particles by counting the ionization clusters created in the gas[34)]. This technique encountered many technical problems due to close distances between successive clusters, even when expanding them by diffusion at low electric fields. We are presently investigating the cluster counting method using low-pressure media: electrons released in an ionization volume drift in a direction perpendicular to the track and are amplified by a LPMSC structure. The "cluster density" is given by the number of activated wires. Such a device can also provide the direction of the track. Another way of measuring the primary ionization is by drifting the ionization clusters in a direction parallel to the track and then counting successive electron avalanches formed in an endcup LPMSC. LPMSCs can also be applied in the same domain as high rate imaging devices for multiparticle detection when preceded by a collection drift region. One or two primary electrons are sufficient for an efficient detection. Furthermore, electrons arising from different particle tracks will be released at various depths. The timing of the successive avalanches may efficiently solve ambiguity problems related to the imaging of many-particle events. The idea of coupling sensing elements based on gaseous detectors to secondary emission electrodes[35)] or to proper photocathodes[16)], could have an important impact in the domain of particle and photon detection. It was previously suggested that low-pressure MWPCs could be used for such applications[7)]. A method consisting of the extraction of secondary electrons in vacuum from low density CsI layers traversed by charged particles and accelerating the electrons through a thin window foil into a LPMSC for accurate time and coordinate determination is presently under investigation[36)]. This technique can also be applied to vacuum photocathodes coupled to LPMSC sensing devices. Heavy particles traversing proper target foils yield a large number of secondary electrons, in proportion to their energy loss in the foil[37)]. Timing devices based on the amplification of such electrons with channel plate detectors are often used in heavy ion physics[38)]. Electrons emitted by very low ionization heavy ions can be measured in LPMSCs directly coupled to the emission foil, providing a new potential tool for timing and imaging of very high or very low energy heavy ions.

4. THE TREC

4.1. Mechanism of Operation

The Tracking Range and Energy Chamber (TREC) is a new concept in heavy ion detection[39]. It is a 4π three-dimensional imaging proportional detector of the Time Projection Chamber (TPC)[40] type. It is capable of following tracks of particles arising from an internal source (fission fragments) or from a target placed at its center (reaction products). The particles are fully stopped in the sensitive volume filled with a low-pressure gas, adapted to each particular application. The parameters measured are the track direction, total range in the gas, total energy and the sampled specific ionization along the track (Bragg curve). A schematic view of the detector is shown in fig. 23. Particles are emitted from a central source or a target and are stopped in a "drift-volume". Ionization electrons produced along their trajectories drift towards end-cap MWPCs and are amplified. A direct charge pulse is produced on the anode wires while pulses of an opposite polarity are induced on the two cathodes. The various parameters are measured as follows:

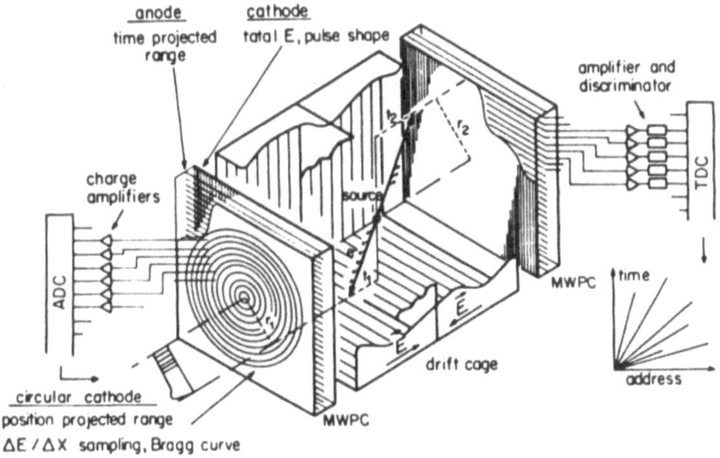

Fig. 23. A schematic view of the TREC principle and its electronics.

- The total energy E is obtained by integrating the charge signal induced on the fromt MWPC cathode.
- The time projected range (t) is recorded from the arrival time of successive electrons, produced along the track, at the anode wires. Signals from individual wires are amplified, shaped and digitized by a system of TDCs. The output of the TDCs versus the wire address allows one to follow each track and to measure the total time projected range (t).
- The position projected range (r) is recorded from the Bragg-like distribution of charge pulses induced on the rear cathode, made of concentric rings on a printed board. Pulses from successive rings are read out through charge-sensitive amplifiers and are digitized by a system of ADCs. The output of the ADCs versus the ring address represents a Bragg Curve, the length of which is the position projected range (r). The total range and the angle of emission are obtained from both r and t.
- The ADC output from each ring provides a sample of the specific ionization over a solid angle of 2π. The surface under the Bragg Curve provides the total energy as well. Other "Bragg Curve parameters" can be measured to provide an efficient particle identification as proposed by Gruhn[3)].

4.2. Applications and Properties

Detectors of this type were designed and built for two different applications. One is used in the search for Heavy Relic particles from the Big Bang by studying the spontaneous fission of ^{252}Cf [41)]. The experiment consists of identifying fission fragments, arising from a source placed in the center of the detector, by measuring their energy and range and searching for anomalous, single fission events that might indicate the presence of a heavy particle inside the fissioning nucleus. The second detector was designed for the detection of fission fragments and heavy recoils in studying heavy ion reactions of medium mass projectiles on heavy targets, in the energy range of 20-200 MeV/u. In this case, the ^{252}Cf source is replaced by a target,

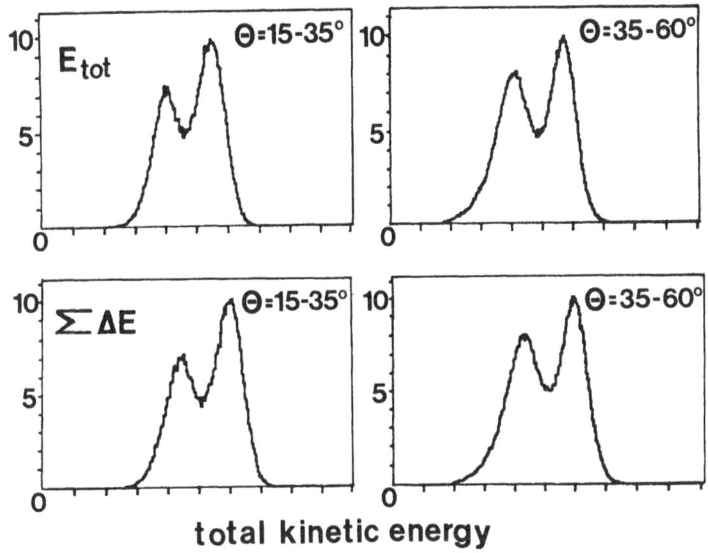

Fig. 24. Energy distributions of ^{252}Cf fission fragments stopped in the TREC volume, emitted at two angular ranges. Two methods of measurement: a) Total induced charge on the front cathode. b) The sum of charges induced on individual circular strips of the rear cathode (surface under the Bragg curve).

and the accelerator beam passes directly through the detector gas. The idea is to have a high efficiency, close to 4π, detector in which particles emerging from the target will be detected without traversing any gas window. It is of particular importance when detecting low energy heavy recoils.

Energy spectra of fission fragments from a ^{252}Cf source placed at the center of the detector are shown in fig. 24 for two angular ranges. The detector is operated at a pressure of 80 Torr of isobutane in a proportional mode with a low gain of about 100, which is sufficient in the present case. The upper part of the figure shows spectra of the total charge measured on the front cathode of the MWPC. The lower spectra are that of the sum of the sampled charges induced on individual radial strips of the rear cathode, thus the area

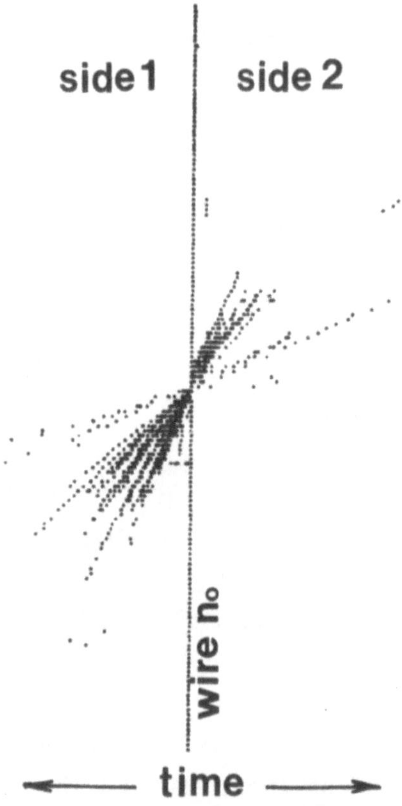

Fig. 25. Example of tracks of fission fragments emitted from a central source into the two sides of the TREC (see fig.23). The tracks are recorded from the arival times of successive electrons to the anode wires of the MWPC. The total time projected range can be measured.

under the Bragg curve. The two distributions have practically the same shape. There is a certain angular correlation, especially at very forward angles, to the measured energy due to space charge effects, that can be corrected.

Fission fragment tracks stopped in the detector volume on both sides of the source are shown in fig. 25. The maximum range of the fission fragments at the present operation conditions is about 60 mm. As can be seen from this projection in time on the anode wires, tracks

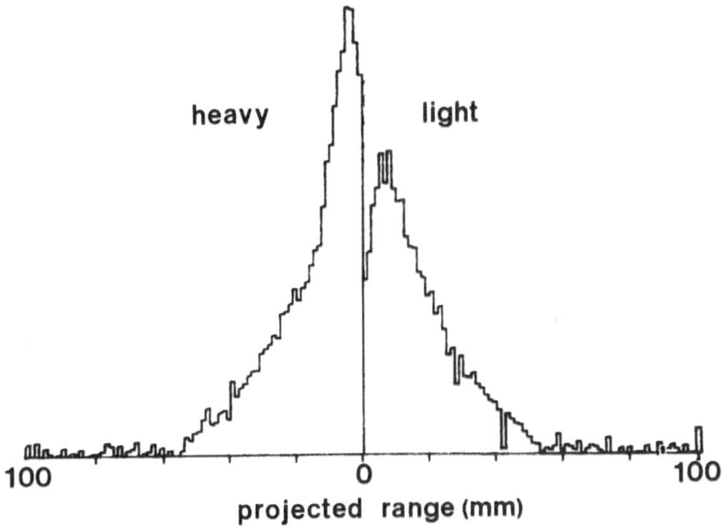

Fig. 26. Bragg curves recorded for the two fragments of a fissioning ^{252}Cf nucleus. The information of each channel corresponds to a sampled charge induced on a cathode strip. The length of the distribution corresponds to the position projected range.

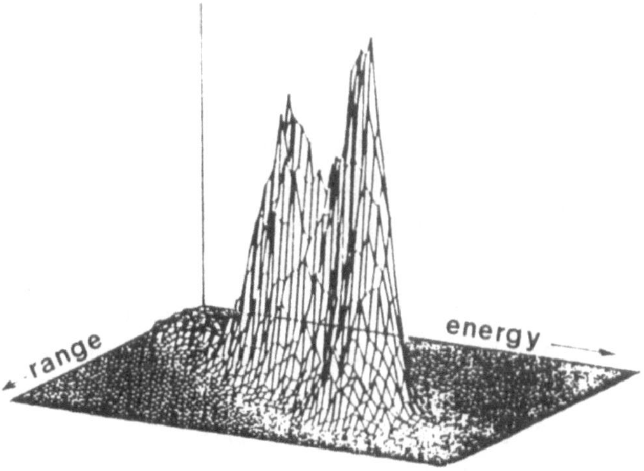

Fig. 27. Three-dimensional plot of the energy versus total range of fission fragments. The absolute range is calculated from the measured time and position projected ranges.

can be followed, wire by wire, until their origin and the total time projected range (t) can be determined.

Typical Bragg curves for a single event are shown in fig. 26. Each channel on the abscissa correspond to the charge induced on one circular strip of the rear cathode. The position projected range (r) of each of the fragments and their total energies can be determined from this figure.

The total range and the emission angle of each track are calculated from the values of the time and position projected ranges. A plot of the total range versus total kinetic energy of fission fragments emitted over the whole solid angle (close to 2π) is shown in fig. 27.

In view of direct beam applications with energetic heavy ions we have replaced the central fission source by an uranium target. A beam of 20 MeV protons from the Pelletron accelerator of the Weizmann Institute was used to induce fission of uranium. The proton beam, collimated to a diameter of 2 mm could penetrate the detector volume through thin Ni foils and was traversing the gas volume at a rate of about $10^{11} - 10^{11}$ protons per second. A ceramic tube (Al_2O_3) of 5 mm diameter was inserted in the drift volume, traversing the endcup MWPC at its center and ending at a distance of about 10 mm from the target, to reduce eventual secondary effects. Construction details will be given elsewhere[42]. The detector was operated at a pressure of 50 Torr of isobutane. An energy spectrum of uranium fission fragments is shown in fig. 28. Despite the high direct beam rate the detector has shown a stable operation and did not suffer from any particular background problems. The detector is presently tested with a reduced ceramic tube length, covering only the width of the MWPC, in order to increase the acceptance at forward angles.

4π TREC detectors seem to have several promising applications in the domain of nuclear physics as versatile and flexible multiparameter tools for the identification of reaction products. Being able to operate at variable pressures the same instrument can be adapted to particular problems of detecting particles over a broad range of energies and specific ionizations.

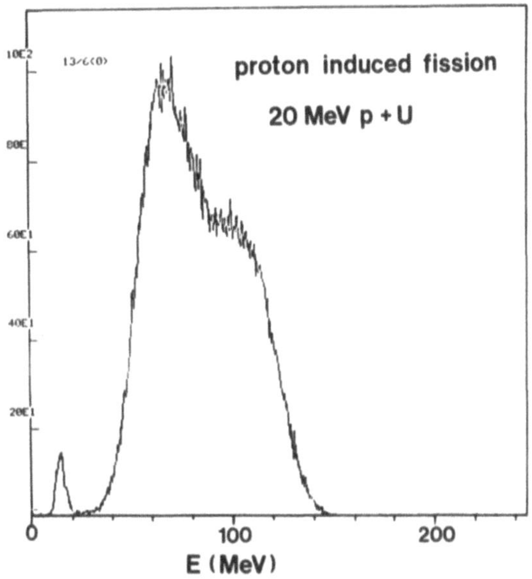

Fig. 28. Energy distribution of fission fragments from an uranium target in a proton-induced fission experiment. The direct proton beam passes through the gas volume at a rate of $10^{10} - 10^{11}$ protons per second.

5. CONCLUSION

Various low mass detectors described here have applications over a broad range of charged particles, from heavily ionizing ions down to single electrons. Among the properties of low pressure timing and imaging detectors such as multiwire chambers and multistep counters are high gain, rapidity, good localization, high rate capability, good multiparticle resolution and low sensitivity to background radiation. The high efficiency of low-pressure multistep chambers for the detection and imaging of single electrons have triggered several potential applications in particle and high energy ion physics, having the advantages of low pressure operation. Low energy heavy ions can be identified over solid angles close to 4π Str. by means of TREC detectors even in the case of low ionization, high intensity, direct beam passing through the sensitive gas volume.

REFERENCES

1. See for example proceedings of the workshop on Detectors for Relativistic Nuclear Collisions, edited by L. Schroeder, LBL-18225, UC-37, CONF-840 3137 (Sept. 1984).
2. H. Sann, these proceedings.
3. C.R. Gruhn et al., Nucl. Instr. Meth., 196, 33 (1982).
4. I. Tserruya, A. Breskin, R. Chechik, E. Duering, S. Kaplanis, N. Trautner and N. Zwang, Nucl. Instr. Meth., 196, 225 (1982).
5. A. Breskin and N. Zwang, Nucl. Instr. Meth., 146, 461 (1977).
6. Y. Eyal and H. Stelzer, Nucl. Instr. Meth., 155, 157 (1978).
7. A. Breskin, Nucl. Instr. Meth., 196, 11 (1982).
8. A. Breskin, R. Chechik, Z. Fraenkel, P. Jacobs, I. Tserruya and N. Zwang, Nucl. Instr. Meth., 221, 363 (1984).
9. A. Breskin, R. Chechik and N. Zwang, Nucl. Instr. Meth., 165, 125 (1979).
10. A. Breskin, R. Chechik and N. Zwang, IEEE Trans. Nucl. Sci. NS-27, 133 (1980).
11. A. Breskin, R. Chechik, I. Levin and N. Zwang, Nucl. Instr. Meth., 217, 107 (1983).
12. See for example A. Breskin, G. Charpak, C. Demierre, S. Majewski, A. Policarpo, F. Sauli and J.C. Santiard, Nucl. Instr. Meth., 143, 29 (1977).
13. D.F. Anderson, Phys. Lett., B118, 230 (1982).
14. M. Brugger, S. Polikanov, G. Herrmann, H. Overbeck, N. Trautmann, A. Breskin, R. Chechik, Z. Fraenkel and U. Smilansky, A Search for Superheavy Relic Particles from the Big Bang by Observing Anomalous Heavy Ion Scattering, GSI proposal No. 341, July 1984.
15. G. Charpak, in Detectors in Heavy Ion Physics, edited by W. von Oertzen, Lecture Notes in Physics (Springer-Verlag, Berlin, 1983), Vol. 178, p.10.
16. G. Charpak, W. Dominik, F. Ssuli and S. Majewski, IEEE Trans. Nucl. Sci. NS-30, 134 (1983).

17. J. Seguinot and Y. Ypsilantis, Nucl. Instr. Meth., 142, 377 (1977).
18. R. Bouclier et al., Nucl. Instr. Meth., 205, 403 (1983).
19. D.F. Anderson, R. Bouclier, G. Charpak, and S. Majewski, Nucl. Instr. Meth., 217, 217 (1983).
20. D.F. Anderson, G. Charpak, W. Kusmierz, P. Pavlopoulos and M. Suffert, CERN-EP/84-82, Nucl. Instr. Meth., in press.
21. See for example A. Breskin, G. Charpak, S. Majewski, G. Melchart, G. Petersen and F. Sauli, Nucl. Instr. Meth., 161, 19 (1979).
22. A. Breskin, G. Charpak, S. Majewski, G. Melchart, A. Peisert, F. Sauli, F. Mathy and G. Petersen, Nucl. Instr. Meth., 178, 11 (1980)
23. A. Breskin, G. Charpak and S. Majewski, Nucl. Instr. Meth., 220, 349 (1983).
24. A. Breskin and R. Chechik, WIS-84/6 Febr-PH, Nucl. Instr. Meth., in press.
25. A. Breskin and R. Chechik, Detection of Single Electrons and Low Ionization Radiation with Low-pressure Multistep Detectors, WIS-84/43 Oct.-PH, presented at the IEEE Nucl. Sci. Symp. Orlando, Oct. 1984, IEEE Trans. Nucl. Sci., in press.
26. W. Dominik, D. Anderson, A. Breskin, G. Charpak, F. Sauli and S. Majewski, Possible Applications of Low-pressure pressure Multistep Chambers to Čerenkov Ring Imaging, CERN-EP/84-29. A contribution to the Int. Conf. on Instr. for Colliding-Beam Physics (Novosibirsk, USSR, March 1984). To be published in the Conf. Proc.
27. I. Tserruya, P. Jacobs, A. Breskin, R. Chechik, Z. Fraenkel, S. Lantzmann, U. Smilansky and N. Zwang, Sub-barrier Fusion of O+Sn Using a New Detection System. Invited talk, Int. Conf. on Fusion Reactions Below the Coulomb Barrier (MIT, June, 1984), to be published in the Conf. Proc.
28. See for example D.S. Gemmel, Nucl. Instr. Meth., 194, 255 (1982).

29. A. Breskin, R. Chechik and N. Zwang, Ann. Israel Phys. Soc., 4, 254 (1981).
30. W. Koenig, E.P. Kanter, A. Faibis, Z. Vager and B.J. Zabransky, A Multiparticle 3D Imaging Technique to Study the Structure of Molecular Ions. A contribution to the Conf. on Application of Accelerators in Research and Industry, Denton, Nov. 1984.
31. See for example O.N.W. Siegmund, S. Clothier, J. Thornton, J. Lemen, R. Harper, I.M. Mason and J.C. Culhane, IEEE Trans. Nucl. Sci., NS-30, 503 (1983).
32. See for example M. Adams et al., Nucl. Instr. Meth., 217, 237 (1983).
33. G. Charpak and F. Sauli, Use of TMAE in a Multistep Chamber for Čerenkov Ring Imaging and Other Applications, CERN-EP/83-128, presented at the 2nd Piza meeting on Advanced Detectors, Castiglione della Pescara, June 1983.
34. A.H. Walenta, IEEE Trans. Nucl. Sci., NS-26, 73 (1979).
35. G. Charpak et al., CERN 78-05 (1978).
36. A collaboration work CERN-Weizmann Institute-Saclay.
37. See for example W. Meckbach, Beam-foil Spectroscopy, edited by L.A. Sellin and D.J. Pegg (Plenum Press, New York, 1976), Vol. 2, p.577 and references.
38. See for example J. Girard and M. Bolore, Nucl. Instr. Meth., 140, 279 (1979).
39. A. Breskin, T. Johansson, S. Polikanov and J.C. Santiard, Nucl. Instr. Meth., 217, 131 (1983).
40. See for example W.W. Allison, C.B. Brooks, J.N. Bunch, J.H. Cobb, J.B. Lloyd and R.W. Plenning, Nucl. Instr. Meth., 119, 499 (1974).
41. G. Barbiellini, A. Breskin, M. Červenka, R. Chechik, G. Hermann, T. Johansson, S. Polikanov and N. Trautmann, On Anomalous Fission as Signature for Heavy Relic Particles from the Big Bang, CERN-EP/83-09.
42. A. Breskin, R. Chechik, Z. Fraenkel, B. Hall and G. Maman, to be published.

PLASTIC SCINTILLATOR DETECTORS FOR THE STUDY OF TRANSFER AND BREAKUP REACTIONS AT INTERMEDIATE ENERGIES

H.R. Schmidt, M.Bantel[a), Y.D. Chan, S.B. Gazes,
R. Kamermans[b), C. Albiston, S. Wald[c) and R.G. Stokstad

a) Now at Max-Planck-Institut fuer Kernphysik, Germany
b) On leave from University of Utrecht, Netherlands
c) Now at Weizmann Institute of Science, Israel

Invited talk presented at the Conference on Instrumentation for Heavy-Ion Nuclear Research, Oak Ridge, October 22-24, 1984 by R.G. Stokstad

The detection of light particles associated with projectile like fragments can be used to separate transfer and breakup reactions provided the detectors cover a large solid angle. Three detection systems are described: (i) a 4π detector in the shape of a cube, 20 cm on a side, (ii) a X-Y position sensitive ΔE-E detector having an area of 20x20 cm^2, and (iii) a multi-element detector consisting of eight position sensitive strips. The latter two detectors are of the phoswich type having the thin element of NE102 (τ=2.5 ns) and the thick element of NE115 (τ=225 ns). The performance characteristics of the three detectors are described.

I INTRODUCTION

The study of heavy-ion reactions becomes increasingly complicated at higher bombarding energies. At energies above 10 MeV/nucleon, the assumption of a two-body final state is no longer valid. A heavy ion in the exit channel may result from different, competing reaction pathways. A ^{12}C ion produced, e.g., in the reaction ^{20}Ne + ^{197}Au could originate via the following processes:

i) transfer of ^{8}Be to the target: $^{197}Au(^{20}Ne,^{12}C)^{205}Bi^{*}$

ii) inelastic scattering followed by the decay of the ejectile: $^{197}Au(^{20}Ne,^{20}Ne^{*})$; $^{20}Ne^{*} \rightarrow {}^{12}C+2\alpha$

iii) transfer of one alpha particle to the target followed by decay of the ejectile: $^{197}Au(^{20}Ne,^{16}O^{*})$; $^{16}O^{*} \rightarrow {}^{12}C+\alpha$

iv) direct fragmentation: $^{20}Ne \rightarrow {}^{12}C+{}^{8}Be$

Reaction pathway i) is easily distinguished from ii)-iv) provided a 4π device enables rejection of all ^{12}C nuclei associated with light charged particles. A measurement of the energy, position, multiplicity and type of the light particles in coincidence with the heavy ion could, in principle, distinguish among reaction types ii) through iv).

Our first approach was to build a 4π array of plastic scintillators, which had the capability to detect charged particles with maximum efficiency and solid angle - the Plastic Box. It permits the above mentioned identification of reaction mechanism i), i.e., it separates transfer and breakup reactions. Later, the Plastic Box was upgraded by replacing individual elements of it by more elaborate detectors that provide more information on reactions ii) and iii). These detectors are referred to as Phoswich One and Two.

II THE PLASTIC BOX - THE POOR MAN'S 4π-DETECTOR

II.1 Design

The Plastic Box consists of six walls of plastic scintillator arranged in a 20 cm cube around the target (Fig. 1). Each of the walls is made up of two parallel scintillator sheets (NE102) of 1 mm thickness. The plastic sheets are coupled individually via adiabatic light guides to RCA 8850 and 8575 photomultiplier tubes. The scintillators are wrapped with 6 microns of

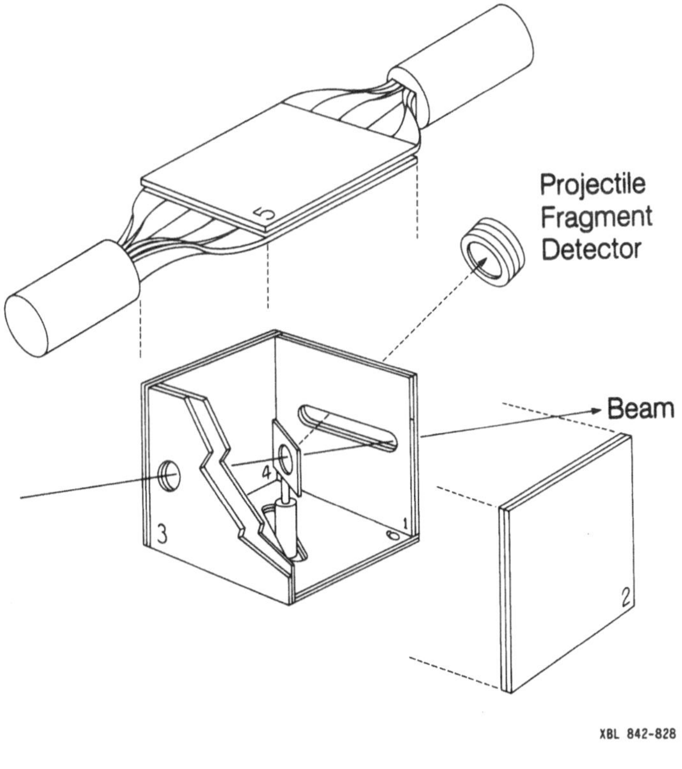

Fig. 1. Schematic diagram of the Plastic Box .

aluminized mylar. Although this is a neglible thickness for light particles, the mylar prevents the detection of heavy residues or fission fragments. The purpose of the double layered wall with an inner (A) and an outer (B) element is to distinguish neutral particles hitting the Plastic Box from charged particles. Because of the close geometry the two layers provide, at best, only a crude particle identification (see Fig. 2). For obvious reasons, several walls must have openings. The extent of the slot in wall number 1 allows a maximum angle of 21° for the trigger telescope. Part of the solid angle lost due to this slot was regained by using another plastic scintillator wall behind the telescope. With all detectors in place, the total solid angle subtended by this arrangement was approximately 97% of the full 4π.

II.2 Operation

Fig. 2 shows a two dimensional plot of the light collected with layer A versus that in layer B for an actual experiment. In tests where light particles were well-collimated, protons and alphas were easily resolved. Here this separation is considerably degraded. This is due both to the averaging over many angles of incidence as well as the position dependence of the light collection. Fig. 2 also shows the portion of neutral particles seen with the plastic scintillators: due to the small absolute efficiency of the 1 mm NE102 for neutrons and gamma rays, the amount of neutrals seen in either one of the two layers is the same. The counts on the vertical line in Figure 2, representing those particles that caused a signal in layer B but not in A (i.e. neutrals) give a measure of the contribution of neutrons and gamma rays. In each event, the above described features enable the unambiguous determination of the presence or absence of a charged particle in coincidence with a projectile-like fragment. For the reaction ^{20}Ne+^{197}Au at 341 MeV, those yields are presented in Figure 3, where the quantity S stands for number of walls which have fired. This number is, insofar as multiple hits of one and the same wall are ignored, a direct measure of the number of light charged particles emitted in coincidence with the observed ejectile of charge Z. A detailed discussion of the operation of the Plastic Box and the results for ^{20}Ne on ^{197}Au at 220 and 341 MeV incident energy can be found in Refs. 1 and 2.

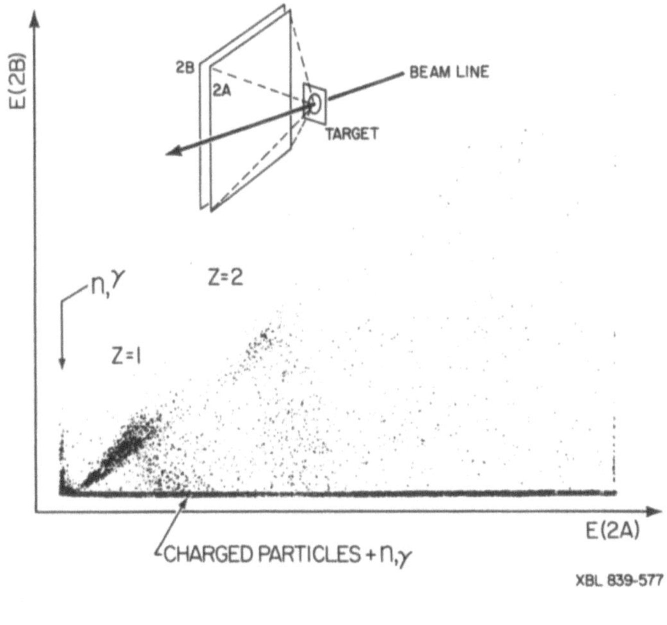

Fig. 2. A two-dimensional singles spectrum of the light output from the inner (A) and outer (B) paddles of the wall. Events appearing on the vertical axis are due to neutral particles.

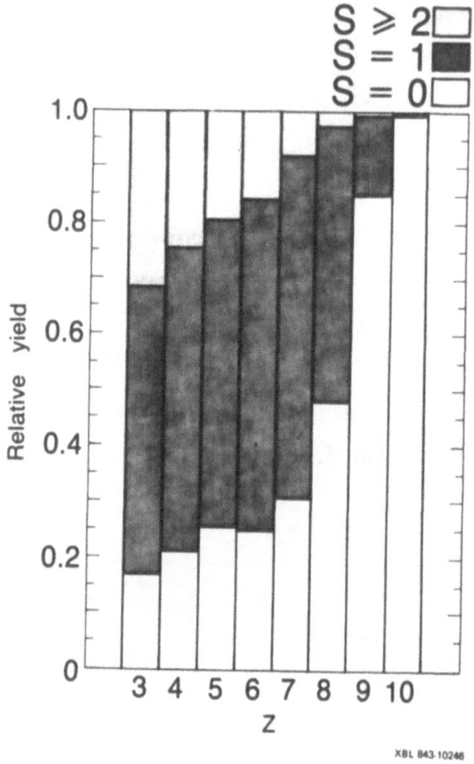

Fig. 3. The relative yield of ejectiles having none, one, or more than one scintillator wall in coincidence for an ejectile of charge Z, at a bombarding energy of 341 MeV.

III PHOSWICH ONE - ADDING PARTICLE IDENTIFICATION AND POSITION SENSITIVITY

III.1 Design

a) The Phoswich

The goal for Phoswich One was to obtain particle identification and position sensitivity and to do this in a size that could replace one of the Plastic Box elements. Fig. 4 shows a sketch of the Phoswich One detector. Particle identification was achieved employing the phoswich technique (Ref. 3). Previous phoswich detectors typically had CaF_2 (which has a long decay time) as the first element ΔE counter. However, production limitations and cost restrict the size of the inorganic crystal and, therefore, the area of the whole detector. To get around these limitations we used a combination of NE102 (decay time

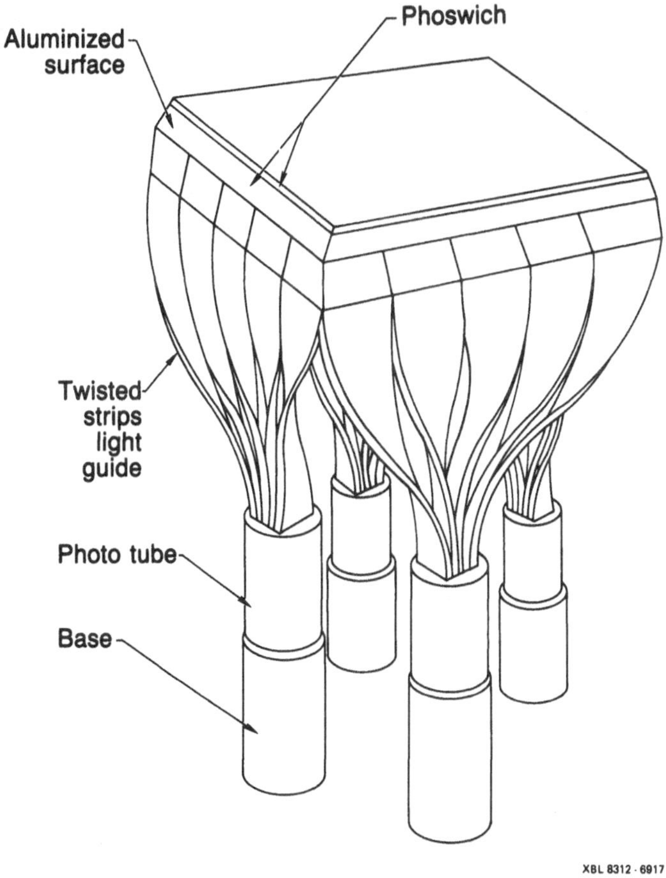

Fig. 4. A sketch of the entire Phoswich One detector, including light guides, phototubes and bases.

τ=2.5 ns) and a new plastic material NE115 which has a long decay time (τ=225 ns). In contrast to other detectors (Refs. 3,4) the fast plastic scintillator is the first element of the phoswich (0.5 mm thickness). Since the light output of NE102 is higher than that of NE115, a reduction of thickness for the ΔE element is possible and, most important, good timing properties are obtained even for particles that do not penetrate into the second element. The first layer stops protons with E < 6 MeV and alpha particles with E < 22 MeV, and still gives a reasonable signal for 40 MeV protons. The second layer (4.5 mm thickness) of NE115 stops protons up to 20 MeV and alpha particles up to 90 MeV. The two layers were glued together with optical cement only at the edges to avoid a dead layer over the entire detector. The light is read out via twisted lightguides at each of the four edges of quadratic layers and fed into RCA 8575 phototubes. A typical signal from the phototubes is illustrated in Fig. 5. The sharp peak is from the ΔE plastic and the tail from the NE115 backing. The combined signal is analyzed by applying a short and a long gate, respectively, as shown in Fig. 5. Results for the particle identification will be shown in the next section on Phoswich Two. A more detailed description of Phoswich One is found in Ref. 6.

b) Position Sensitivity

To achieve position sensitivity, we made use of the natural light loss occurring at reflection from the scintillator surface due to surface irregularities (Fig. 6). We observed that, for a sufficiently thin sheet of scintillator the number of reflections that the light has to undergo to travel to the edge of the sheet is large enough to result in a dependence of the collected light

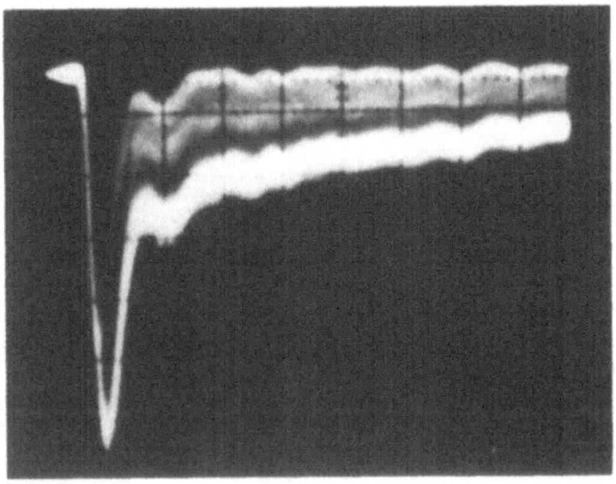

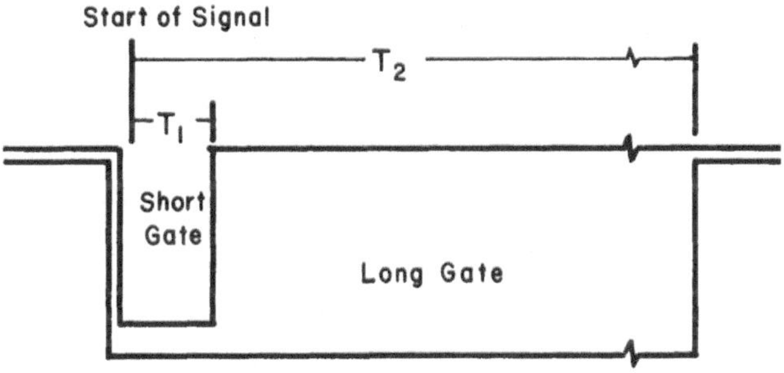

Fig. 5. Anode signals for reaction products from the reaction of 60 MeV alpha particles with Au nuclei. The gate signals used for the two QDC's were T_1 = 30 ns and T_2 = 600 ns.

on the distance of the light source to the photomultiplier. This behavior is shown in Fig. 7 together with Monte-Carlo simulations of light reflection from a partly irregular surface. Comparing the signals seen with the four phototubes allows the determination of the position where the particle hits the detector in both the X and Y directions. Fig. 8 shows the result of a test in which the detector was irradiated with scattered alpha-particles of 60 MeV beam energy. The figure

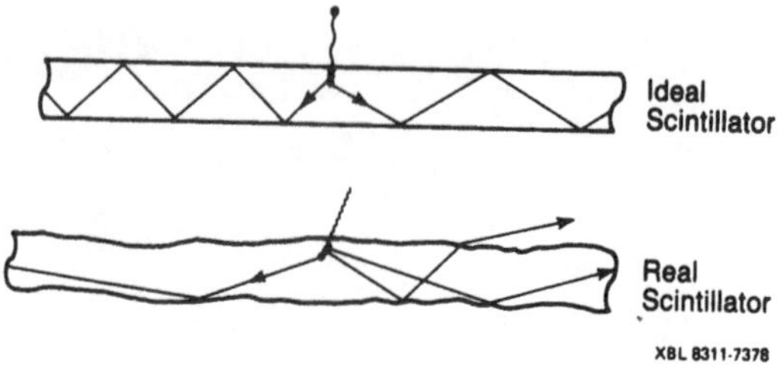

Fig. 6. Schematic illustration of light transmission by internal reflection in an ideal thin scintillator. The irregular surface of the real scintillator causes light loss for some of the reflections because the critical angle for internal reflection is exceeded.

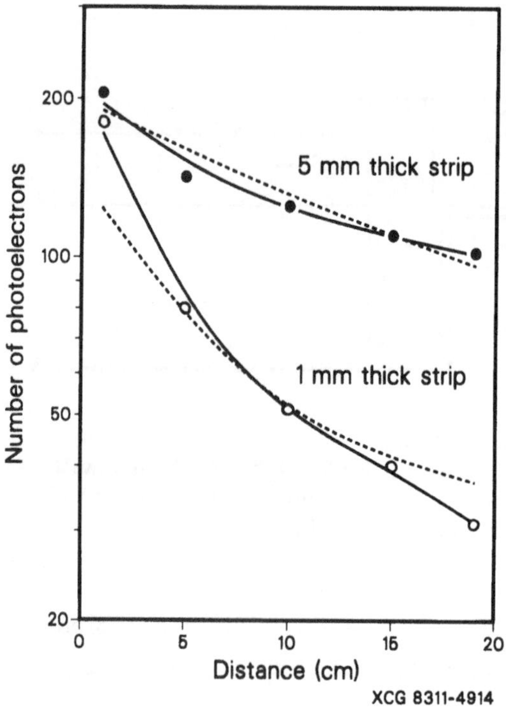

Fig. 7. The light collected (in units of photoelectrons) at one end of a thin strip 20 cm long and 2 cm wide. Results for two thicknesses, 1 mm and 5 mm are given as a function of source distance along the strip. Solid lines guide the eye to the experimental data points. The dashed lines are results of a calculation described in the text.

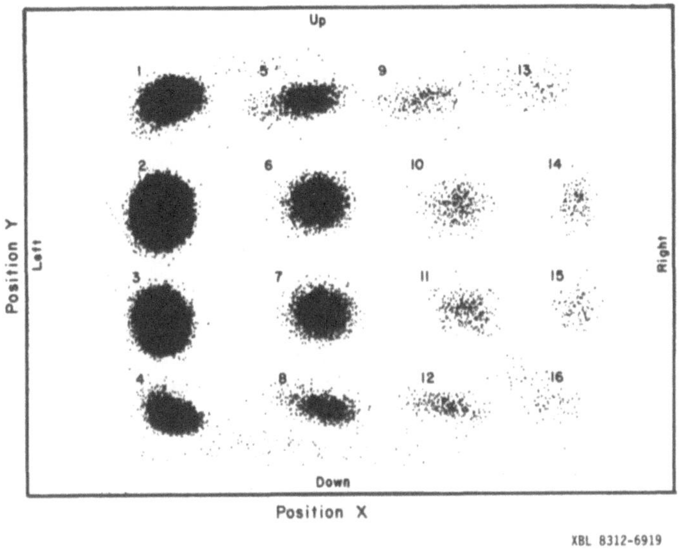

Fig. 8. A scatter plot showing the results obtained for alpha particles passing through the 16 holes of a mask. Holes 9-16 are beyond the grazing angle.

displays the image of a mask with 16 holes installed in front of the detector. Position and energy resolutions at 60 MeV were determined to be about 10 mm (FWHM) and 5-6% (FWHM), respectively. More details about reconstruction of position and energy will be presented in the next section.

IV PHOSWICH TWO - ADDING MULTIPLE HIT REGISTRATION

IV.1 Design

Although Phoswich One has several desirable properties, it still lacks the capability of multiple-hit registration. Therefore we gave up the continuous position information in one of the dimensions by segmenting the 20 by 20 cm^2 plane into 8 separate strips (Fig. 9). A strip has one phototube at each end, which allows continuous position determination along the strip. This design has the following major advantages: the detector can be operated at substantially higher integral count rates (which is of importance at forward angles where elastic scattering dominates), and it allows multiple hit registration.

IV.2 Operation

The strips of Phoswich Two have been tested using an 87 MeV alpha beam from the 88-Inch cylotron. The beam current was reduced sufficiently so that the direct beam could be used to bombard the strips. Moving the horizontally positioned strip through the beam determined the position response. For each position the 87 MeV alpha beam was degraded in energy with aluminum foils of different thickness, providing alpha energies of

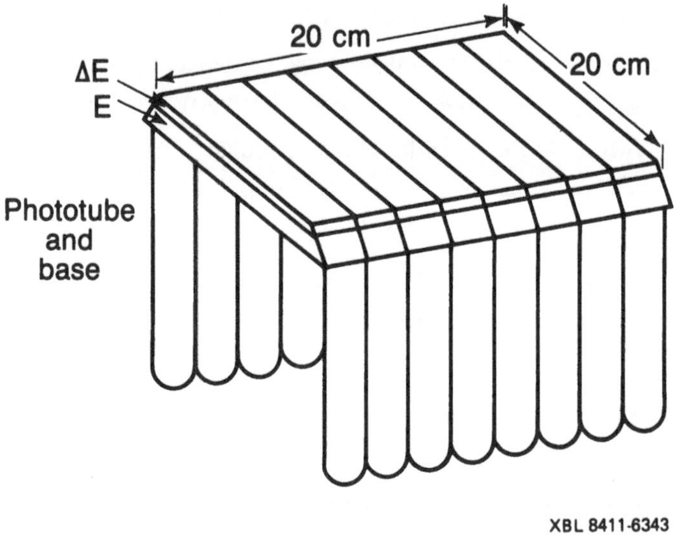

Fig. 9. A sketch of the segmented Phoswich Two detector.

87, 79, 70, 60, 49, 35.5 and 14.5 MeV. Fig. 10 shows the spectrum of these energies measured with a strip. The figure does not yield the intrinsic energy resolution of the the phoswich detector directly since, for all but the highest energy, the energy spread in the degrader foils has to be taken into account. After correction for energy straggling in the degrader foils (Ref. 5) the energy resolution was determined to be 2.2, 2.4, 2.7, 3.1, 3.7, 5.8 and 32.8 % (FWHM) for the various alpha energies. The lowest energy alpha particles (14.5 MeV) were stopped in the thin ΔE layer, which results in a poorer resolution due to the stronger light attenuation in the layer. As shown in Fig. 11, the energy resolution is independent of position. The position was reconstructed according to equation (2) below. The position resolution is energy dependent and - after correction for beam spread due to multiple scattering in the degrader foils - is 0.4, 0.6, 0.8, 1.0, 1.3, 1.6 and 4.2 cm for

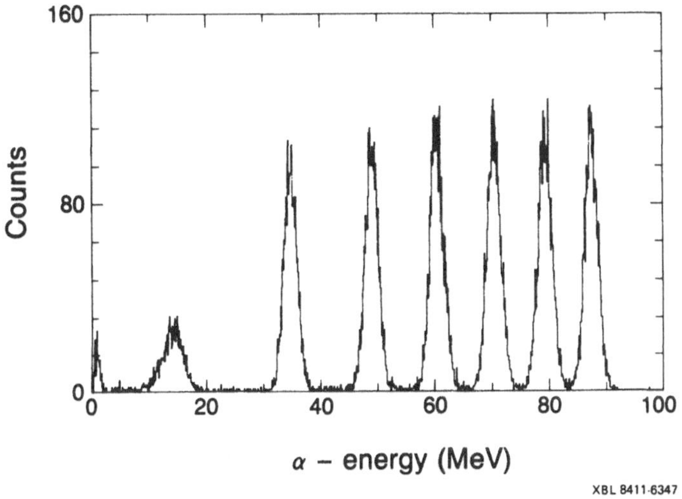

Fig. 10. Energy spectrum of alpha particles. The various alpha energies were generated with the help of aluminum degrader foils of different thickness.

the different alpha energies.

If N stands for the number of light quanta produced by the ionizing particle, the light output I measured by the phototubes at the left and right edges of the scintillator can be described empirically by

$$I_{left} = N \cdot \exp(-ax+bx^2) \qquad (1a)$$
$$I_{right} = N \cdot \exp(+ax+bx^2) \qquad (1b)$$

where x is the position on the strip defined as

$$x := (I_{left} - I_{right})/(I_{left} + I_{right}) \qquad (2)$$

Therefore, the energy $E \sim N^{0.7}$ is given by the relation

$$E \sim \left(\sqrt{I_{left} \cdot I_{right}/\exp(2bx^2)}\right)^{0.7} \qquad (3)$$

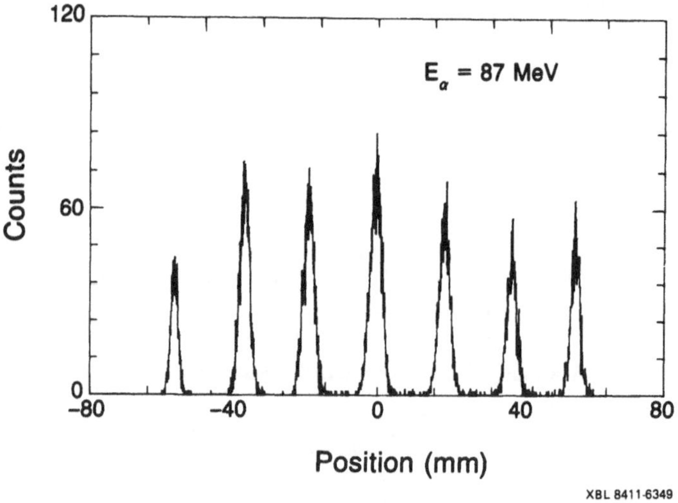

Fig. 11. Position spectrum of an 87 MeV alpha beam hitting a strip at different positions.

Finally, a test was performed by scattering alpha particles from a carbon target with the middle of the strip located at 43^{0} . The result of this test is illustrated by a plot of the energy loss in the first layer versus the total energy deposited in the scintillators. Two bands are clearly defined (Fig. 12) and represent alpha particles and hydrogen isotopes. The lower band contains mainly protons, but it shows also a separated group of deuterons and tritons. A projection of alpha particles on the energy axis is given in Fig. 13. It indicates the resolution of the ground state and the first excited state of the ^{12}C target.

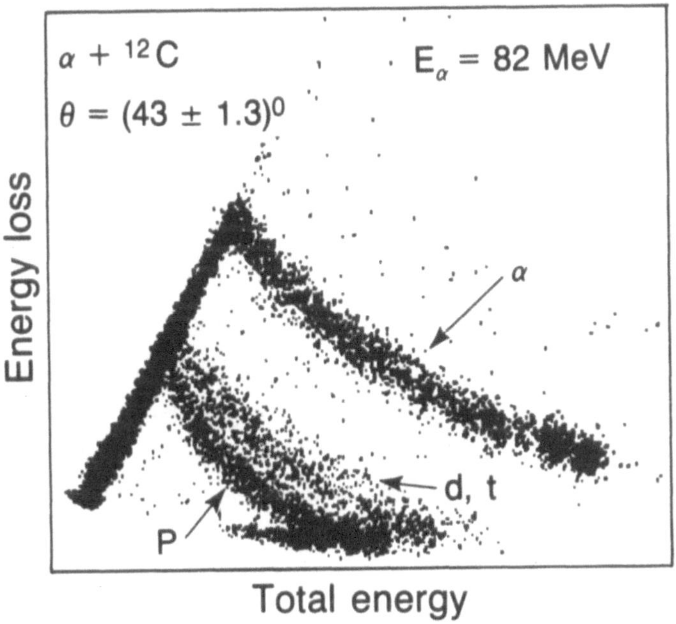

XBL 8411-6344

Fig. 12 ΔE-versus-E_{total} plot of the reaction of 87 MeV alpha particles with ^{12}C. The plot is generated by gating on a narrow region of the position spectrum of one strip of Phoswich One.

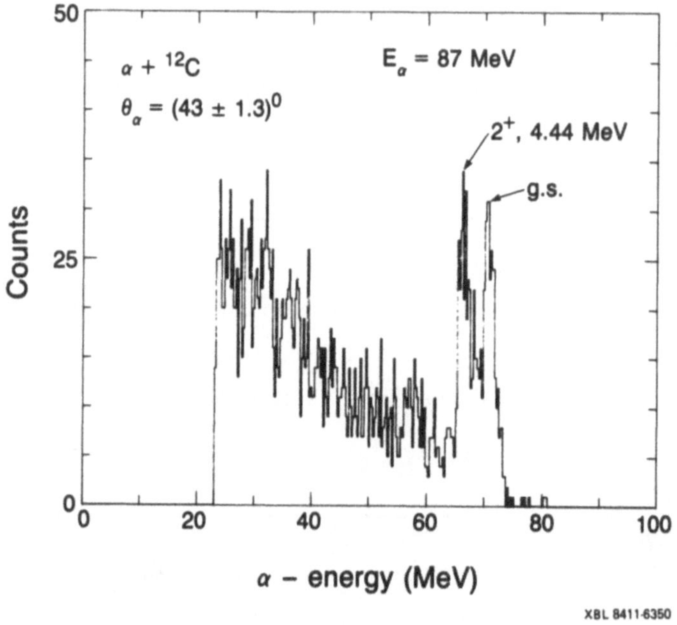

Fig. 13. Projection of the alpha particles of Fig. 12 on the energy axis.

Given the thickness of the ΔE layer, the spectrum is cut at about 22 MeV for alphas. The good timing properties of plastic scintillator, however, should enable the separation of alphas and protons stopped in the first layer from other particles stopped there by time of flight. A time resolution of 580 ps (FWHM) was determined with a source and timing the two phototubes versus each other.

This work was supported by the Director, U.S. Office of Energy Research, Division of Nuclear Physics of the Office of High Energy Nuclear Physics, and by the Nuclear Sciences of Basic Energy Sciences Program of the U.S. Department of Energy under contract Nos. DE-AC03-76SF00098 and DE-AM03-76SF00326.

REFERENCES

1) K. Van Bibber et al., IEEE Trans. Nucl. Sci. NS-31 35 (1984).

2) R.G.Stokstad et al., Proceedings of the International Symposium on Heavy-Ion Physics, Mt Fuji, Japan, (1984). To be published by the Physical Society of Japan.
S. Wald et al., to be published.

3) A. Baden et al., Nuc. Inst. and Meth. 203 (1982) 189.

4) C. Pastor et al., Nuc. Inst. and Meth. 212 (1983) 209.

5) J.B. Marion et al., Nuc. Inst. and Meth. 51 (1967) 93.

6) M. Bantel et al., Nucl. Inst. and Meth. 226 (1984) 394

PROPERTIES OF LARGE BaF_2 CRYSTALS. APPLICATION AS FAST AND EFFICIENT GAMMA-RAY DETECTORS IN THE 4π-γ CRYSTAL CASTLE ARRAY.

F.A.BECK
Centre de Recherches Nucléaires
F-67037 Strasbourg Cedex, France

New properties of large BaF_2 crystals are reported. The fast component of their light emission allows as well for very good timing, comparable to those of plastic scintillators but with a high γ-ray efficiency, as for high counting rates. This properties encourages the use of BaF_2 detectors in 4π-γ arrays. The present realization of such a device, the crystal castle, is described : its versatility will allow particle γ-work or γ spectroscopy with 12 Ge detectors surrounded by BaF_2 Anti-Compton shields. The response of BaF_2 crystals versus the detection of light particles or heavy ions is also discussed.

1. INTRODUCTION

In γ-ray detection the best counter should have at the same time good performances in energy resolution, efficiency and timing. For the moment such a counter do not exist. The best energy resolution is achieved with Ge detectors ($\sim$ 0.2 % for 662 keV γ-rays) which are much less powerfull in efficiency (no large volumes available) and in timing resolution ($\sim$ 5 ns). If ultimate performances in timing are needed, plastic scintillators are used but they have no γ-ray energy resolution and exhibit a low efficiency.

In the use of $4\pi\gamma$ arrays, where the total energy and the γ multiplicity produced in a nuclear reaction are the main quantities to be measured, efficiency, timing (e.g. for neutron-γ discrimination by the time of flight technic), capability of high counting rates are more important characteristics of the detector that the energy resolution even if some applications would also need a good value for this parameter. These considerations have lead in a first time to the use of NaI crystals (energy resolution $\sim$ 8 %, timing $\sim$ 3 ns) for the construction of the Oak Ridge (72 counters) and Darmstadt-Heidelberg (162 counters) crystal balls. More compact systems (e.g. TESSA in Daresbury) have since been built by the use of BGO crystals. Indeed such a crystal is much more efficient than a NaI crystal with a similar timing response but its energy resolution is approximatively twice that of a NaI detector.

We will report here on the properties of BaF_2 crystals which exhibit an energy resolution only 20 % worser that those of NaI detectors, an efficiency between NaI and BGO and a timing ($\sim$ 0.5 ns) of a plastic scintillator. Keeping in mind the characteristics recently pointed out on little samples of BaF_2 crystals [1], we have developped, together with the company MERCK, larger and larger crystals which enable now their use as detectors for 4π-γ arrays. Such a device, the crystal castle, will work in France beginning 1985. Hexagonals (Ø = 10 cm, L = 14 cm) BaF_2 crystals are used to built a crystal castle of 74 counters or a crystal castle of 38 counters and Anti-Compton shields. Following the results reported here, a Karlsruhe group is planing the construction of a crystal ball of 42 BaF_2 detectors.

The properties of BaF_2 crystals versus the detection of light particles and heavy ions have also been studied with a small sample.

2. CHARACTERISTICS OF BaF_2 CRYSTALS

Recently it has been pointed out [1] that BaF_2 crystal emit two light components : a slow one around 320 nm ($\tau \sim 620$ ns) and a fast one around 220 nm ($\tau \sim 0.6$ ns) in the ultra-violet region. If the amplitude ratio of the light emission for the two components is largely in favor of the slow one the situation is not the same for the height ratio of the two components for the pulses observed at the anode of the photomultiplier. The schematic design of this pulse (Figure 1) show a very narrow needle due to the fast

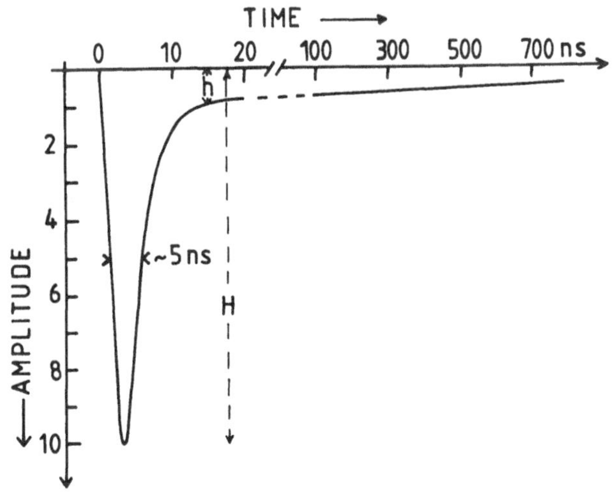

FIGURE 1 Anode pulse observed with an EMI 9821QB photomultiplier.

component while the slow component gives a wide pulse of low height. The height ratio R = H/h will depend on the ultra violet collection and on the photomultiplier rise time. For slow rise time (∿ 11 ns) photomultipliers with quartz windows a ratio value ∿ 2.5 has been observed. For small BaF_2 crystals (Ø = 4 cm, L = 6 cm) this ratio has been increased till 20 by the use of a RTC 2020 Q photomultiplier (diameter photocathode 44 mm, rise time 1.5 ns). With large BaF_2 crystals (Ø = 10 cm, L = 14 cm) ratios ∿ 14 have been observed with EMI 9821 QB photomultipliers (diameter photocathode 68 mm, rise time 2.1 ns). Such a fast pulse will allow very good timing and, when the best energy resolution is not required, very high counting rates. Its detection will need the use of fast photomuliplïers with quartz windows allowing a good light transmission near 200 nm.

2.1. Energy resolution

The energy resolution was previously reported as near of 30 %. With small crystals a value near 10 % has been observed recently [1]. Several kinds of photomultipliers with quartz windows have been used to study the energy resolution variation versus the photomuliplier type. With conventionnal slow tubes values of 9 % have been obtained while a 9.3 % value has been obtained with the best of the fast EMI 9821QB photomultipliers.

Another interesting aspect is the variation of the energy resolution when the light emitted is only partly collected to avoid pile-up events at high counting rates. In order to study this effect we used a charge sensitive ADC Le Croy 4300 for which the gate width was reduced by

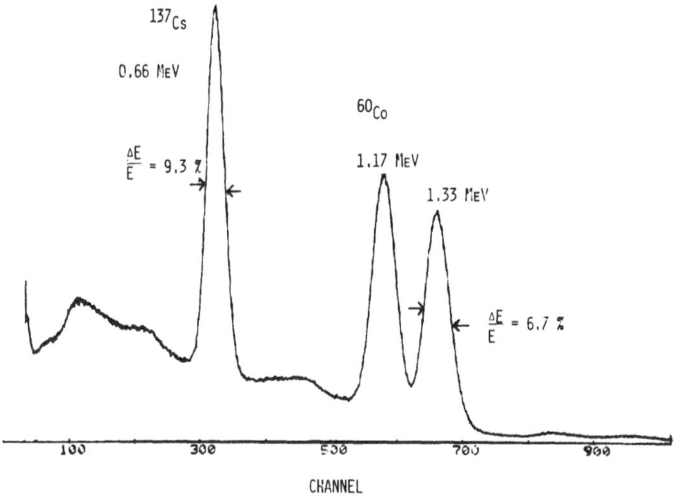

FIGURE 2 Gamma Spectrum observed with a BaF_2 crystal (Ø = 10 cm, L = 14 cm) and a EMI 9821QB PMT

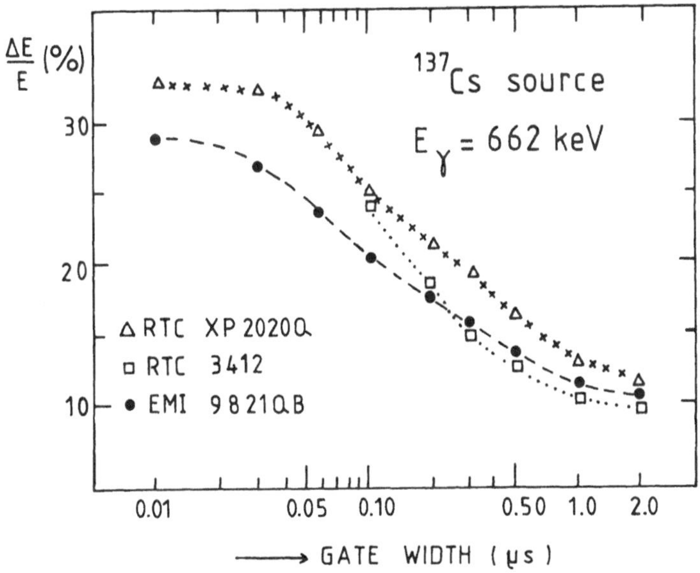

FIGURE 3 Energy resolution at 662 keV as a function of the Δt time of the light collection for different types of photomultipliers.

steps from 2000 ns to 10 ns. With the EMI 98210B we observed an energy resolution of 28 per cent for a gate width of 10 ns. In this case very high counting rates can been achieved with a reasonable individual energy resolution for many applications of 4π-γ arrays (sum energy, multiplicity).

2.2. Gamma efficiency

The absorption coefficients for NaI, BaF_2 and BGO show that the efficiency of BaF_2 crystals is less than that of the BGO crystal and greater than that of the NaI crystals. As shown in Figure 4 a 1 MeV γ ray will be absorbed by 95 per cent in 6 cm thick BGO, 10.5 cm thick BaF_2 and 14 cm thick

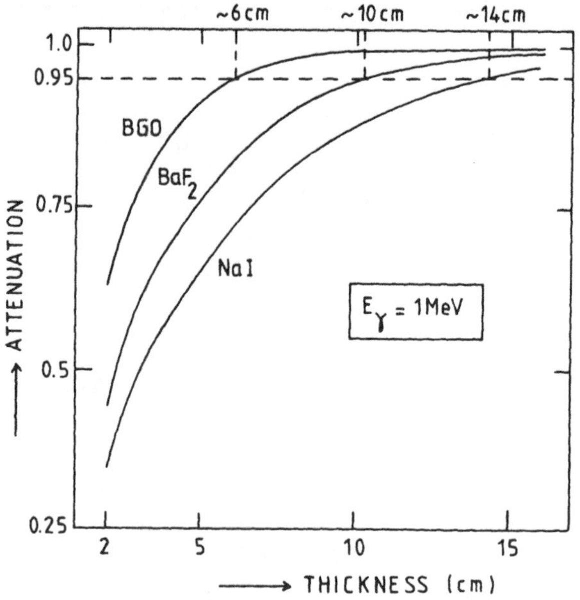

FIGURE 4 Attenuation of 1 MeV γ-rays by BGO, BaF_2 and NaI crystals.

NaI crystal. This means that the BaF_2 can be very useful in respect to efficiency : For the same absorption value they are smaller than the NaI crystals, but they are large enough to allow for a sufficiently large inner diameter of the 4π-γ array without having to increase too much the number of crystals, as compared to the BGO systems.

2.3. Timing

The fast component of the BaF_2 light emission, illustrated by the anode pulse shape (Figure 1), will allow for very good timing. For two small samples of BaF_2 (Ø = 4 cm, L = 6 cm) a timing resolution of less than 400 ps was achieved for a ^{60}Co source, with a threshold at 100 keV. For larger volumes (Ø = 10 cm, L = 14 cm) table I gives the timing resolution values obtained in coincidence with a small sample of BaF_2 under the same conditions, i.e. for a ^{60}Co source, with a threshold at 100 keV, for different types of photomultipliers. If the threshold is raised up to 1 MeV, values of 420 ps and 920 ps are obtained respectively for

TABLE I Timing resolutions with large BaF_2 crystals (^{60}Co source, threshold at 100 keV)

PMT	Rise time	FHWM(ns)	FTWM(ns)
Hammamatsu 1035Q	11 ns	1.9	3.8
RTC 3412	7 ns	1.7	3.4
RTC 2020Q	1.5 ns	0.7	1.5
EMI 9821QB	2.1 ns	0.52	1.1

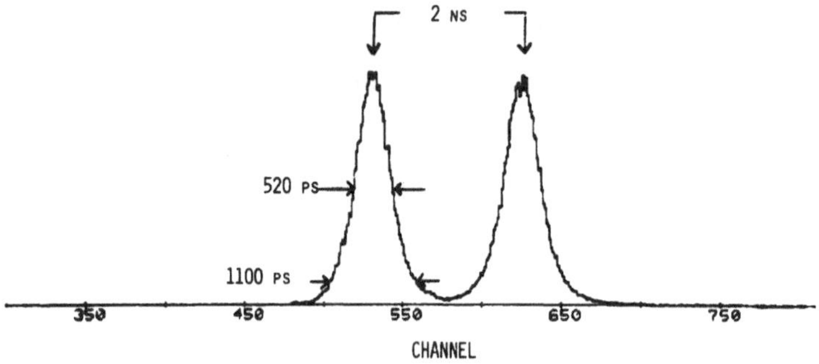

FIGURE 5 Time spectrum obtained between a large BaF_2 crystal and a small one with a ^{60}Co source (threshold at 100 keV)

the time resolution at the half maximum and at the tenth maximum.

Besides the fact that the time resolution of BaF_2 is much better that those obtained with NaI or BGO crystals note the value measured at 1/10 of the maximum (FWTM) which will enable very good neutron discrimination by time-of-flight measurements. Another interesting aspect for arrays with a large number of detectors is the fact that in order to obtain these values no critical adjustments are needed.

Of course these timing properties become even more important when the energy of the bombarding particles is inreased. A time-of-flight spectrum obtained at Ganil with a 44 MeV/nucleon ^{40}Ar beam show a clear neutron-γ separation by use of a BaF_2 detector for which the time reference was given by the RF of the cyclotron.

2.4. Neutron response

The response to neutrons emitted by a ^{252}Cf source for BaF_2 and NaI crystals has been compared in a time-of-flight experiment where the γ rays were detected by a small BaF_2

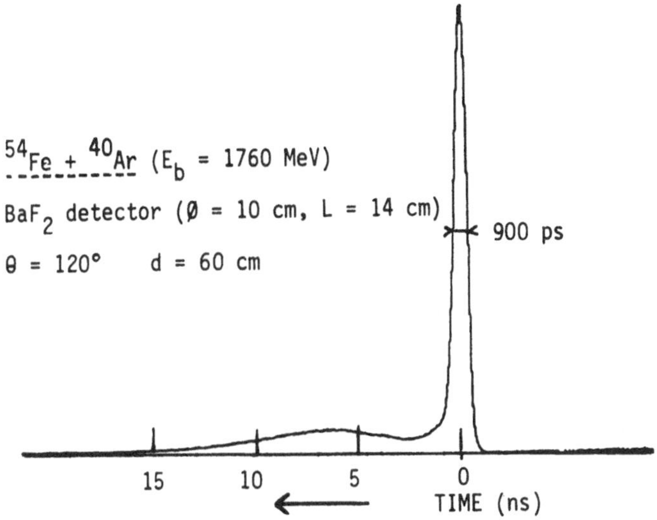

FIGURE 6 Time of flight spectrum observed at Ganil

crystal located near the source and the neutrons by larger BaF_2 or NaI counters shielded against γ rays by 10 cm of lead. For the same γ efficiency the number of neutrons detected in the BaF_2 counter was approximatively 2.6 time less than those measured in the NaI counter.

To check the dammages due to radiations a little sample of BaF_2 has been irradiated by 10^9 neutrons at a reactor. Beside short lifetimes radioactivites due to ^{20}F and Ba isotopes no change in the BaF_2 properties has been observed.

2.5. Comparison of the properties of the NaI and BaF_2 crystals in a heavy ion reaction.

The properties of the BaF_2 and NaI crystals have been compared in the heavy-ion reaction $^{127}I(^{37}Cl,xn)^{164-xn}Yb$ for a bombarding energy E_B = 160 MeV. The BaF_2(Ø = 10 cm,

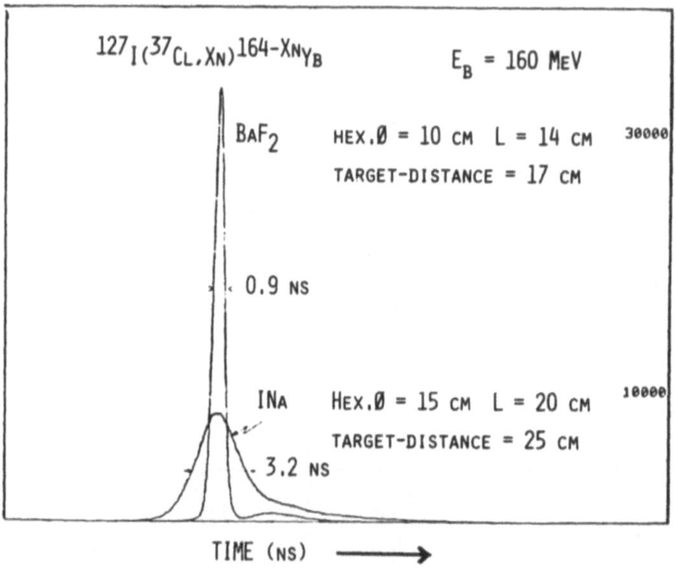

FIGURE 7 Comparison of the time of flight spectra observed in a H.I. reaction with NaI and BaF_2 detectors.

L = 14 cm) and NaI (Ø = 15 cm, L = 20 cm) crystals were located respectively at 17 cm and 25 cm from the target in order to have approximatively the same solid angle for an identical γ efficiency and for a geometry commonly used in 4π-γ arrays. Although the distance to the target was shorter for the BaF_2 crystal it was alone capable of a good neutron-γ discrimination. The measurements have been repeated at larger distances to have a better neutron-γ separation in the NaI crystal : for an equivalent γ counting rate in BaF_2 and NaI detectors the neutrons detected were 2.3 times more important in the NaI than in the BaF_2 counter.

It is also interesting to compare the energy spectra obtained with this two types of counters. The spectra associated to γ rays and neutrons were separated by setting windows on the time-of-flight spectrum : of course the γ

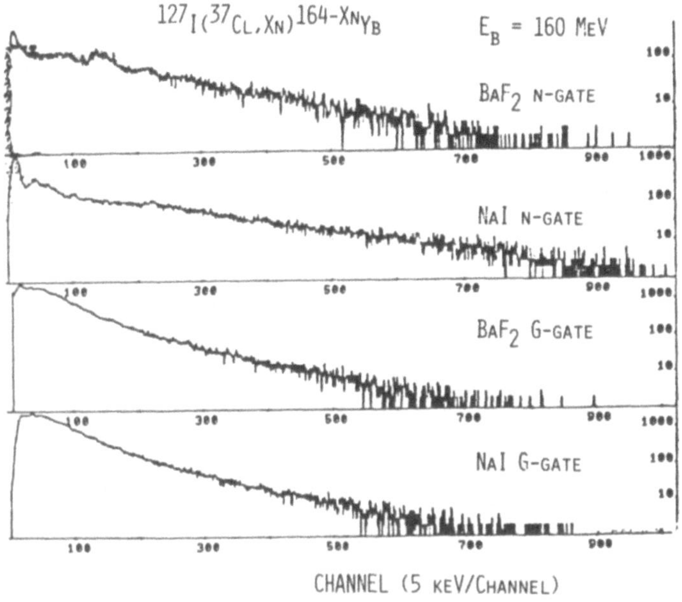

FIGURE 8 Comparison of the energy spectra associated respectively to γ-rays and neutrons produced in BaF_2 and NaI crystals for a H.I. reaction.

spectra are similar but the mean energy deposit by neutrons in the BaF_2 crystal is less important than the one in NaI. This may also be an interesting aspect in sum energies measurements or in studies of high energy γ rays of weak intensity.

2.6. Natural radioactivity contamination

Beside the usual natural radioactivity contaminations detected as well by other γ counters several peaks appear especially in BaF_2 detectors. It has been established that these lines are due to α particles emitted in the natural radioactivity of Radium which belongs to the same chemical family as Barium.This α lines are assigned by their pulse

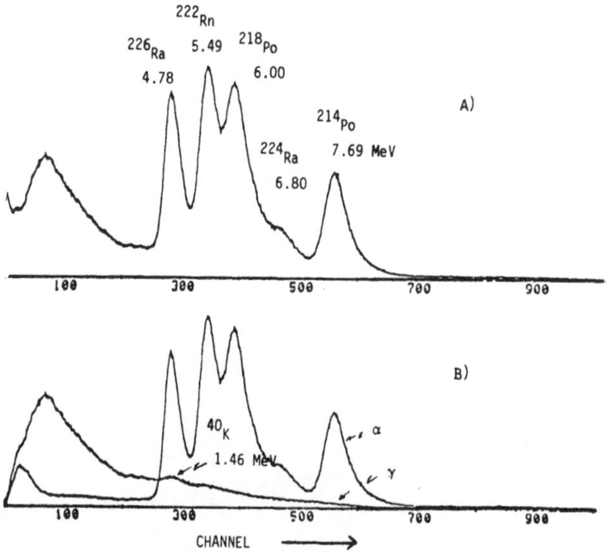

FIGURE 9 Natural radioactivity of BaF_2 crystals,
a) total spectrum observed
b) α and γ unfolded spectra

shape and their energy conversion factor different from these of γ-rays as pointed out in section 4. A pulse shape discrimination system has allowed for separation of γ-rays and α particles (figure 9) leading to an estimation of $\sim$ 200 counts/sec for α particles emission in 1000 cm^3 of BaF_2 crystal. This 0.1 per billion contamination of Ra can be disturbing for low-background γ ray studies, but will not affect studies were high counting rates are expected.

3. USE IN 4π-γ ARRAY : THE CRYSTAL CASTLE

3.1. Description

In the first 4π-γ arrays a crystal ball geometry was used.

Its advantage is that all detectors have approximately the same solid angle ; it is not possible, however, to change the number of detectors or to use the same counters for other purposes e.g. counter walls, Anti-Compton shields, etc.. This was the main reason why for the French 4π-γ array (a collaboration between Bordeaux, Grenoble, Lyon, Orsay, Strasbourg) a more versatile construction has been choosen : a crystal castle. The hexagonal BaF_2 counters ($\emptyset$ = 10 cm, L = 14 cm) are located in five rings of 1, 6, 12 and 18 counters above and belove the reaction plane leading to a number of 74 detectors, with an inner diameter of $\emptyset \sim 34$ cm. One ring less would yield to an array of 38 counters ; just as well, it is possible to add more rings leading to number of detectors 122, 182,...

Of course in such a geometry all counters will not

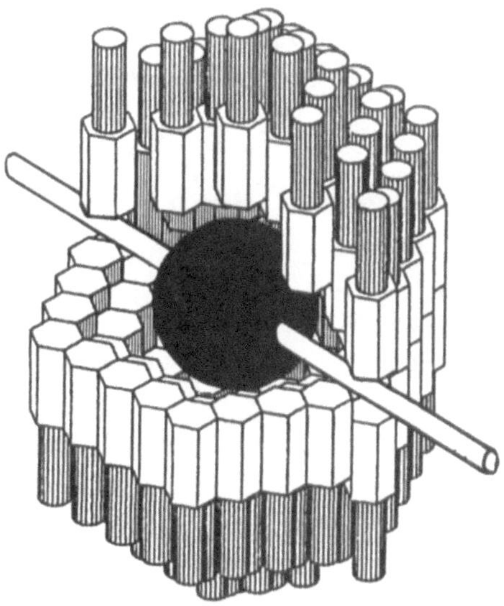

FIGURE 10 Schematic view of the crystal castle

have the same solid angle, leading to a loss in the total energy and multiplicity resolutions. According to a Monte-Carlo calculation [2], however this loss can be limited to less than 10 percent of the values expected for a crystal ball with the same number of detectors.

3.2. Anti-Compton shields

Numerous projects are now under way to perform γ-ray spectroscopy with a large number of Ge detectors surrounded by Anti-Compton shields (N = 6-21) ; within a multidetector array allowing for a selection in the [E_T, M_γ] entry region in heavy ion reactions. The first such device was TESSA 2. In most of these realizations or projects BGO crystal is used, since its high γ absorption allows construction of compact systems as requested for Compton suppression shields. The crystal castle being versatile, it

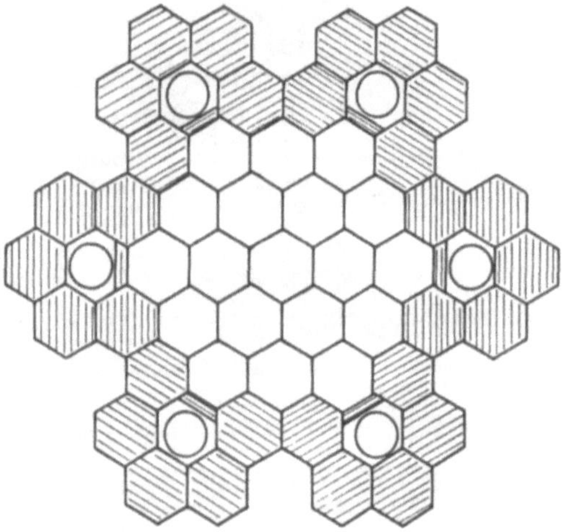

FIGURE 11 Crystal castle of 38 detectors surrounded by 12 Ge detectors with BaF_2 Anti-Compton shields (6 above and 6 belove the reaction plan).

was interesting to try to suppress the Compton events by using the hexagonal BaF_2 detectors of the crystal castle with some small additionnal BaF_2 shields. The suppression factor observed encourages the idea to use for γ-spectroscopy a crystal castle of 38 detectors surrounded by 12 systems Ge-BaF_2 Anti-Compton shields as shown in Figure 11.

4. PARTICLE DETECTION

The fact that BaF_2 crystal is non-hygroscopic makes its use very easy. Thus it was interesting to study its application in particles detection ; in particular for the measurements of the E energy in ΔE/E telescopes. Several checkings have been done with a small sample, 5 mm thick, of BaF_2. To first if the anode pulse shape (see Figure 12) associated to protons is similar to those observed for γ-rays, it changes drastically for alpha particles. Besides the light output for alpha particle is reduced by a factor of 3.7 as compared to γ rays, value measured with an americium source (5.5 MeV alphas). The properties of the small BaF_2 detector have also been studied for alpha and

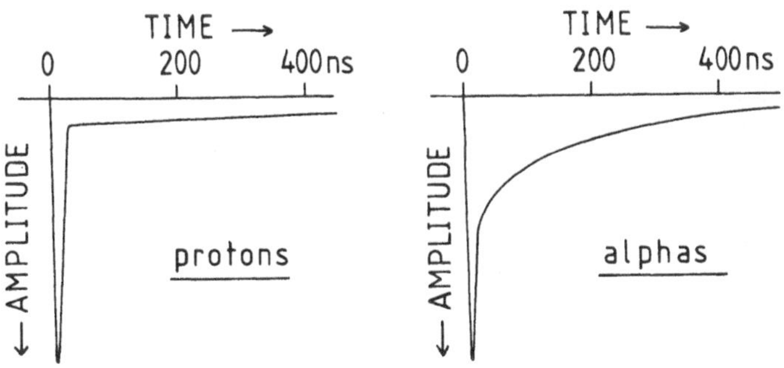

FIGURE 12 Anode pulse shape observed with a small BaF_2 crystal for protons and alpha particles.

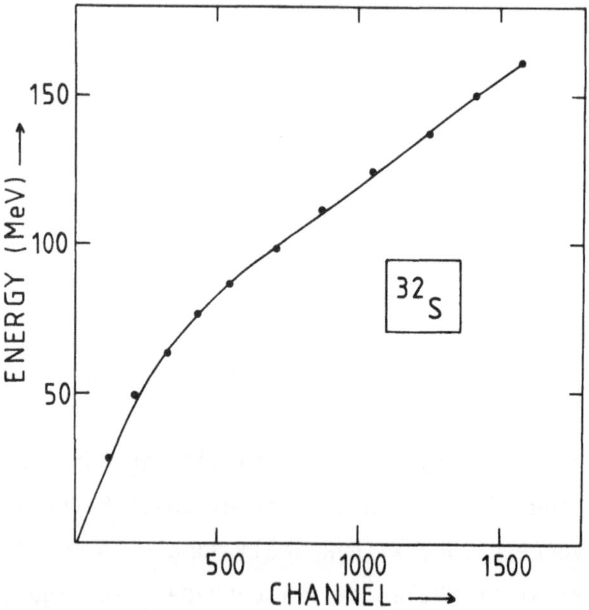

FIGURE 13 Light output versus particle energy for sulfur particles

^{32}S particles by elastic scattering on carbon, gold and lead for different bombarding energies. In the case of the reaction α + ^{12}C the E/ΔE telescope show [3] a very clear separation of the light particles like p, d, t and α.

5. CONCLUSION

At our intigation large BaF_2 crystals have been produced for the first time. Very encouraging performances have been obtained : the energy resolution is only 20 % worser than that of a NaI for a much better efficiency and almost a gain of an order of magnitude in timing. Some contamination by the radium radioactivity advice against their use in low background studies. Nevertheless they seems to be very powerful everywhere where efficiency, timing and not too bad energy resolution are required. This were the main reasons for their choice as detectors for the French crystal castle. For some applications which do not require the best energy resolution very high counting rates can be achieved by the use of the alone fast component of the light emission of the BaF_2 crystal.

REFERENCES

1. M.LAVAL, M.MOSZYNSKI, R.ALLEMAND, E.CORMORECHE, P.GUINET, R.ODRU and J.VACHER, Nucl. Inst. Meth., 206, 169 (1983)
2. J.P.VIVIEN, Annual Report, CRN-Strasbourg,(1982)
3. V.RAUCH, F.A.BECK, D.DISDIER and J.P.VIVIEN, poster session and to be published.

THE GSI MAGNETIC SPECTROMETER

D. Schüll
Gesellschaft für Schwerionenforschung
D-6100 Darmstadt, Germany

The detection of very heavy ions (A>20) using magnetic spectrographs is described. The specific problems and the requirements for the particle identification are discussed. A brief review of timing and focal plane detectors for heavy ions in spectrographs is given. Then the concept and the performance of the GSI-magnetic spectrograph is presented.

1. INTRODUCTION

Magnetic spectrographs have been used since many years in nuclear physics. Especially for the spectroscopy of lighter ions they can be attributed to be the "standard instruments" of a heavy ion laboratory. Originally the ion optical resolution was the most important design goal. Line widths of 50 μm occasionally have been resolved but only with nuclear track detectors along the focal line. The rapid development of various types of ionization and proportional detectors led to the design of a new group of spectrographs with excellent ion optical focusing, a large solid angle and a large dispersion to meet the properties of the available detectors. But very often the detectors could not completely make use of the full focal length of the spectrographs. Later on when heavier ions with good beam quality became available, more sophisticated detectors with ΔE and E measurements were added. Meanwhile many existing instruments are equipped with detection systems which in addition measure the orthogonal coordinate and the angle of incidence which makes experimental ray tracing possible and provide the opportunity for correcting aberrations. Under those conditions some of the older broad range spectrographs with low dispersions but a large momentum range reached new interest.

At a growing number of accelerator facilities meanwhile heavy ion beams with masses above A=30 up to Uranium are becoming available. With those beams the particle identification becomes an important task. The detector systems may require a very precise energy loss and energy measurement or a fast timing system or all of them. I will discuss the special requirements and problems, and after a short review of several detectors present a spectrometer concept which from the very beginning was designed for the work with the very heavy ions.

2. SPECIAL REQUIREMENTS AND PROBELMS FOR HEAVY MASSES

2.1 Identification of particles

For nuclear reaction studies with heavy mass beams usually a great number of different nuclei is emerging from the target. In order to obtain a unique determination of the nuclear charge Z, the mass number A and the atomic charge state q several detectors have to be combined. The position spectra in the focal plane are represented by a superposition of different charge states, masses and excitation energies of the products.

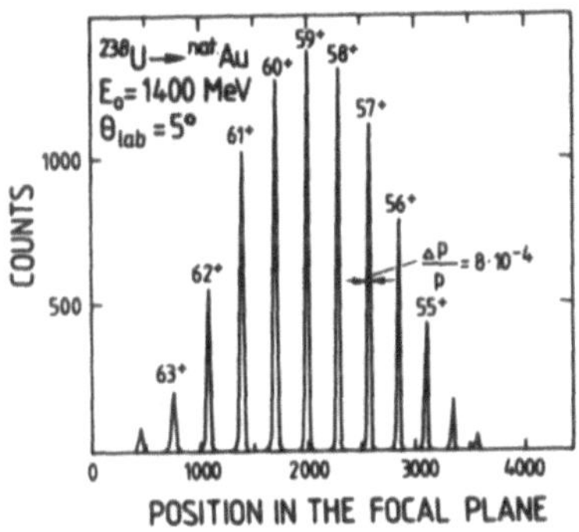

FIG 1
Charge state distribution of elastically scattered U-ions

In fig. 1 a charge state distribution of elastically scattered Uranium ions is shown. The average charge state for this energy is 59 and the width amounts to about 5.5 units (FWHM). If one intends to obtain reliable numbers for relative cross sections a reasonable part of the charge state distribution has to be covered or at least measured sequentially by several magnet settings. The average charge state depends mainly on the Z of the fragment but as well on velocity and target thickness. The consequence is that a spectrograph has to cover a large momentum byte of the order of 20 %

or more in order to avoid arbitrary cuts in the spectra for each nuclide.

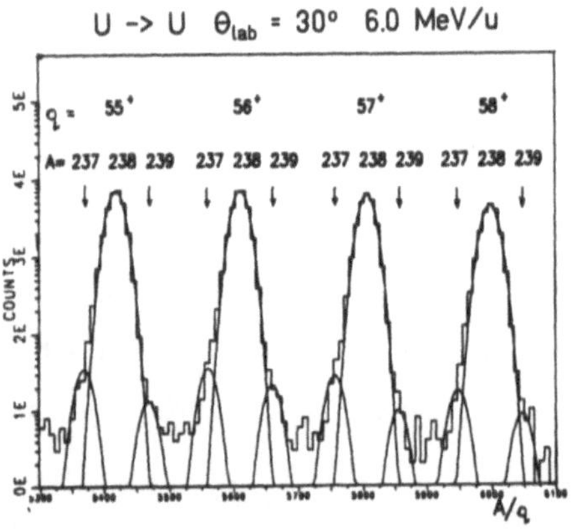

FIG 2
A/q spectrum for Uranium-like transfer products. The mass resolution in the spectra is 1/400.

The second problem is the ambiguity in the mass determination since a transfer product three or four masses away from the projectile has the same A/q as the projectile with the neighbouring charge state. Therefore an independent measurement for determination of the atomic charge state is necessary. A rather high accuracy is required since tails from elastic lines may cause contaminations of other A/q spectra.

The particle identification is usually performed by a $B\rho$, ΔE and E-technique. This method works well for masses up to about 60 and small reaction Q-values as long as the charge states do not overlap the $B\rho$-spectra. For heavier masses a time-of-flight measurement (TOF) has to be added to supply sufficiently good accuracy. The accuracy which is needed to study e.g. transfer reactions from Uranium is demonstrated in fig. 2 by a A/q-spectrum of Uranium like products. Even with a mass-resolution of 400 it is very hard to resolve the one-particle transfer products from the elastically scattered Uranium ions.

2.2 Kinematical Broadening

One of the most serious difficulties arises from the kinematical broadening i.e. the strong variation of the momentum with the opening angle in the reaction plane. It is usually expressed by the kinematical factor $k = 1/p \cdot dp/\Theta$. Values for k of -1 are typical and they can go up to -2 especially if one uses inversed reaction kinematics. The kinematical broadening leads to a shift of the focal plane of the spectrograph. For the large k-factors the shift may amount to several meters which can practically not be handled by a shift of the detectors. Therefore many instruments have multipole elements which can bring the focal plane back to the detector position. Another possibility is not to compensate for the kinematics and determine the focal plane by a reconstruction of the particle trajectories. In this case the accuracy of the energy and TOF measurement is strongly limited by the determination of the entrance angle Θ into the spectrometer. The energy width is influenced through

$$(\delta E)/E = 2 \cdot k \cdot \delta(\Theta) \qquad (1)$$

2.3 Target Effects

To study reactions with very heavy ions with a high energy resolution is the most difficult problem because there exist some severe limitations from the process in the target. Ford et al. (ForE77, Eng79) have made an analysis of system resolving powers for a number of heavy ion reactions. The main contributions for the energy broadening are the energy loss straggling, target inhomogeneities and multiple scattering in the target in conjunction with a large kinematical factor. The multiple scattering in the target disturbes the angle-energy correlation and can therefore not be compensated by ion optical methods. In essence one can conclude that an energy-resolution of several hundred keV ($\delta E/E \approx 10^{-3}$) seems to be a realistical limit for the experiments. Compared to lighter ion work this is a rather moderate value and the momentum resolution features of a spectrograph are therefore relatively simple.

In conclusion one can say that a spectrograph for very heavy ions should have a broad momentum range of about 20 %, kinematical compensation features for $|K| \geq 1$, small flight path differences on the order of $\delta s/s \approx 5 \cdot 10^{-3}$ and a moderate energy resolution of $\delta E/E \approx 10^{-3}$.

3. TIMING DETECTORS

As mentioned above time-of-flight measurements seem absolutely necessary for mass determination in the range of A>100. Usually this requires signals from a start and a stop detector. A simpler method can be used when the beam from the accelerator has a time structure which can be tuned by RF-cavities such that the time spread at the target position is less than 0.5 ns. In such a case one does not need a timing start detector which may disturb the energy resolution because it cannot be made infinetely thin. The other advantage is that there exist no count rate limitations if one wants to go to small forward angles. One disadvantage comes up in the particle identification when the flight path is rather large and the momentum byte accepted by the spectrometer is large as well. The reaction products from different micro-bunches may surpass each other in the spectrograph. The ambiguity in the mass determination has to be solved then by a comparison of energy and TOF.

Transmission start detectors ought to be as thin as possible and should have an excellent homogeneity. Besides that they have to be rather compact since usually there is very little space between the target and the first ion optical element. If the detector is position sensitive it can provide a signal for the entrance angle Θ which is needed for trajectory reconstruction.

The group of start detectors which can be made very thin are secondary emission channel plate detectors (CPD). They consist of a self supporting carbon foil of about 10-20 $\mu g/cm^2$ thickness. The electrons being emitted when a heavy ion passes through are accelerated and bent to a micro-channel plate. The output signals can be directly fed into a trigger unit. The timing resolution is excellent. Values of 150 ps can be easily achieved. A detector using a magnetic field to bent the electrons is described by Renfordt et al. (RenN81). Using a resistive anode a position signal can be derived by a charge division method. Position resolutions of 0.7 mm have been reached. At the GSI-spectrometer (see section 5) we are using a detector which is much smaller (BusP80) because no magnetic field is applied. An electrical mirror is used instead. Both detectors need grids for the acceleration of the electrons which may cause scattering and energy loss. A detector which avoids this disadvantage is reported by Odenweller (OdeN82). The electrons are deflected by crossed electric and magnetic fields using a special arrangement of acceleration electrodes. The active area of CPD's cannot be made too large since channel plates are available only up to a size of about 3". Those plates become rather expensive. For the heavy ions the efficiency is usually close to 100 % since the number of emitted electrons is high enough. For lighter ions however it may drop down. A new idea is reported from Dubna (KavL83) using porous dielectric layers of MgO evaporated onto a carbon foil. The number of secondary electrons can be increased up to a factor of 100, however, the homogeneity of these layers is still poor.

No restriction in the size of the sensitive detection area exists for scintillation detectors. Scintillator foils can be made as thin as $\approx$ 60 $\mu g/cm^2$ with a good homogeneity. They have no grids and they can be placed into a tilted position - as reported by Garron et al. (GarJ82) - to compensate for the flight path differences in the spectrograph. However no position signal can be derived. The timing resolution - depending on the size - can reach 200 ps. Another scintillation detector being used as transmission stop detector is described in section 5.2.

Similarly good timing properties are obtained by parallel plate avalanche counters (PPAC) and low pressure multiwire proportional counters (MWPC), see e.g. the article of A. Breskin in this conference. Both detector types can be manufactured by low costs and to any size. The thickness is however restricted by the entrance and exit foils. Two foils are the minimum but four are usual for PPAC's. Therefore the total thickness may not be below 150 $\mu g/cm^2$. The MWPC need at least one grid. A PPAC-detector with excellent position and timing properties being used as active aperture in a Q3D-spectrograph is reported by Lettau et al.(LetB83).

4. FOCAL PLANE DETECTORS

It is not the aim to give a complete review of all spectrometer detection systems and discuss the figures of merit for each of the different designs. But I would like to offer some personal remarks on the applicability of these detectors with respect to the detection of very heavy ions. Not too many developments with new design ideas have been published since the excellent compilation of particle detectors edited by A. Bromley in 1979 (Brom79). Many of the detectors decribed there were prototypes or in a test phase at that time. They are now routinely used.

Three different groups may be found: the multiwire proportional counters (MWPC), the single wire proportional counters (SWPC) and the hybride-counters. MWPC's have a good linearity and high count rate capabilities. The position resolution for normal incidence is basically the wire spacing. Chambers with 0.5 mm wire distance are meanwhile used in high energy physics and in some spectrographs (Gläs77). For inclined tracks the resolution can be improved by an interpolation between wires. A further development (HynF84) of those detectors is the vertical drift chamber (VDC). I must not go into details of this detector since T. Sjoreen has presented it already within this conference. The position resolution amounts to ≤ 0.5 mm and the Θ-resolution is about 7 mrad. For low dispersion instruments this excellent position resolution is very important, a much better resolution however seems less important since the target spot size and the angular divergence at the target or in the focal plane become dominant.

SWPC's are much simpler in their mechanical construction. The readout can be performed using the charge division or the rise time technique or an induction readout with a delay line. The position resolution of ~1 mm is slightly worse than the MWPC's. Both detector types have to be backed by an additional detector for the ΔE and/or E which is usually a scintillation counter. For the very heavy ions the energy loss in the entrance and exit foils or in the light covers introduces detection thresholds which may not be tolerable.

Therefore the hybrid-counter (ShaV75, ErsB76) became the most popular detector at nearly all spectrometer systems for heavy masses. This detector consists basically of one or two SWPC integrated into an ionization chamber which acts as a ΔE-E counter. Sometimes it may be backed by a scintillator for particles with a larger range. A multiple measurement of the position allows the determination of the angle Θ and therefore a track reconstruction with accuracies of about $\delta\Theta \sim 5$ mr. In principle one can derive a timing signal from the cathode or from an extra grid. But the signals are rather small and only for higher absolute energies sufficiently good timing results are reached.

5. THE GSI-SPECTROMETER

5.1 Design concept

The spectrometer was designed for nuclear reaction studies with heavy ions in the mass range A>40. To identify those particles the detection system has to meet some requirements. Since a trajectory reconstruction seems absolutely necessary the detectors have to provide in addition to the position, the ΔE and the E signals information about the entrance angles Θ and Ø. If these quantities are measured it has an impact on the ion optics. Higher order aberrations can be corrected by software in the data analysis and need no complicated field configurations. On the other hand the detector construction is much simplified, when e.g. the angle of incidence into the focal plane is close to 90°.Therefore an integration of ion optics and detector system properties was considered from the very beginning. A modular design with simple magnets and no fixed imaging characteristics, together with a variable dispersion provides enough flexibility to adapt to the experimental program.

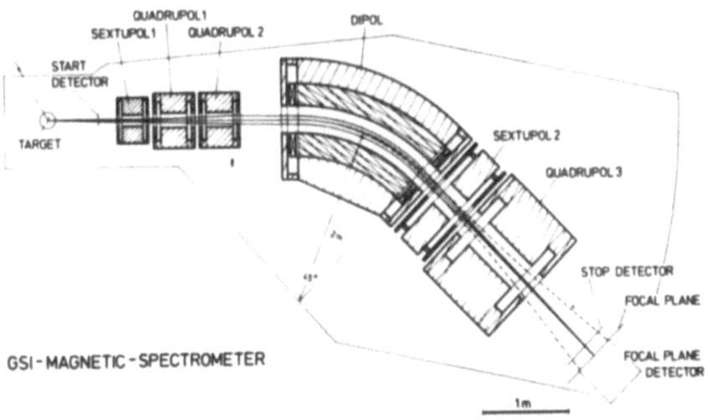

FIG 3: Schematic view of the spectrometer

Fig. 3 shows the layout of the spectrometer. The optical elements are:

a) a 45°-bending magnet with straight field boundaries perpendicular to the centre trajectory. The radius of curvature is 2 m, the maximum field 1.7 T.

b) the two quadrupols Q1 ad Q2 for the focusing in both planes,

c) a large aperture (30 cm) quadrupole Q3 by which the dispersion can be varied within a range of 2 to 7 cm/%,

d) the sextupole Sex2, which allows to rotate the tilt angle of the focal plane to be perpendicular to the centre trajectory.

e) the sextupole Sex1 is an attempt to provide some second order correction.

I don't want to go into the details of the ion optics, it is described by Th.Walcher (Wal74). But I should like to make some remarks which are connected with the total detection accuracy. We have chosen the relatively small deflection angle in order to keep the flight path differences small. It is known (Bro70) that the momentum resolution and the flight path differences are inversely proportional. So one has to find a compromise between the necessary momentum resolution and the mass resolution. $\Delta p/p$ is limited in our case to about 1/3000. Because of the large kinematical effect the opening angle in the reaction plane has to be kept small (±15 mr). In order to have a reasonable total solid angle the aperture in the ϕ-coordinate has to be larger (±50 mr). Double focusing as in most other instruments implies that the geometrical aberrations which arise from a large ϕ cannot be corrected by help of a proper measurement of the vertical image size y in the focal plane. Most gas counters can measure this coordinate without much effort. The basic features of the instrument are compiled in table 2.

type	SQQDSQ
deflection angle	45°
Bρ	3.2 Tm
solid angle	3msr
dispersion variable	2....7.5 cm/%
focal plane	50 cm, ⊥
energy acceptance	50.....13 %
energy resolution	$5 \cdot 10^{-4}$ FWHM k = 0
	10^{-3} FWHM k = 1
flight path differences	$6 \cdot 10^{-3}$
kinematical correction	$\lvert k \rvert \leq 1.5$
rotation angle	-45° to 135°

Table 1: Spectrometer Characteristics

5.2 Detector system

The detector system consists of a start detector at the entrance of the spectrometer, a scintillator stop detector and a hybrid ionization chamber as focal plane detector. These detectors measure the time-of-flight through the spectrograph, the entrance angle Θ, the horizontal and vertical positions in the focal plane, the energy loss ΔE and the total energy E.

The start detector (BusP80) is a secondary emission channel plate detector with a carbon foil of about 20 μg/cm² mounted about 50 cm apart from the target. The electrons emitted from the carbon foil into the backward hemisphere are accelerated to 2.2 keV and then imaged to a channel plate with an active area of 20 x 40 mm² by an electrical mirror. The catcher anode consists of 2 mm wide strips connected through a resistor chain. The position signal is determined by the charge division method. A schematic drawing of the detector is shown in fig. 4. The timing resolution is about 200 ps, the position resolution 1.2 mm which corresponds to about 2.5 mr angle resolution.

The stop detector consists of a thin self supporting scintillation foil (~ 150 $\mu g/cm^2$) with dimensions 50 x 7 cm^2 placed close before the entrance window of the focal plane detector. The light emitted from the surface when a heavy ion passes is detected by four 3" phototubes mounted in series on top of the vacuum chamber. A cylindrical reflector improves the detection efficiency. A cut through the detector is shown in fig. 5. The output signals from the four phototubes are summed by a cable and put on the input of a constant fraction trigger. The timing resolution is about 400 ps. The light emission from the different foil positions has been simulated by Monte Carlo calculations. They showed differences in the path length for the light given by the detector geometry corresponding to about 380 ps.

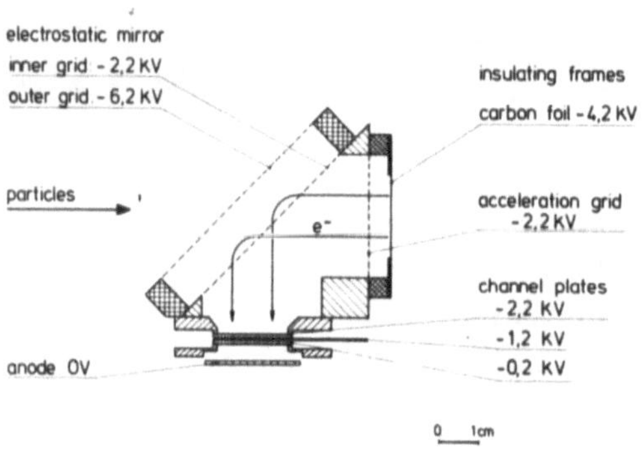

FIG 4
Cross section of channel- plate start detector

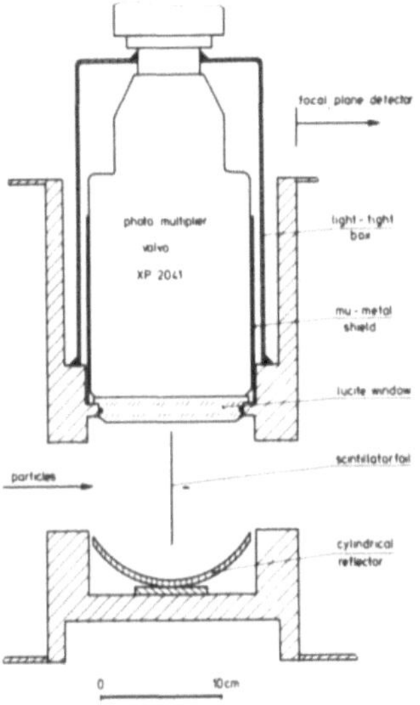

FIG 5
Cross section of scintillation stop detector

The focal plane detector is from the hybrid type. It has two proportional counters integrated into the anode group similar to the detectors used in Rochester and Argonne (ShaV75, ErsB76). The length is 50 cm . The anode is segmented into 9 stripes for the ΔE-E measurements. The length of the ΔE-section can be externally chosen. It was found that an energy loss of about 40% of the total energy gives the best Z-resolution. The entrance window is supported by a chemically etched mesh with a transmission of 75%. The detector is 50 cm deep in order to stop all ions up to 20 MeV/u. Other arguments for a "deep" detector are: the good charge collection near the entrance window since the field distortions are relatively small, a large dynamical range and nearly negligable energy losses in the proportional wire sections. The ΔE-resolution for 1000 MeV Kr ions is 1.1%. An energy resolution of 0.5% was reached. The detector is usually operated with pure methane gas. For higher energies tetrafluoromethane (CF_4) with a four times higher stopping power is used. The proportional counters consist of a high resistive carbon coated Si-wire of 25 μ thickness. The position signal is taken from both sides and processed by the charge division technique. The resolution is 1.2 mm.

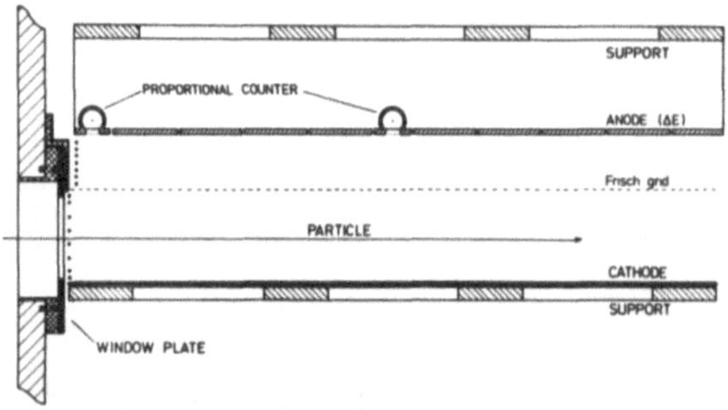

FIG 6: Schematic view of the focal plane counter

5.3 Data analysis

The electronic for the different signals is straight forward. All signals are fed into Analog to Digital Converters and event by event read from a PDP 11/45 computer. This computer is coupled via a fast link to the IBM 3081 computer where the online and offline data processing is performed. For the control of the experiment a powerful online computer is essential since the interesting physical quantities cannot simply be derived from the measured signals. By a combination of some of them the atomic charge state, the mass and the Z can be calculated. A preliminary set of corrections has to be applied. Very helpful are some two dimensional scatter plots like position vs. θ (see fig.7), position vs. TOF, or ΔE vs. E especially during the tuning procedure. The adjustment of the right field settings of the quadrupoles or sextupoles can be done very quickly by looking to slopes or curvatures in the scatter plots and try to correct them. When the intrinsic resolution is good one can decide to do it later by software in the final analysis.

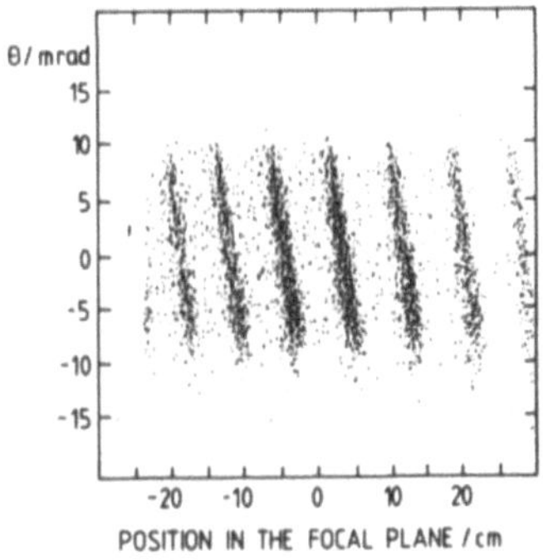

FIG 7
Two dimensional scatter plot of position in the focal plane vs. entrance angle Θ. The lines correspond to different charge states of elastically scattered Xe-ions.

Fig. 7 shows as an example the scatter plot horizontal position x vs entrance angle θ. The inclined lines correspond to different charge states of elastically scattered Xe-ions. The slope directly measures the kinematical factor k. By changing the fields in the quadrupoles Q1 and Q2 these lines rotate. A complete kinematical compensation is achieved when these lines are vertical which is equivalent to a shift of the focus into the focal plane. It should be noted that all these tuning procedures and the experimental ray tracing become much simpler for the heavy masses because there are always several atomic charge states in the focal plane who correspond to the same energy and kinematics, thus providing a set of calibration points along the focal plane by which all nonlinearities can be evaluated.

For the particle identificaton the four basic relations between the measured quantities ΔE, energy E, flight time T, position in the focal plane x, the particle characteristics nuclear charge Z, mass number A, velocity v and atomic charge state q are used:

$$\begin{aligned} &a)\ \Delta E \sim Z^2 \cdot A/E \sim Z^2/v^2 \\ &b)\ T = s(\Theta)/v \\ &c)\ B\rho \sim x \sim A \cdot v/q \\ &d)\ E = A \cdot v^2/2 \end{aligned} \tag{2}$$

A combination of three signals gives the charge state

$$q = 2/s(x, \Theta) \cdot E \cdot T/B\rho(x, \Theta) \tag{3}$$

The resolution for q is determined mainly by the uncertainty of the energy measurement since the other signals can be determined much more accurately. The resolution for 1500 MeV Kr-ions goes up to 1/200 but for 400 MeV Sm-ions it may decrease to 1/80. This seems sufficient for the whole mass range but it should be considered that the tails of strong elastic lines sometimes may cause contaminations in the mass and energy spectra.

Having gates on the charge states the mass is obtained by

$$A/q = B\rho(x, \Theta) \cdot T/s(x, \Theta) \tag{4}$$

Since the $B\rho$ can be measured better than 1/1000 the accuracy is mainly limited by the time of flight measurement. The flight path differences however and the variation of the dispersion along the focal plane have to be taken into account to reach a mass resolution of 1/400. When q and A are fixed the energy of the particle is calculated from the relation

$$E_p = (B\rho)^2 \bullet q/(A/q) \tag{5}$$

Since A and q are discrete values the energy resolution is mainly due to the intrinsic $B\rho$-resolution and the target effects. In standard mode operations resolutions of $\delta E/E= 1/1000$ are typical. I should mention that the maximum mass resolution and energy resolution for large k-factors can only be achieved when θ is measured to a proper accuracy. This can be understood by looking to formulas. Formula 5) holds only for a kinematical factor k=0. If the spectrometer is operated in the uncompensated mode the lines are broadened in the focal plane. In equation 4) the velocity cancels out so that θ contributes only via the flight path differences ($\leq 10^{-3}$)to the mass resolution. In order to get a kinematically corrected energy spectrum equ. 5) has to be replaced by

$$\begin{aligned} &\text{a) } E_p \sim (B\rho)^2 \bullet (1-K\Theta)^2 \\ &\text{b) } A/q = B\rho \bullet T/s(\Theta) \end{aligned} \tag{6}$$

for spectrometer mode: "uncompensated"

For the "compensated" mode equ. 2c) must be modified since not the particles with the same velocity but with the same velocity- dependence are focused to the same position. In the time measurement the actual velocity is determined. Equ. 4) and 5) will then be for the "compensated"-mode:

$$\begin{aligned} &\text{a) } E_p \sim (B\rho)^2 \\ &\text{b) } A/q = B\rho \bullet T/\left(s(\Theta)\cdot(1-K\Theta)\right) \end{aligned} \tag{7}$$

for spectrometer mode: "compensated"

The comparison of equations 6) and 7) show that the uncertainty in θ which is on the order of 3 mr contributes in the first case preferably to the energy width, in the second mode it contributes to the mass resolution. So we can choose an operation mode which optimizes either the mass or the energy resolution or find a mode between these extremes. For large k-values on the order of -1 and a $\delta\theta \geq 3$ mr the mass and the energy resolution are limited by this effect to 1/300 and 1/150 respectively.

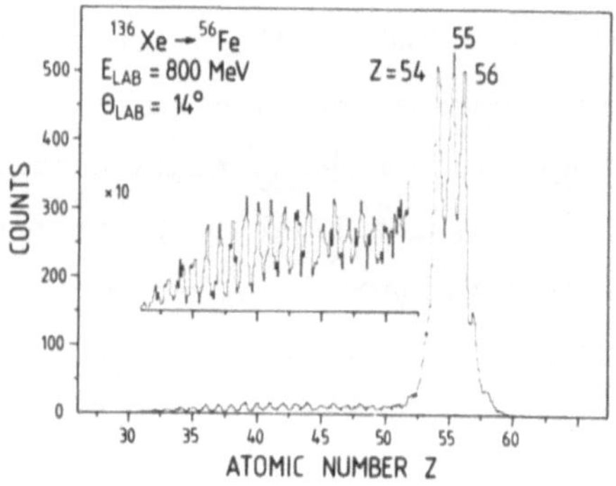

FIG 8
Z-spectrum of Xe-like reaction products

In order to get the nuclear charge Z one has to linearize the ΔE-T functions or the ΔE-E functions for each mass.

$$Z \sim f(\Delta E, E/A) \sim f(\Delta E, T) \tag{8}$$

The Z-resolution is $\Delta Z/Z \sim 1/90$ for 800 MeV Xe-like products. In fig. 9 the Z-distributions for quasielastic Xe-like ions is shown.

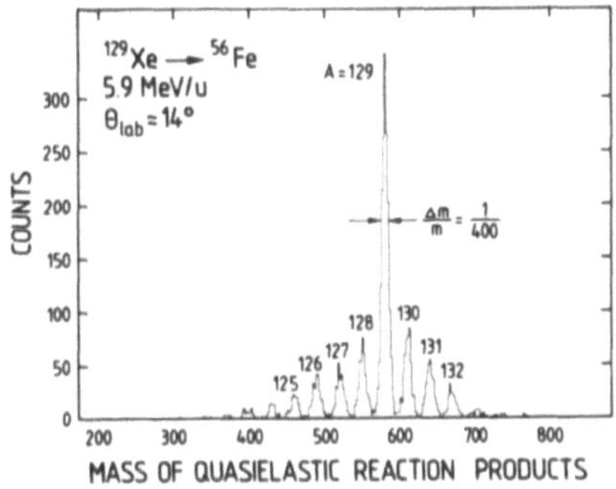

FIG 9
Mass spectrum of quasielastic products from Xe-induced reactions.

A mass spectrum is displayed in fig. 10 which demonstrates the excellent mass resolution of $\delta A/A = 1/400$.
Three energy spectra of ^{86}Kr from the reaction ^{86}Kr + ^{208}Pb at three different bombarding energies is shown in fig. 11. The resolution of about 3 MeV can only be visualized on the high energetic tail. The low energy tail is due to many inelastic levels. A structure at excitation energies of 23 MeV at 10 MeV/u bombarding energy and at 38 MeV for 18 MeV/u is clearly visible. By means of a Monte Carlo simulation it could be shown (SohF84) that these structures are due to the decay of a primary formed ^{87}Kr emitting a neutron. A comparison of these structures in all neighbouring nuclei allowed the determination of the energy partition between the reaction partners which is strongly channel dependent.

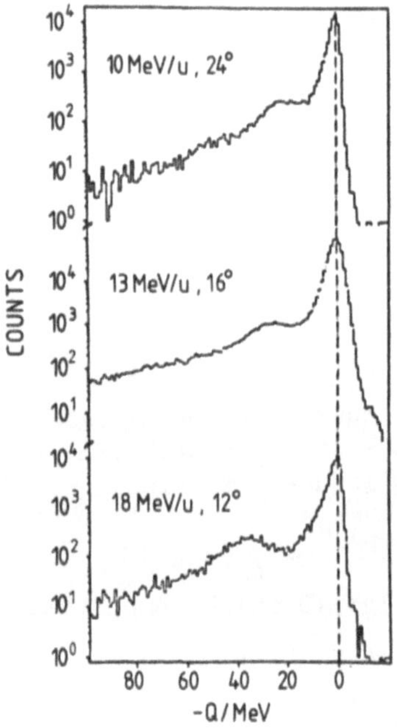

FIG 10
Energy-spectra of ^{86}Kr from the reaction ^{86}Kr + ^{208}Pb for three different bombarding energies.

6. Conclusion

I have presented a view on an instrument which was designed for the detection of heavy reaction products (A>40). To take advantage of the recent detector developments an integration of the detector properties with the features of the ion optics was intended. The energy resolution is moderate. But for heavy masses it is limited by target effects anyhow. The good particle identification properties make it a versatile instrument for nuclear reaction studies. It is used for the investigation of elastic and inelastic scattering (JiaC81), the domain of application however is the study of quasielastic transfer reactions (SchS81, BreC82). Large area coincidence detectors can be mounted for complete decay studies.

But other applications like accelerator mass spectroscopy experiments are as well under way.

The spectrometer was proposed by F. Pühlhofer and set up by groups from the University of Marburg, Max-Planck-Institut für Kernphysik Heidelberg and GSI. Th. Walcher has suggested the ion optical layout. I like to thank all group members especially F. Pühlhofer, W. Pfeffer, F.Busch, Th. Walcher, B. Langenbeck, K.Blasche and R. Bock for their creative engagement and the continuous help. The work was supported by the Bundesminister für Forschung und Technologie.

REFERENCES

BreC83 C. Brendel, J. Carter, F. Beck, A. Richter, G. Schrieder, D. Schüll,G. Delic and R.H. Lemmer, Phys.Lett. 124B(1983)474

Bro70 K.L. Brown, Stanford Linear Acc. Centre, Stanford, SLAC report No. 75

Brom79 Nucl.Instr. Meth. 162 (1979) 263 ff, A. Bromley ed.

BusP80 F. Busch, W. Pfeffer, B. Kohlmeyer, D. Schüll, and F. Pühlhofer, Nucl. Instr. and Meth. 171 (1980) 71

Eng79 H.A. Enge, Nucl. Instr. Meth. 162 (1979) 161

ErsB76 J.R. Erskine, T.H. Braid, and J.C. Stolzfus, Nucl. Instr. and Meth. 135 (1976) 67

ForE77 J.L.C. Ford, Jr., H.A. Enge, J.R. Erskine, D.L. Hendrie, and M.J. LeVine ORNL/TM-5687

GarJ82 J.P.Garron, J.C.Jacmart, N.Frascaria, C.Stephan and L.Tassan- Got, Nucl.Instr.Meth. 196(1982)431

Gläs77 P.Glässel ,Nucl. Instr. Meth. 140(1977) 61

HynF84 M.V.Hynes, J.L.C.Ford, T.P.Sjoreen, J.L.Blankenship and F.E.Bertrand Nucl.Instr.Meth. 224(1984) 89 T.P.Sjoreen, J.L.C.Ford, J.L.Blankenship, R.L.Auble, F.E.Bertrand, E.E.Gross, D.C.Hensley, D.Schüll and M.V.Hynes , Nucl.Instr. Meth. 224(1984)421

LetB83 H.Lettau, H.G.Bohlen, M.R.Glover, H.Ossenbrink and W.von Oertzen, Nucl.Instr.Meth.184213(1981)461

JiaC81 Jiang Cheng-Lie, P.R. Christensen, O. Hansen, S. Pontoppidan, F. Videbaeck, D. Schüll, Shen Wen-Qing, A.J. Baltz, P.D. Bond, H. Freiesleben, F. Busch, and E.R. Flynn, Phys. Rev. Lett. 47 (1981) 1039

KavL82 R.L.Kavalov, S.M.Luk'Yanov, Yu.L.Margaryan, Yu.Eh. Penionzhkevich, S.G.Gasparyan, G.G.Chubarryan, JINR 13-83-188, Dubna 1983

OdeN82 T.Odenweller,H.Noll, K.Sapotta, R.E.Renfordt and R.Bass, Nucl.Instr.Meth. 198(1982)263

RenN81 R.E.Renfordt, H.Noll and K.Sapotta, Nucl.Instr.Meth.185(1981)157

SchS81 D. Schüll, W.C. Shen, H. Freiesleben, R. Bock, F. Busch, D. Bangert, W. Pfeffer, and F. Pühlhofer Phys. Lett. 102B (1981) 116

ShaV75 D. Shapira, R.M. DeVries, H.W. Fulbright, J.Toke and M.R. Clover, Nucl. Instr. Meth. 129(1975) 123

SohF84 H.Sohlbach, H.Freiesleben, P.Braun-Munzinger, W.F.W.Schneider, D.Schüll, B.Kohlmeyer, M.Marinescu, F.Pühlhofer, to be published

Wal74 Th. Walcher, MPI-Sonderdruck MPI-4-1974-V25, Heidelberg 1974

THE MSU REACTION PRODUCT MASS SEPARATOR

J.A. NOLEN, JR., L.H. HARWOOD, M.S. CURTIN, E. ORMAND
AND S. BRICKER
National Superconducting Cyclotron Laboratory
Michigan State University
East Lansing, Michigan 48824-1321

In this session of this conference on Instrumentation for Heavy-Ion Nuclear Research we will hear about several devices which can be grouped under the general topic of "mass separators". The Reaction Product Mass Separator (RPMS) which is now operational at MSU was designed and constructed to separate and permit the study of isotopes produced by the heavy ion beams from the superconducting cyclotrons at the NSCL. At the present time nuclear research is being carried out at the NSCL with the k=500 superconducting cyclotron which was completed in 1982. This cyclotron currently utilizes an internal Penning ion gauge (PIG) ion source. A k=800 superconducting cyclotron is under construction and a project to build an electron cyclotron resonance (ECR) ion source has been recently initiated. Plans are to operate the K800 cyclotron with the ECR source in 1986 and the coupled cyclotrons with the ECR source in 1988. The heavy ion beams which will be available from various combinations of the cyclotrons and ion sources are indicated in Figure 1. It was our goal to build a device which would permit useful measurements to be made using the wide variety of heavy ion beams which will be available at the NSCL.

MSU-84-067

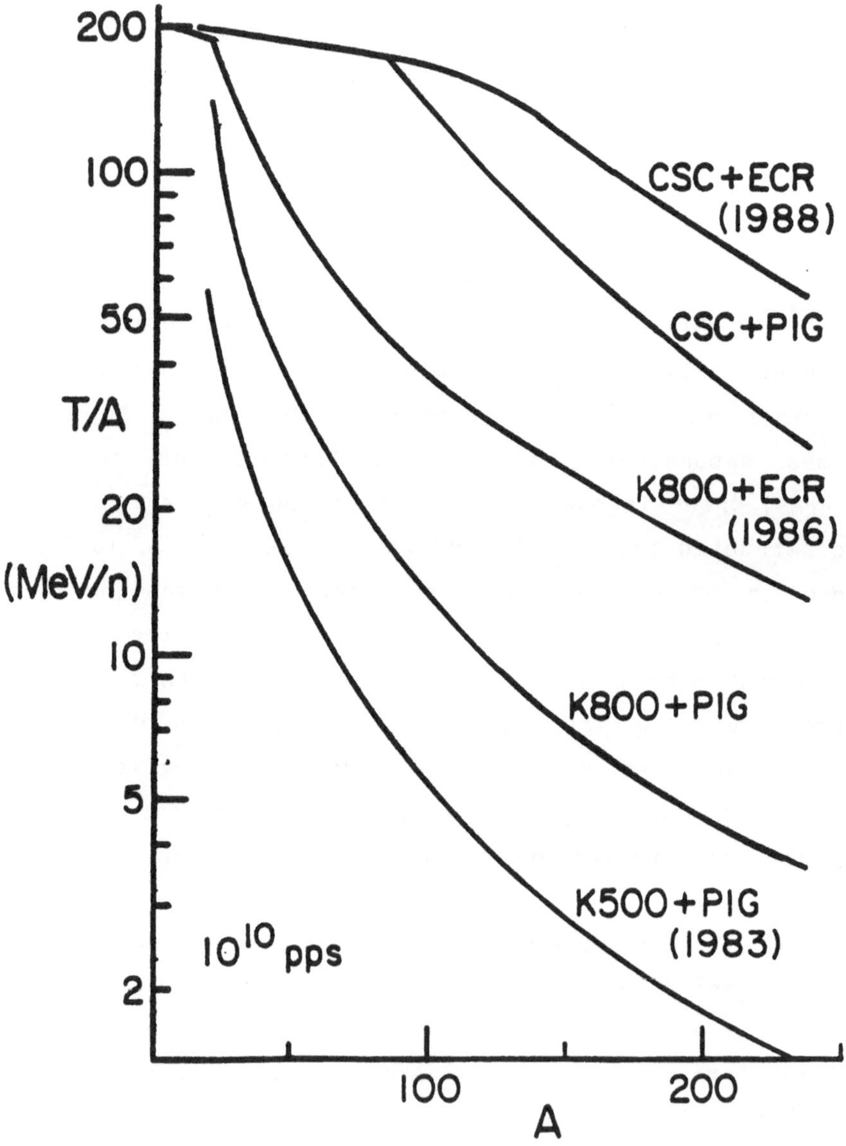

Figure 1. NSCL heavy ion beam energies per nucleon for various combinations of the two cyclotrons and ion sources. The years of initial operations of three of the combinations are indicated.

With heavy ion beams a variety of reaction mechanisms are available to make and study rare and new isotopes. At low energies (5-10 MeV/u) fusion and deep inelastic reactions have been used, at low to medium energies (5-25 MeV/u) direct and quasi-elastic reactions have been exploited, while at intermediate to high energies (40 MeV/u to over 200 MeV/u) the fragmentation mechanism has been prolific. A dramatic illustration of the potential of the fragmentation mechanism is given by the work of Westfall, _et al_[1] in which the particle stability of 14 new isotopes was demonstrated in one 8 hour run of a 200 MeV/u ^{48}Ca beam on a ^{9}Be target.

Hence, it was decided to design and build at MSU a device which would directly separate heavy ion reaction products and would be useful over a broad energy range, ~10-100 MeV/u. The reaction products would pass through the separator with flight times ~1 μsec so that isotopes with very short lifetimes could be studied. There would be no intermediate ion source so that there would be no chemical sensitivity in the isotope yields. The device would focus all isotopes with a range of energies but a selected value of m/q to a small spot in the focal plane. This is to permit studies of lifetimes and decay schemes of rare isotopes by physically separating them from abundant species and concentrating them in small detectors.

Thus the proposed device was an electromagnetic particle separator which focussed in x, y, and velocity space and was dispersive in m/q. In mass spectrometer language it was to be a "triple-focussing" device. Such a device must involve a combination of electric and magnetic dispersion adjusted to have zero velocity dispersion, $D_v \equiv$

$(x \mid \delta_v) = 0$, but non-zero mass dispersion at the focal plane, $D_{m/q} \equiv (x \mid \delta_{m/q}) \neq 0$. This is possible because electric dipoles disperse according to the ions' kinetic energy per unit charge, $T/q \sim 1/2\ (m/q)\ v^2$, whereas magnetic dipoles disperse according to the momentum per unit charge, $p/q \sim (m/q) \cdot v$. The mass spectrometer condition requires the magnetic dispersion to be of the opposite sign and twice the magnitude of the electric dispersion.

Three simple geometries which satisfy this requirement are shown qualitatively in Figure 2. Figure 2a) illustrates the classic mass spectrometer arrangement known as the Aston spectrometer, an E x B device followed by a magnetic dipole. Figure 2b) is the classic Mattauch spectrometer, an electric sector followed by a magnetic dipole. A third possibility, shown in Figure 2c), is an electric sector with the same bend angle as in 2b) but with twice the magnetic field used in a conventional Wien filter as in 2a). All three devices as shown have the same net bend angle, same electric flux, the same magnetic flux (with the magnetic flux equal to twice the electric), and, therefore, the same m/q resolving power. The focussing indicated by the quadrupole doublets in Figure 2 is only schematic; it could be integrated into the dispersive elements via appropriate edge angles or field gradients.

To our knowledge there are only three devices of the specific "triple-focussing" types discussed above currently in use in heavy ion nuclear science and one additional one nearing completion. (Due to lack of space other closely related devices such as the SHIP at GSI, the EMS built by the MIT group at BNL, and the MIT/ORNL velocity filter cannot be discussed in this paper. A review of all types of

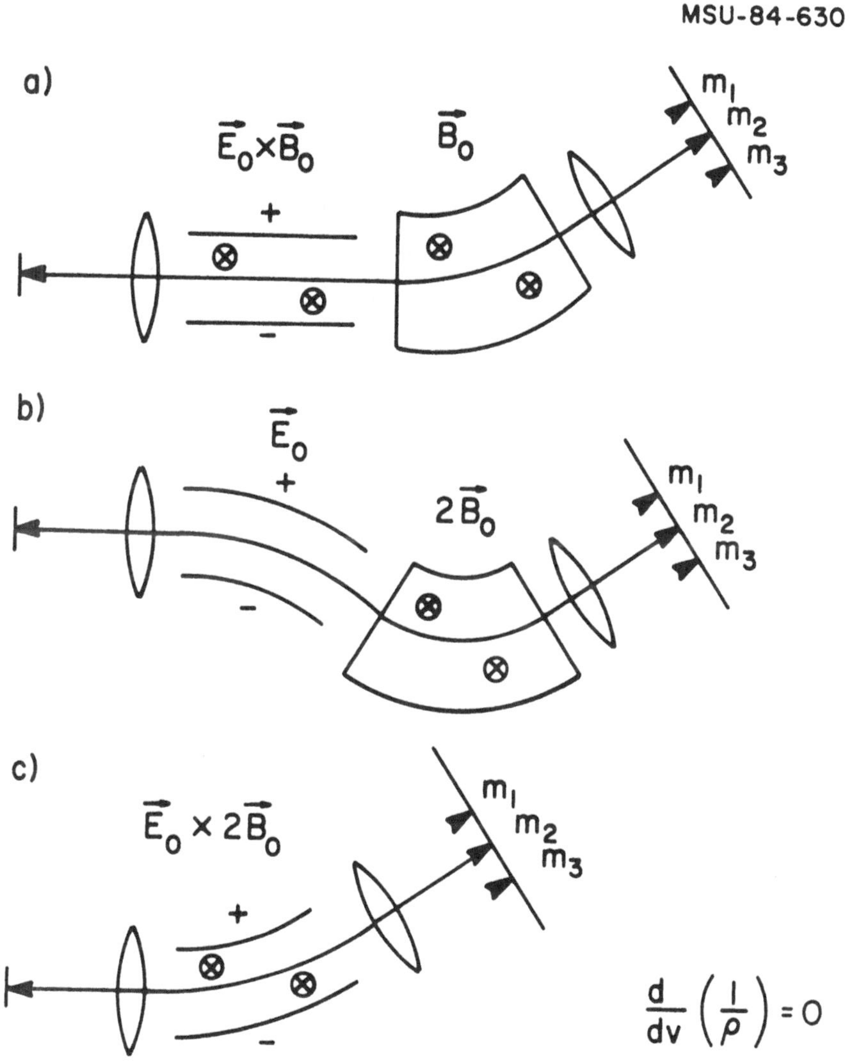

Figure 2. Schematic examples of three possible combinations of electric and magnetic dispersion which can be adjusted to be "triple-focussing" with mass dispersion.

recoil separators was presented recently by H. Enge.[2]) The Rochester RMS (Recoil Mass Separator) has been operating for about two years and is described in some detail later in this same session. The MSU RPMS was first tested in early-1984 and used for research in mid-1984. A device recently completed at Osaka (CARP) has by far the most sophisticated optical design of these triple-focussing devices.[3] It features a 55° magnetic bend followed by a 32° electric bend, includes 2 large quadrupoles, 3 sextupoles (multiples), and variable dispersion via moving the focal plane. It is to have a large solid angle (~15 msr) and large mass resolving power $(m/q)/\Delta(m/q)$ ~1000. The fourth device, nearing completion at Daresbury, will probably be operational in 1985.

In the early stages of the MSU design, K. Brown and H. Enge were consulted to help get the project started. We gratefully acknowledge their help in this area. Our original consideration of the "Aston" geometry, a Wien filter followed by a magnetic dipole, was stimulated by an early proposal which the MIT group had made for the Holifield facility at ORNL. The MSU design was also influenced by the availability of certain surplus equipment. The heart of the RPMS, a 5 m Wien filter, is a "meson separator" on loan from the Bevalac at LBL. The large 8" quadrupole doublet at the front of the RPMS was originally at SREL in Virginia, the 4" quadrupole doublet at the back of the device came from the University of Maryland, and the 6" gap magnetic dipole is from Cornell. These components arranged as shown in Figure 3 now form the 14 m long RPMS at MSU.

The most unique feature of the RPMS is that it is the only one of the triple-focussing mass separators designed for heavy ion energies above about 5 MeV/u. The layout as indicated in Figure 3 is "broad band" since it has no fixed bend angles to create an upper energy limit. The Wien filter is intrinsically straight and, hence, can be tuned for any velocity, with velocity dispersion decreasing with energy. With this configuration it is only the focal length of the first quadrupole doublet which sets the high energy limit at about 30 MeV/u for N=Z isotopes at this time. It may be necessary in the future to either extend the target-to-quadrupole drift length to more than the present 2 m value or change to a higher gradient doublet to increase this energy limit.

A second unique feature of the RPMS is that it is not fixed at 0°; the scattering angle can be varied from 0° to 30°. However, since the device is 14 m long this angle change is accomplished by pivoting about a point under the magnetic dipole (as indicated in Figure 3) and not around the target as in most magnetic spectrometers. This necessitates a lateral motion of about 1 m at the target and an "inflection" dipole capable of bending the beam by ±15°. A bellows system and "floating" scattering chamber arrangement have been devised to accommodate the variable lengths and directions of the beam taken in the vicinity of the target chamber. This system works well and was a cost-effective way to implement the 30° angular range within the space limitation of our experimental area.

The optical design of the MSU RPMS was presented at an earlier conference[4]. The calculations for the design were carried out to second order with a modified version of the

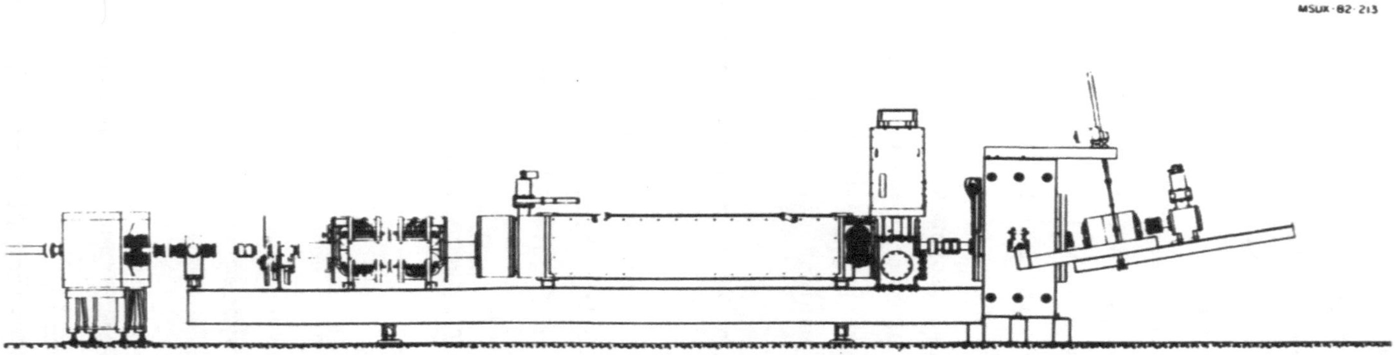

Figure 3. Pictorial drawing of the RPMS. The beam inflector magnet is the first component at the left, followed by the bellows system and target chamber, aperture holder, quadrupole doublet, Wien filter (5 m long) with high voltage supplies on the top right end, magnetic dipole and pivoting "tail" with quadrupole doublet and focal plane detectors at the right end.

widely-used code TRANSPORT. The modification involved removing all "path-length" terms in the first and second order matrices. The "path-length" positions in the beam vector and matrices were then taken by "m/q" terms. The fifth column is thus used for mass dispersion in the same way the sixth column is used for momentum dispersion. The first and second order matrix elements of the Wien filter were calculated in a separate program using the formulation of Ianoviciu[5] and used as input for the modified code TRANSPORT. For these design calculations the 5 m Wien filter parameters were assumed to be: maximum electric field of 40 kV/cm across a 10 cm gap and a maximum magnetic field of 600 gauss.

After the design was completed using the second order matrix calculation, its higher order performance was calculated with well-known code RAYTRACE developed by H. Enge and S. Kowalski of MIT. Figure 4 illustrates the performance of the RPMS as predicted by the RAYTRACE calculations. These results are also presented in Table I. The angular acceptance is 1° in the dispersive direction and 4° in the non-dispersive resulting in a solid angle of 1 msr.

In the initial design calculations for the RPMS the mass resolving power was severely limited by the (momentum) chromatic aberrations of the quadrupoles. For the central mass the device transmits a velocity or momentum range of about 10%. Even though the first order momentum dispersion, $(x|\delta)$, at the focal plane is zero the variation of the quadrupole focal length with momentum leads to a very large second order term, $(x|\Theta\delta)$. The effect of this term was

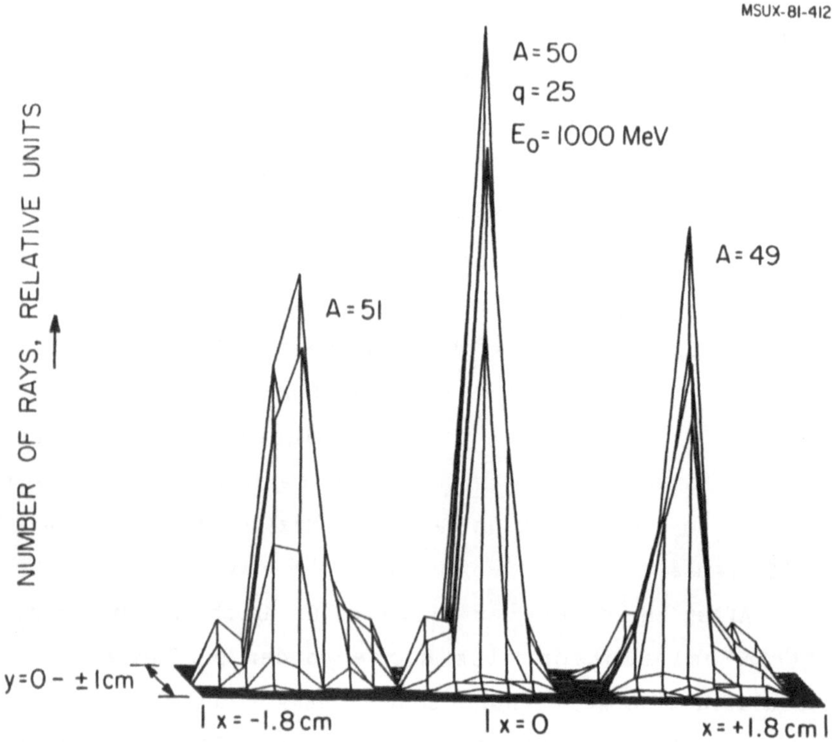

Figure 4. Two-dimensional histogram calculated with the program RAYTRACE, illustrating the RPMS mass separation at 20 MeV per nucleon.

Table I

RPMS	Parameters	(Prototype)
Layout:	QQWDQQ	(W Wien filter)
Length:	14.3 m	
Max Energy:	30 MeV/u	
Angular Range:	0° - 30°	
Resolution:	$\frac{M}{\Delta M}$ = 100 *(base-to-base)	@ 1 msr and 16% energy range.
	= 200**	@ 1 msr and 8% energy range.

* Calculated for E/A = 20 MeV and M = 2Q with TURTLE.

** Calculated for $^{40}Ca+^{40}Ca$ fusion products at E_{Beam}/A=5MeV.

reduced by noting that the displacement at the exit of the Wien filter is given by:

$$x = (x|\Theta)\ \Theta + (x|\delta)\ \delta + (x|\Theta\delta)\ \Theta\delta.$$

Hence, for positive Θ the second two terms add, whereas, for negative Θ they tend to cancel. This cancellation tended to increase the momentum bandpass of the device and broaden the calculated line width at the focal plane. By offsetting the first quadrupole doublet so that only positive Θ's are accepted this effect was reduced greatly and the mass resolving power increased. This, of course, resulted is a reduced momentum bandpass but the greatly increased resolution was considered a more important consideration. The offset did not reduce the geometric solid angle because the 8" aperture quadrupole doublet was large enough to match the 4" gap Wien filter even with the displacement of its axis. The mass resolution as calculated with RAYTRACE, with and without the offset, is indicated in Figure 5. The momentum bandpass calculated for the central mass with the quadrupole offset is shown in Figure 6.

An interesting result encountered during these design calculations is illustrated in Figure 7. For devices with large momentum bandpass the third and higher order terms can become as large or larger than the second order ones. In such cases raytracing is essential to predict system performance.

The initial experiments with the RPMS have used a 30 MeV/u ^{18}O beam on targets of ^{9}Be and ^{181}Ta to measure the yields and lifetimes of neutron rich light isotopes. The feasibility of such experiments with higher energy heavy ion

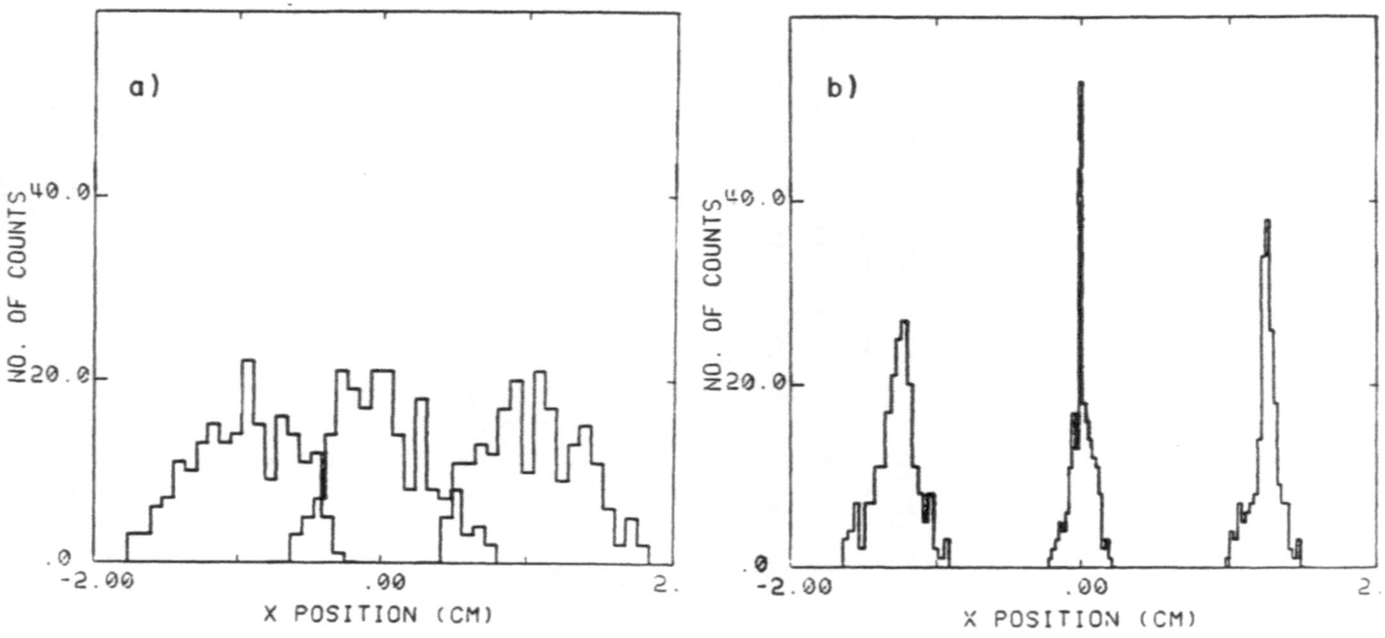

Figure 5. One-dimensional histograms calculated for three mass groups with q=25 and A=49,50 and 51 at 20 MeV per nucleon. a) With quadrupole axis the same as the Wien filter axis. b) With axis displaced by 4 cm to reduce the effect of chromatic aberrations.

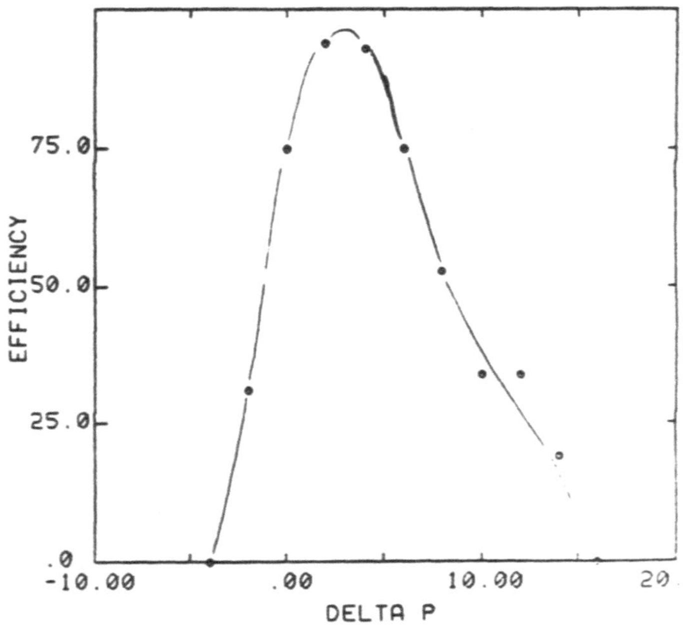

Figure 6. Calculation of the momentum bandpass of the RPMS with the axis offset.

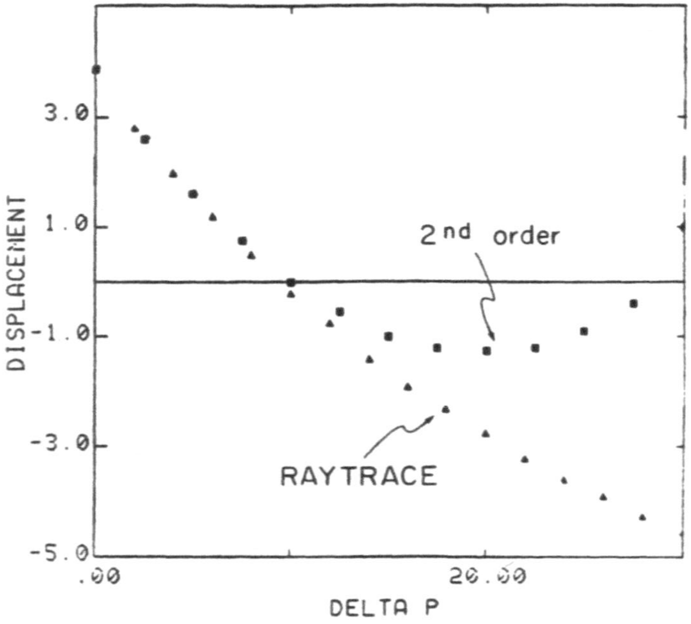

Figure 7. Calculation of the ray displacement at the exit of the Wien filter as a function of momentum. This illustrates the large difference between second order calculations and ray tracing for large momentum ranges.

beams was demonstrated originally by Murphy, et al.[6] at the LBL Bevalac.

The experimental setup in the focal plane of the RPMS for the present work is illustrated in Figure 8. For tuning the RPMS the 2-dimensional position sensitive gas proportional counter is most important. Initially the device was tuned and the velocity filter calibrated by putting the beam straight through it at 0° and focussing on a scintillator in the focal plane. Further tuning is done on light beam fragments such as $^{4}He^{2+}$ and $^{6}Li^{3+}$ for m/q=2 and $^{6}He^{2+}$ and $^{9}Li^{3+}$ for m/q=3. The Si-detector telescope is used for particle identification and the position sensitive counter to tune the magnetic dipole deflection angle to obtain a velocity focus. This tuning can be done with any beam such as ^{14}N before changing to the more expensive isotopic feed gas, ^{18}O. In the experiments done to date the beam has been either stopped or degraded to very low energy by the target. This is possible because the light neutron-rich fragments have more range than the beam. The reaction products are also penetrating enough to permit operating the gas counter and Si telescope in air beyond a vacuum window at the focal plane as indicated in Figure 8. Also, at these energies (≧ 10 MeV/u) the reaction products are mostly fully stripped which concentrates the yields of rare isotopes into one charge state. Lower charge states of more abundant isotopes are seen but they are easily identified by the ΔE-E telescope.

In the first experiments the neutron-rich isotopes ^{11}Li, ^{14}Be and ^{15}B, and ^{17}C have been observed and their lifetimes measured[7]. For ^{14}Be and ^{17}C these were the first lifetime measurements, and it appears that ^{14}Be is now the

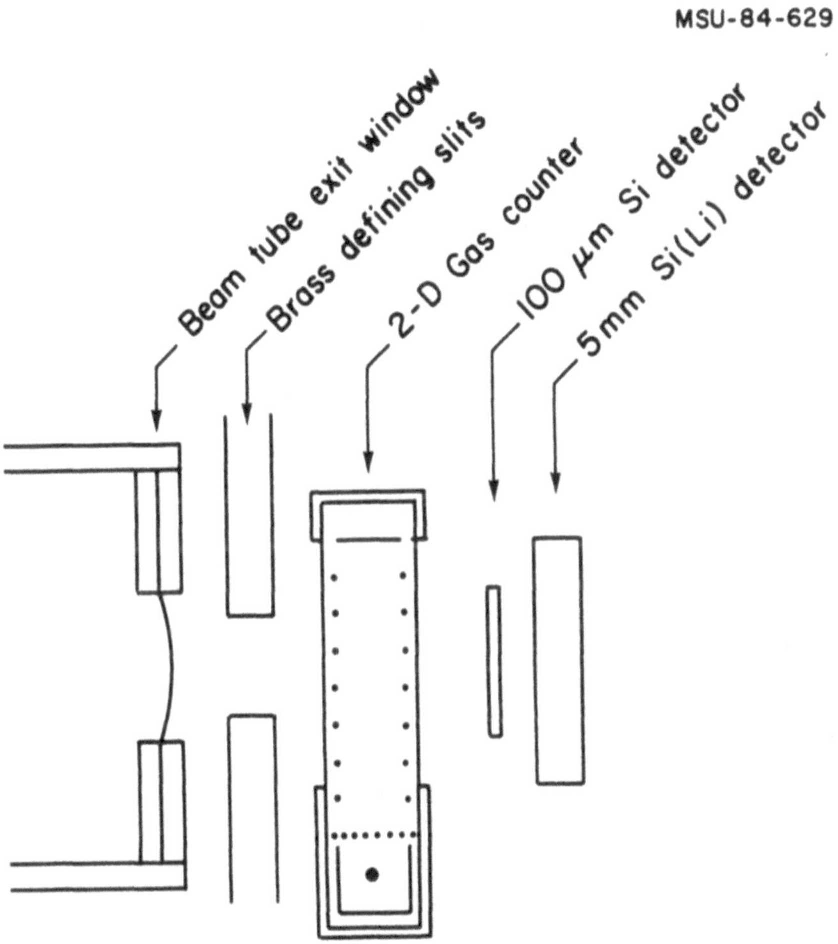

Figure 8. Schematic drawing of the focal plane detector arrangement for lifetime measurements.

shortest known half-life for beta decay, ~4 msec. During these measurements the "zoom-lens" feature of the RPMS was found to be very useful, as illustrated in Figure 9. In Figure 9a a very low dispersion was used, centered on m/q=3.0, permitting the measurement of the yields of many isotopes simultaneously. In Figure 9b higher dispersion was used to eliminate some of the more abundant isotopes such as ^{16}C while permitting the simutaneous measurement of the lifetimes of ^{15}B and ^{17}C. At even higher dispersion it was possible to eliminate the m/q=3 line and further clean up the measurement of ^{17}C (m/q=2.83).

In these experiments the decay curves were obtained by turning off the beam every time a heavy ion stopped in the second Si detector. A logic signal from the computer to the cyclotron rf phase turned the beam off in 20 μsec so that lifetimes of a few msec were easily measured. The beta decay signals were observed in the same Si detector in which the ions stopped. The beam off times were long enough (~1 sec) to prevent building of spurious activities. Hence, each beta decay event could be associated with the ion which was detected in the telescope, and decay curves for many isotopes could be accumulated separately during a single cyclotron run. A typical "on-line" computer display is shown in Figure 10, where the decay curves of six isotopes from Figure 9b were measured simultaneously.

In the future, with more elaborate focal plane detector arrangements many other experiments are possible. As higher energy heavy ions become available, yields will increase and more detailed decay scheme measurements can be carried out on these rare isotopes. For isotopes closer to stability the yields are larger and it should be possible to

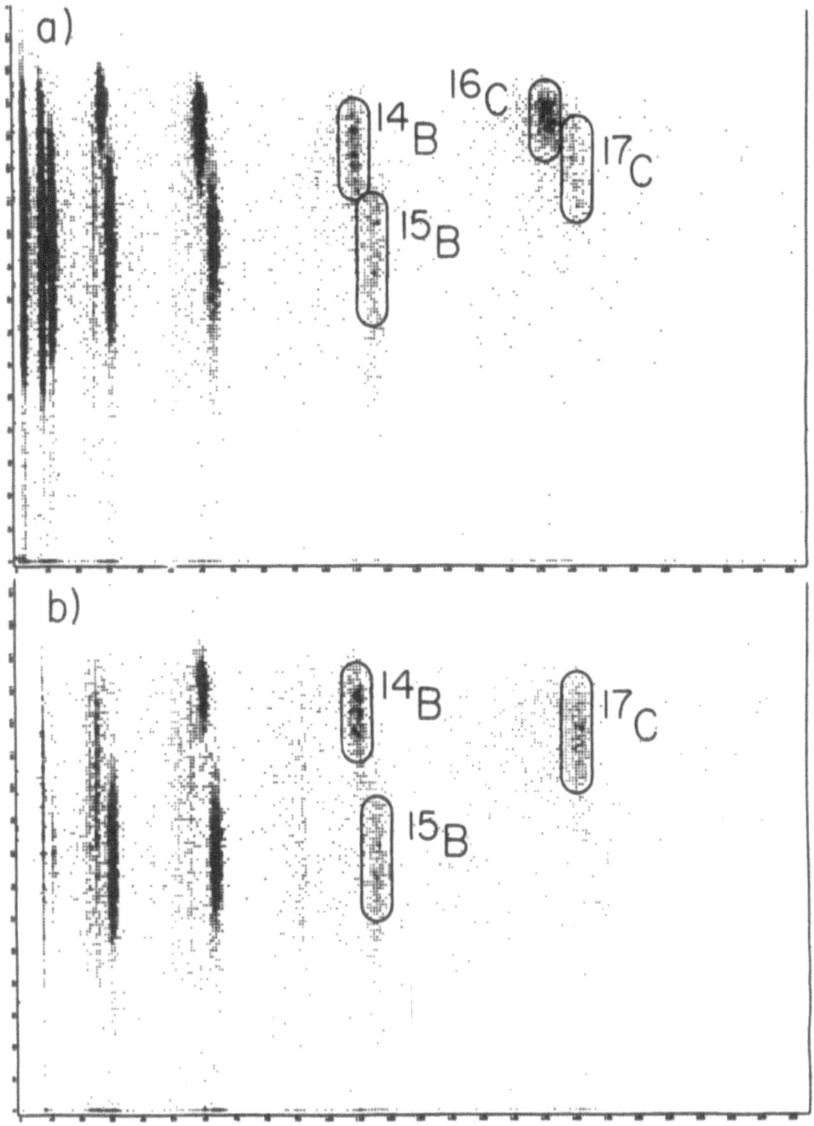

Figure 9. Vertical position vs particle ID signal for a) low mass dispersion and b) higher mass dispersion. In a) Z=1-6 groups are seen with ^{3}H, ^{6}He, ^{9}Li, ^{12}Be, and ^{15}B forming the m/q=3 horizontal line. Above this line ^{8}Li, ^{11}Be, ^{14}B, and $^{16,17}C$ are visible. In b) the Z=1 and 2 groups were mostly eliminated by electronic gates and the dispersion increeased to eliminate ^{16}C.

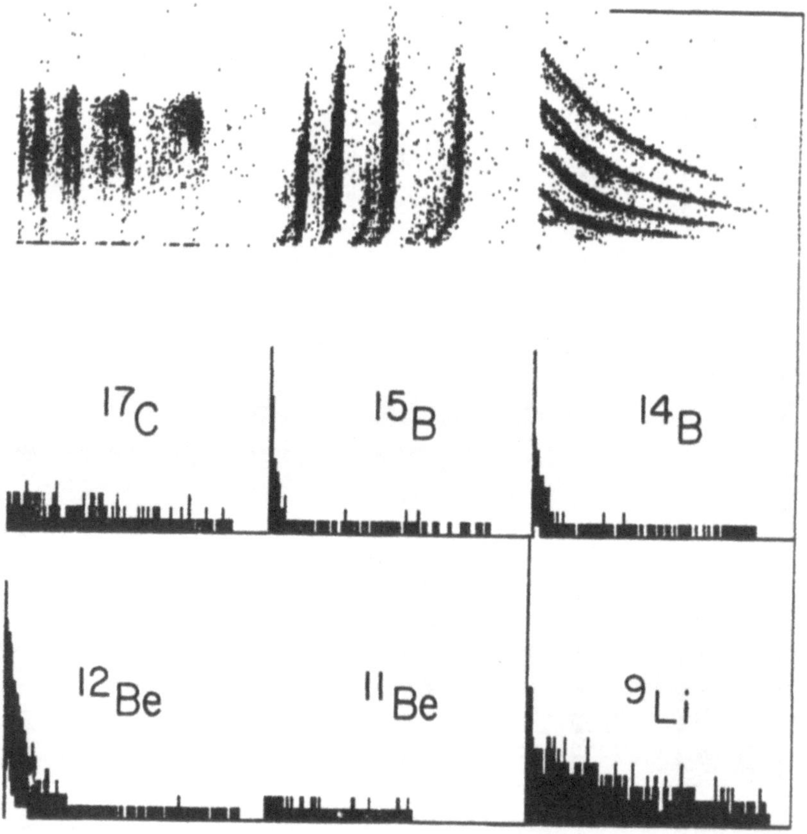

Figure 10. A typical on-line computer display during a run like that shown in Figure 9. The top three displays, left to right are: vertical position (m/q) vs particle ID, E-detector vs particle ID, and ΔE vs E. The other six displays are decay curves for the isotopes indicated. ^{9}Li, ^{11}Be, and ^{17}C are long-lived (≥ 100 msec) while ^{12}Be, ^{14}B, and ^{15}B are short-lived (≤ 20 msec).

carry out very precise measurements in cases of special interest.

This work was supported by Grant No. PHY-83-12245 from the U.S. National Science Foundation.

References

1. G.D. Westfall, et al., Phys. Rev. Lett. 43 (1979) 1959.
2. H.A. Enge, Nucl. Instr. and Meth. 186 (1981) 413.
3. S. Morinobu, I. Katayama and H. Nakabushi, Proc. Int. Conf. on Nuclei Far from Stability, CERN 81-09 (1981) 717.
4. L.H. Harwood and J.A. Nolen, Jr., Nucl. Instr. and Meth. 186 (1981) 435.
5. D. Ioanoviciu, Int. J. Mass Spect. and Ion Phys. 15 (1974) 89.
6. M.J. Murphy, T.J.M. Symons, G.D. Westfall and H.J. Crawford, Phys. Rev. Lett. 49 (1982) 455.
7. M.S. Curtin, L.H. Harwood, B. Sherrill, Z.-Q. Xie and J.A. Nolen, Jr., to be published.

SECONDARY BEAMS AT GANIL

M. LANGEVIN
Institut de Physique Nucléaire
BP 1
91406 ORSAY Cedex (France)

and R. ANNE
GANIL
BP 5027
F 14021 Caen (France)

The LISE facility at the French national heavy ion accelerator GANIL was designed for the production of secondary beams of good optical quality. The first experimental results obtained in atomic physics, radioactive beams production and search for new nuclear species are briefly reported.

1. INTRODUCTION

The new accelerator facility GANIL at Caen is designed to deliver heavy ion beams in the intermediate energy-range 20-100 MeV/u. It consists of a compact injector cyclotron associated with two isochronous separated sectors cyclotrons coupled together by intermediate charge stripping. Therefore rather highly stripped ions are delivered and additional stripping can produce fully stripped or hydrogen and helium like heavy ion species.

These beams are of great interest in atomic physics but

their practical use needs a beam of high optical quality which can only be obtained with magnetic separation and an achromatic beam transport device. This gave origin to the project "LISE" (Ligne d'Ions Super-Epluchés : Highly Stripped Ions Beam Line). Replacing the stripper foil by a thick target enables to transport the projectile fragmentation reaction products and, from the beginning, it was specified that this facility owing to some minor modifications would serve as well for research on nuclei far from stability. LISE was designed in 1981-82 and constructed in 1983.

2. GENERAL FEATURES OF THE LISE SPECTROMETER

LISE is installed in two adjacent concrete-shielded experimental halls (fig. 1). It is composed of :

a) Target or stripper

The incoming beam is directed and focused onto the target location by means of four magnetic quadrupole lenses. Up to ten different targets can be mounted on a water-cooled copper wheel. Easy target change is available by remote control of the wheel position.

b) Analysing section

The first magnetic dipole D1 (45° deflection, max. rigidity $B\rho \cong 3.2$ T.m, central trajectory radius $\rho = 2$m) serves as a dispersive element. The four quadrupole lenses adjacent to the dipole enable the adjustement of the dispersion on the intermediate dispersive focal plane. Remote controlled variable aperture slits are located at this position. An additional stripper foil, an energy degrader or a thin position-sensitive detector can be placed behind these slits. With a dispersion D = 1.84 cm/%, an object size at the target X_o = 0.15 cm and an image magnification G = 0.75 the nominal

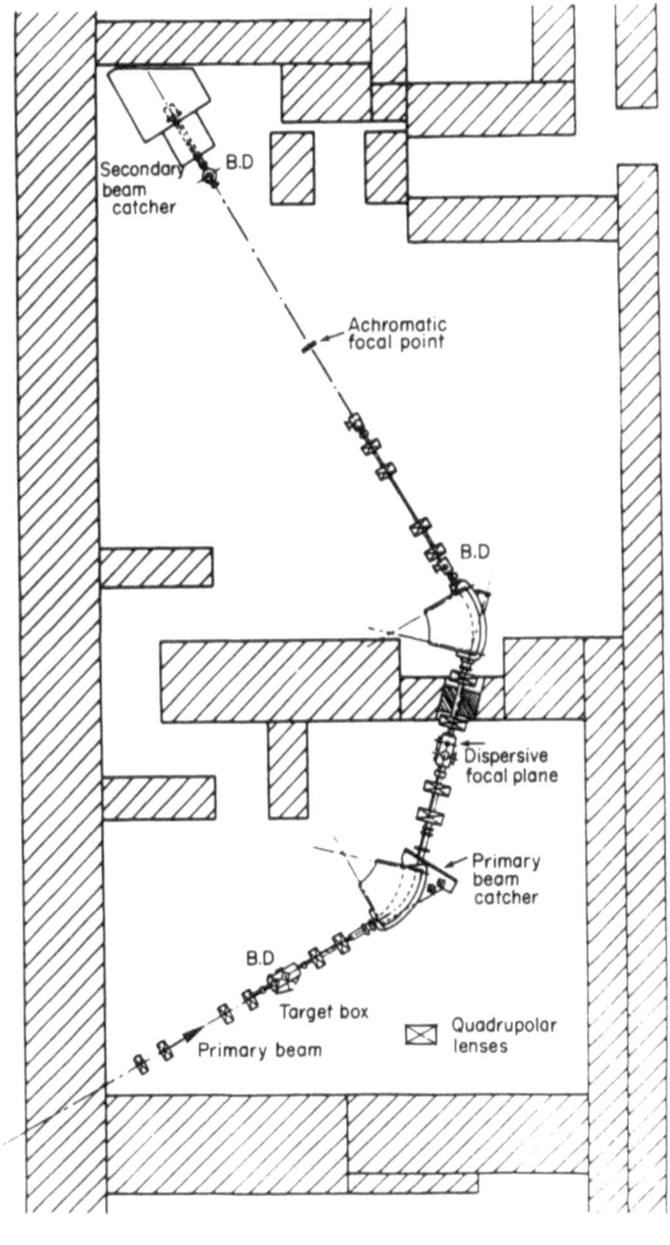

FIGURE 1 General lay-out of the LISE spectrometer
B.D. indicates the main locations of the beam diagnostics

$B\rho$ resolution of the system is determined to be $\frac{\Delta B\rho}{B\rho}=1.2\ 10^{-3}$

c) Dispersion compensating section

The achromatism in angle and in position is assumed after the dispersive focal plane by two quadrupoles and the second magnetic dipole D2. They allow for adjusting the chromatic terms in angle and in position for different uses of the LISE analyser. In the typical configuration the double achromatism is obtained at the exit of D2.

These conditions are normally obtained with the same magnetic field value in D1 and D2 (coils mounted in series). With the use of an intermediate energy degrader the double achromatic operation can be restored at the first order by the use of the additional excitation coils mounted on both dipoles.

d) Focusing system

The four last quadrupole lenses guide the beam to the experimental area. Due to the double achromatism assumed in front they deliver only focusing terms which permit two main configurations at the experiment location : beam focused on the horizontal and the vertical plane or focused in one direction and parallel in the other one .

c) Beam diagnotics

At each place of interest a modular beam diagnostic device allows the remote controled insertion of a Faraday cup and a beam profile monitor. The location of these beam controls are given in fig. 1. The beam profile and position is obtained for high intensities (> 10^8 pps) with a multiwire secondary emission monitor. On the other hand low intensity secondary beams are monitored by Ar-CO_2 filled multiwire proportional chambers used in the ion chamber mode ($\sim 10^8$ pps)

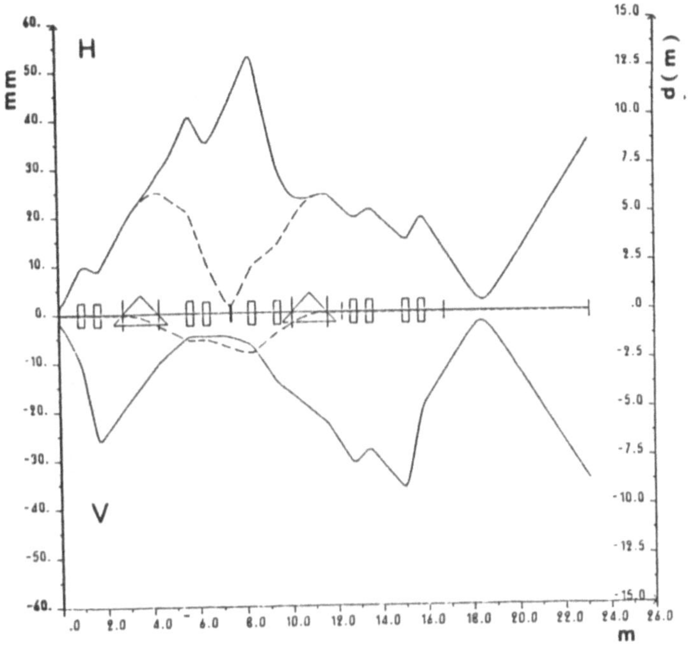

FIGURE 2 Beam transport schematics
Full lines are the beam horizontal (up) and vertical (down) envelopes for $x_o = 0.15$ mm $\varepsilon_o = 15\ \pi$ mm.mr and $\delta p/_p = \pm\ 2.5\ \%$
Interrupted lines are horizontal beam envelope with $\delta p/_p = 0$ (up) and horizontal dispersion coefficient (down). Total solid angle is 1 msr.

or in the proportional gas multiplication mode (down to 10^3 pps).

Fig. 2 gives the schematics of the beam transport and fig. 3 some of the beam profiles obtained with an ^{40}Ar (44 MeV/$_u$) primary beam. The distance between two adjacent wires is 1 mm.

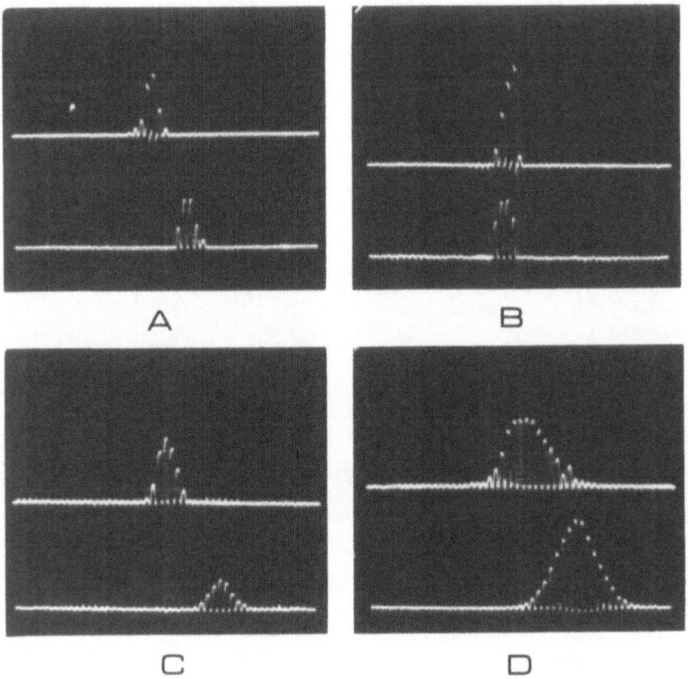

FIGURE 3 Beam profiles

a) Primary beam ($^{40}Ar^{16+}$ 44 MeV/n) at the target location

b) Ar^{18+} beam at the dispersive focal plane

c) Projectile fragmentation products at the final achromatic focus point

d) Reaction products at the achromatic focus point with a 40 mg/cm^2 Al degrader behind the dispersive focal plane and corresponding achromatic tuning of the spectrometer

a) and b) were obtained with the secondary emission profile monitors, c) and d) with the gas multiplication profile monitor

All the elements of the spectrometer : magnets, slits and beam diagnostics are controled and actuated by an on-line computer. In the usual working conditions a very weak intensity beam is transmitted in presence of a high intensity primary beam. Therefore largely-dimensioned C-type magnetic dipoles allow primary beam catching far away from the transmitted beam. In order to protect radiation sensitive beam profile monitors or experimental devices any spurious change in the dipole magnetic fields is detected by pick-up coils and instantaneously stops the accelerator beam. More technical informations about the LISE spectrometer is given in ref. [1] .

3. FIRST EXPERIMENTAL RESULTS

a) Atomic physics

During the ^{40}Ar (44 MeV/$_u$) beam tests a preliminary study of the stripping processes was performed. The experimental results for Carbon and Gold foils are given in tables I and II

TABLE I Carbon stripper

Ar charge state	16^+	17^+	18^+
100 μg/$_{cm^2}$	61 %	34 %	5 %
250 μg/$_{cm^2}$	27 %	45 %	28 %
450 μg/$_{cm^2}$	5 %	39 %	56 %

TABLE II Gold stripper

Ar charge state	16^+	17^+	18^+
160 $\mu g/cm^2$	2 %	17 %	81 %
300 $\mu g/cm^2$	<0.5 %	10 %	90 %

These experimental values are in excellent agreement with theoretical predictions [2] taking into account a mean free path $\lambda = 350\ \mu g/cm^2$ for carbon and $\lambda = 53\ \mu g/cm^2$ for gold.

In the first beam foil spectroscopy experiment the achromatic parallel beams of hydrogen-like and helium-like krypton (30 MeV/u) were used for the lifetime measurement of 2^1S_0 and $2^2S_{1/2}$ atomic excited states. The decay curve of the 2E1 decay of the $^1S_0$ level is shown in fig. 4 [3]. The precision of a few percent is outstanding and proves that LISE is an exceptional instrument in that field.

Other atomic physics experiments performed with LISE are in way of data analysis.

b) Nuclear Physics

The Bordeaux-GANIL collaboration [4] operates LISE as a Projectile Fragmentation Isotope Separator (PFIS). In addition to the Z/A selection assumed by the magnetic analysis the isotopes undergo a $Z^{1.5}/A^{2.5}$ selection due to a differential slowing down provided by an energy degrader instal-

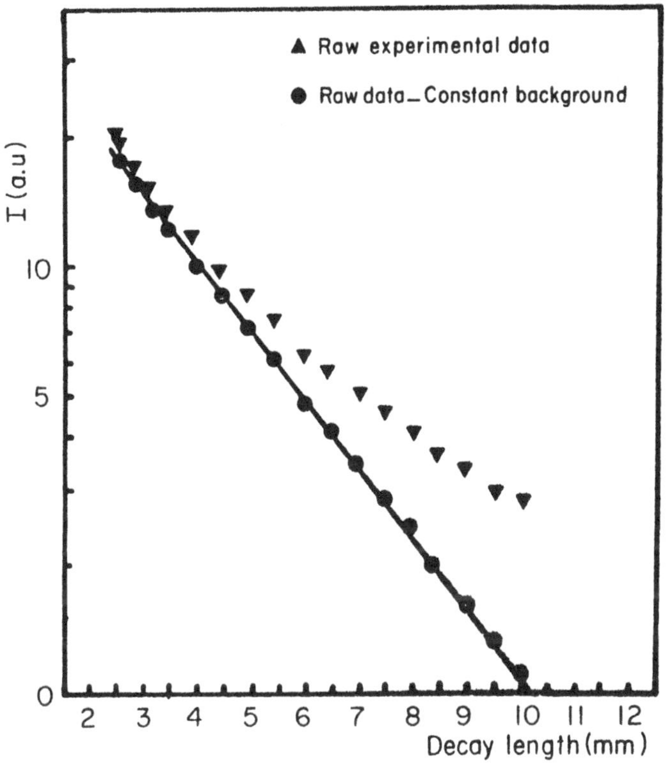

FIGURE 4 Decay of the Kr 2^1S_0 excited state
The corresponding half-life is 3.3 ± 0.2 10^{-11}s

led behind the dispersive focal plane. This effect can be reinforced taking into account the different range of the fragments in gaseous or solid materials. In a first recent run using a 44 MeV/u ^{40}Ar beam on beryllium targets the two isotopes ^{34}P and ^{35}P were collected. They were identified through β-γ coincidences and a mass resolution $\Delta A/_A$= 1% was observed.

For the fragments rather close to the projectile mass

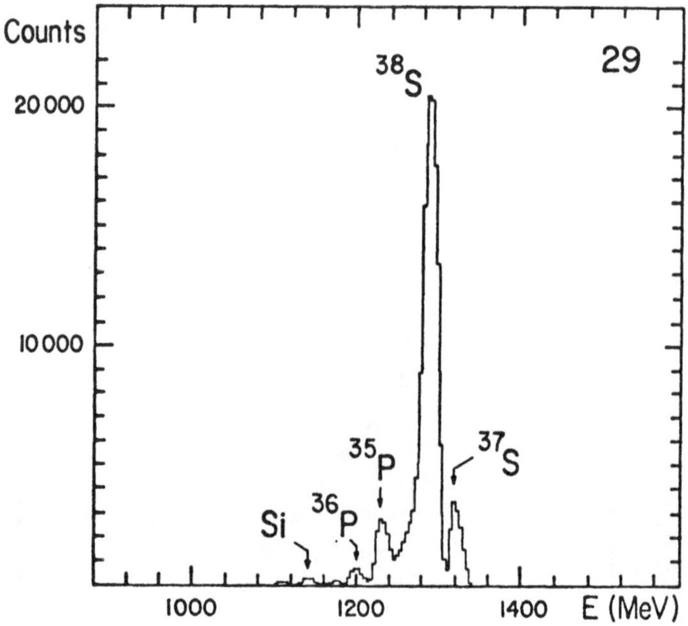

FIGURE 5 Energy spectra of the ^{38}S radioactive beam analysed by a solid state ΔE - E telescope

the LISE spectrometer can deliver radioactive beams with a good energy resolution and a small amount of contamination. Table 3 gives the main characteristics of the beams obtained in a recent test [5]. Figure 5 illustrates the energy resolution and the small amount of contaminating beams for a ^{38}S radioactive beam.

The collaboration IPN-Orsay - GANIL [6] has developed a system for identification in Z and A of the produced fragments. The immediate aim is to identify new exotic nuclei and possibly reach the limit of particle stability up to $Z \cong 18$ for the neutron-deficient and $Z \cong 10$ for the neutron-rich isotopes.

TABLE III Secondary beams produced by interaction of a 44 MeV/u ^{40}Ar beam with a 100 mg/cm^2 Be target. The symbol E/A denotes the secondary beam energy per nucleon, I and I_o are the secondary and primary beam intensities, respectively, I_{max} the secondary beam intensity for a standard GANIL ^{40}Ar beam (I_o=3.10^{11}pps), ΔE/E the FWHM of the secondary beam energy distribution.
All these values were obtained with a ±9mm slit aperture in the dispersive focal plane.

Ion	E/A (MeV/u)	I/I_o	I_{max} (pps) (I_o=3.10^{11}pps)	ΔE/E	Contaminants
^{39}Cl	33.9	(2±1)10^{-5}	6.10^6	±0.9%	^{40}Cl (5%)
					$^{36+37}$S (10%)
					others ($\lesssim$1%)
^{38}S	33.8	(0.5±0.2)10^{-6}	1.5.10^5	±0.8%	^{37}S (10%)
					^{39}S (3%)
					$^{34+35}$P (10%)
					others ($\lesssim$1%)

The produced fragments are detected in a telescope consisting of two ΔE Si surface barrier detectors (thickness 500 μm and 1000 μm), a 4000 μm thick Si-Li E detector and a veto detector. The time-of-flight is measured between the target and the telescope providing a basis of 18m. Hence, the fragments are identified doubly in Z (ΔE_1, ΔE_2 and time-

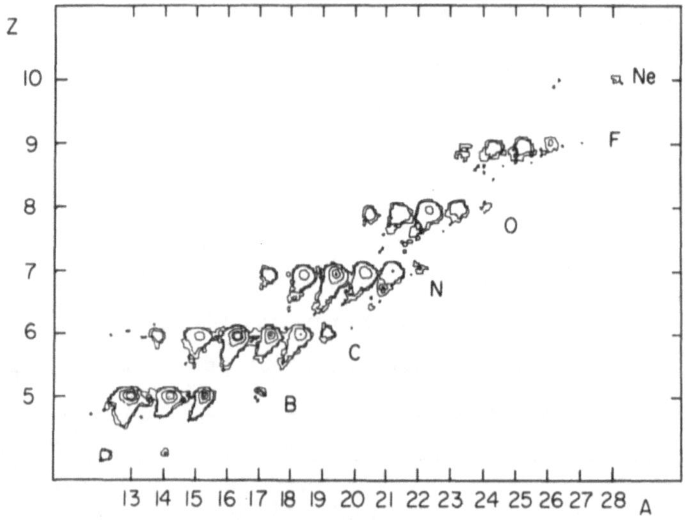

FIGURE 6 Z and A identification diagram.
The levels are chosen to emphasize the ΔE_1 detector defects.

of-flight) and in A (total energy and time-of-flight). In a first test of the method, a 85 mg/cm^2 tantalum target has been irradiated by the 44 MeV/A ^{40}Ar beam with an intensity of 1.5×10^{11} particles/s.

A 20 μm aluminum stripper foil was placed at the dispersive focal plane between the two dipoles to avoid the high counting rate due to the remaining 13^+ and 14^+ charged states of the primary beam. Under these conditions the reaction products are completely stripped and with the magnetic analysis the time of flight of about 200 ns of the detected fragments from 100 ns spaced beam pulses does not give rise to any ambiguity in the Z and A determination.

This assignation was performed by double Z identification (ΔE_2 and time of flight) and single A identification

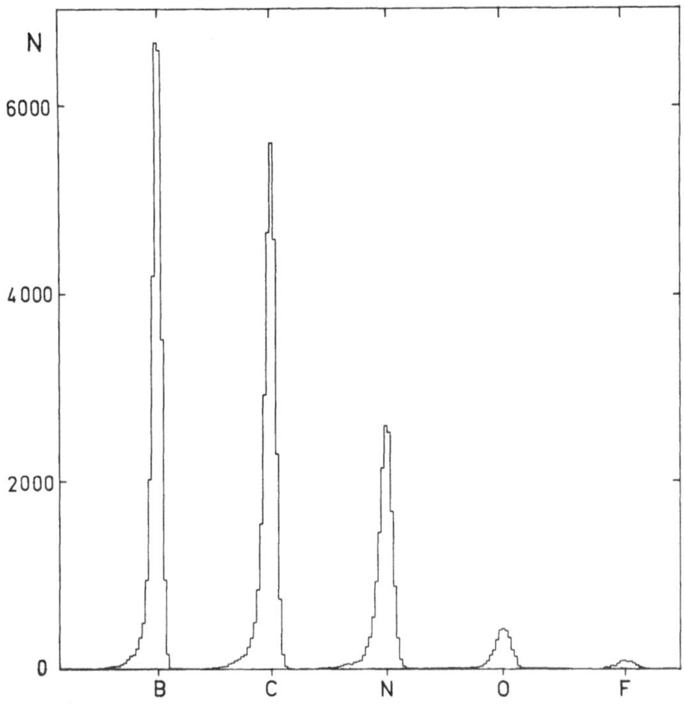

FIGURE 7 Z histogram for all the fragments collected at $\overline{B\rho}$ = 2.66 T.m. N is the number of counts per bin

(total energy and time of flight). The background was furthermore reduced by rejecting all events for which the Bρ calculation did not fall within the known acceptance of the spectrometer. The identification diagram is given fig. 6. The Z identification (fig. 7) is excellent ($\Delta Z/_Z \cong 1.3\%$) without any overlap between adjacent elements but with a slight defect due to the poor resolution and asymetric behaviour of the first 500 μm totally depleted Si surface barrier detector (fig. 8). This detector was still selected amongst many commercially available 300 μm and 500 μm Si

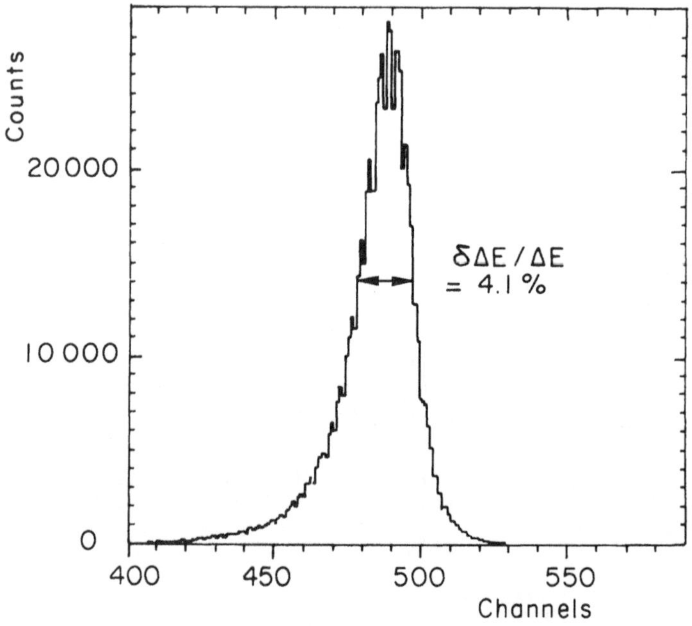

FIGURE 8 500 μm thick ΔE_1 detector response to 44 MeV/u ^{40}Ar ions

(SB) detectors hence the results obtained recently (fig. 9) with a planar ion-implanted Si detector [7] show that further improvement can be obtained in the near future.

The mass separation is good ($0.95\% \leq \frac{\Delta M}{M} \leq 1.45\ \%$) and is limited by the present width of the GANIL beam pulse.

The optimum setting of the LISE spectrometer for a given reaction fragment (Z,A) relies on the reaction mechanism which defines its most probable energy and on the energy loss effects in the target. The spectrometer was tuned for the nuclei of interest taking into account the experimental results of ref. [8] and the energy loss values of ref. [9]. This tuning was experimentally checked by the Bρ distribu-

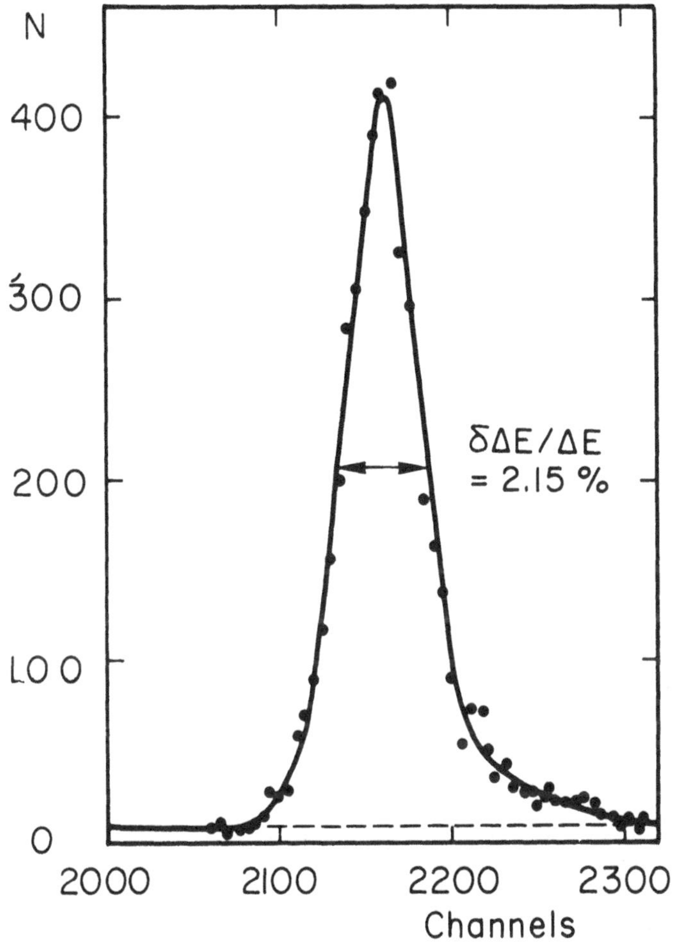

FIGURE 9 ΔE response of a 300 μm planar totally depleted Si detector to 35 MeV/$_u$ ^{40}Ar ions

tion of the events for each observed isotope. For $5 \leqq Z \leqq 10$ at least three isotopes are collected altogether with a relative efficiency $\geqq 75$ %. Due to the small angular acceptance and the lack of knowledge of the angular distribution of the projectile fragments close to 0° at these interme-

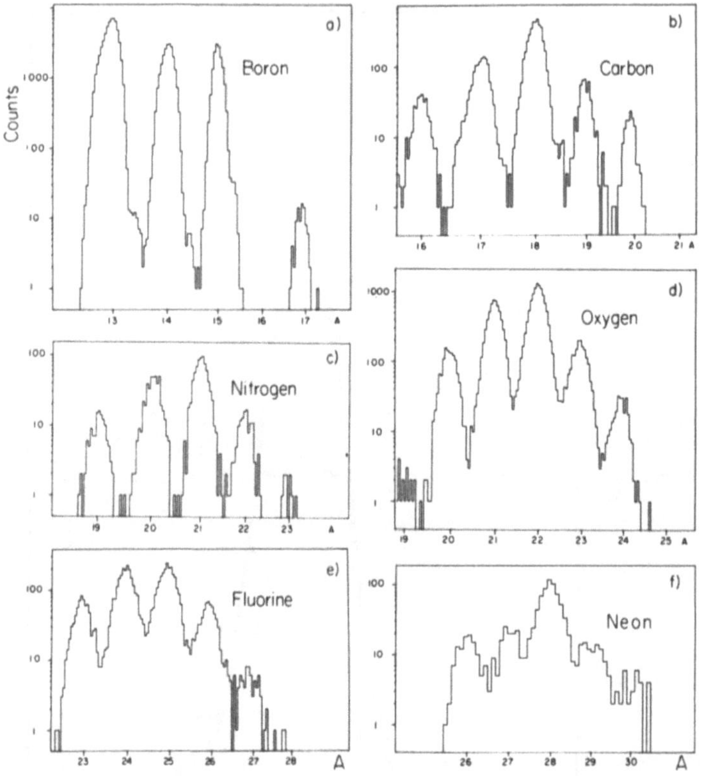

FIGURE 10 Mass histograms of neutron rich isotopes for Boron (a), Carbon (b), Nitrogen (c), Oxygen (d), Fluorine (e) and Neon (f).
Histograms (a) (d) (e) (f) were obtained at $\overline{B\rho}$ = 2.66 T.m (optimal setting for $^{15-16}B$, $^{23-25}O$, $^{26-28}F$ and $^{29-31}Ne$ respectively), (b) and (c) at $\overline{B\rho}$ = 2.92 T.m (optimal setting for $^{19-21}C$ and $^{21-23}N$ respectively.

diate energies only relative yields can be obtained.

Two settings of the spectrometer ($\overline{B\rho}$ = 2.66 T.m and $\overline{B\rho}$ = 2.92 T.m) were used throughout the experiment both with an exposure time of about ten hours. The experimental results are given in fig. 10.

No background is observed as can be seen in particular at the ^{16}B and ^{18}B locations in fig. 10[a]. With the $\overline{B\rho}$ = 2.92 T.m setting which favours masses 17 and 18 instead of masses 15 and 16 more than 300 ^{17}B events were observed without a single event recorded at the mass 18 location.

The first observation of ^{23}N, ^{29}Ne and ^{30}Ne and their particle bound character result clearly from the mass histograms of fig. 10[c] and 10[f]. The experimental results for carbon (fig. 10[b]), oxygen (fig. 10[d]) and fluorine (fig. 10[e]) confirm the existence of ^{20}C and ^{27}F but do not reveal any evidence for the possible existence of ^{21}C and ^{25}O claimed in ref.[10]. The non-observation of an isotope cannot carry the absolute proof of its unbound character. However in the present work as well as in the high energy fragmentation process (fig. 2 and 3 of ref. 11) the yield decrease from A to A + 1 for very neutron rich isotopes averages 7.5. Odd-even effects seldom bring it to more than a factor of ten. Hence a factor more than 100 can be considered as a good criterion for the particle unstability of the A + 1 isotope. Under this rule ^{18}B, ^{21}C and ^{25}O can be considered as particle-unstable.

The preceeding results are proof of the interest of medium energy heavy ions beams for the search of new exotic isotopes. All the results reported here are also proof of the outstanding performances of the LISE facility in the field of intermediate energy heavy ions physics.

REFERENCES

1. R. Anne and C. Signarbieux GANIL Report RA/$_{NJ}$ 278-82
2. J.P. Rozet private communication and to be published
3. R. Marrus et al., 10th Conference in Atomic Physics, Seattle 1984
4. J.P. Dufour, R. Del Moral, H. Emermann, A. Fleury, F. Hubert, C. Poinot, M.S. Pravikoff and H. Delagrange private communication and to be published
5. R. Bimbot, S. Della-Negra, P. Aguer, R. Anne, H. Delagrange and F. Hubert, private communication and to be published
6. M. Langevin, E. Quiniou, M. Bernas, J. Galin, F. Naulin F. Pougheon and R. Anne, C. Détraz, D. Guerreau, D. Guillemaud-Mueller, A.C. Mueller, submitted to Physics Letters
7. J. Kemmer, P. Burger, R. Henck and E. Heijne, IEEE Trans. Nucl. Sci. NS-29, 733 (1982)
8. V. Borrel et al., Z. Phys. A 314, 191 (1983)
9. F. Hubert et al., Ann. Phys. Fr. 5, 1 (1980)
10. J.D. Stevenson and P.B. Price, Phys. Rev. C 24, 2102 (1981)
11. G.D. Westfall et al., Phys. Rev. Lett. 43, 1859 (1979)

ON-LINE ISOTOPE SEPARATION AT CHALK RIVER AND ELSEWHERE

J.C. HARDY, H. SCHMEING and E. HAGBERG
Atomic Energy of Canada Limited
Chalk River Nuclear Laboratories
Chalk River, Ontario, Canada KOJ 1J0

Isotope separators are installed on-line to accelerators and reactors at many laboratories worldwide. Critical aspects of such facilities are described, including isotope-production techniques, types of ion sources and separation quality. Some examples are given of experiments performed with on-line separator beams.

1. INTRODUCTION

An on-line isotope separator (ISOL) is any device that accepts directly the products of a nuclear reaction and produces with them an accelerated beam that is magnetically separated into individual beams, each characterized by a unique atomic mass. A block diagram of the functional components of an ISOL appears as Figure 1. The dashed boxes indicate that the functions are not always distinct: in some cases, the product nuclei are stopped in the target itself, while in others they are allowed to recoil into the ion source where they stop; in still

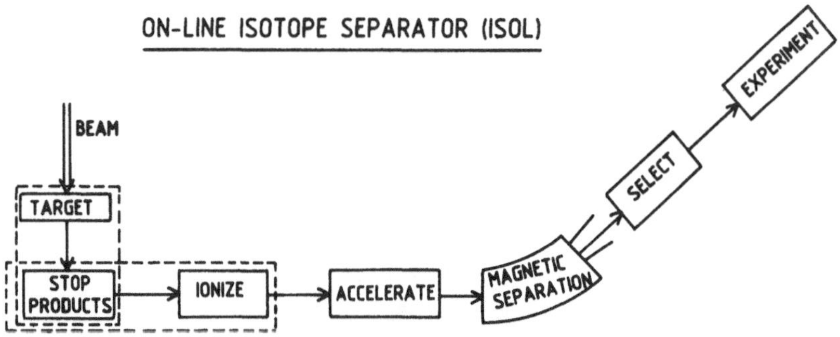

FIGURE 1 Functional Block Diagram of an on-line isotope separator

another they are stopped in a gas where they remain singly ionized, thus obviating the need for an ion source at all. However, in all cases product nuclei are singly ionized and accelerated; and it is ultimately these features that distinguish an ISOL from a recoil spectrometer, which employs the charge state and direct recoil momentum that the products acquire in the initial reaction. This simplification makes the recoil spectrometer faster but, by requiring thin targets, it usually produces weaker beams, and these are less well resolved from contaminants.

An ISOL is only rarely an experimental end in itself. By now, such instruments have been available for two decades; they have fed and are feeding[1] an enormous variety of experimental devices, often simultaneously through a beam transport network that could rival in itself the capability of some small accelerator facilities. The beam currents, too, are impressive: Figure 2 shows what is actually possible with existing technology[2]. Notice that, in all, beams of 39 isotopes of caesium have been produced, 12 of them at more than one nanoampere! Many other elements are available nearly as prolifically.

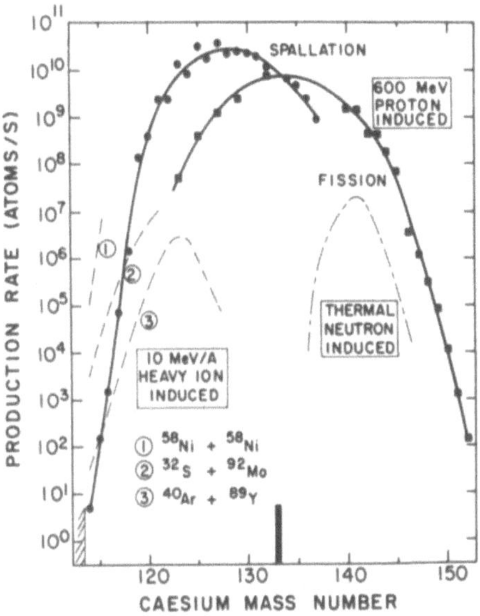

FIGURE 2 Production rates for mass separated beams of caesium isotopes generated by reactions induced with 600 MeV protons at ISOLDE[2] (heavy solid lines and points), by thermal neutron fission of uranium at OSIRIS[3] (dot-dashed line at the right) and by 10 MeV/A heavy ions (dashed lines at the left). The latter curves are the results of calculations with the computer code ALICE[4], which have some experimental support (eg. ref. 5).

2. PRODUCTION TECHNIQUES

While the separator beam currents illustrated in Figure 2 do depend upon the quality of the ion source and separator, ultimately it is the nuclear reaction used to produce them that sets the scale. It does so not only through the reaction cross section but through the effective target thickness afforded by the projectile used. As suggested clearly by the Figure, intermediate energy protons are a highly productive tool. Cross section measurements illustrate part of the reason why. In Figure 3 the results[6] from proton bombardment of a lead target at energies from 0.3 to 3.0 GeV show that many nuclides are strongly produced, particularly as the projectile energy is increased. With protons, such ≈10 mb cross sections can be combined with targets of ≈15 g/cm^2 thickness to yield the enormous ISOLDE[2] production (see Fig. 2) at the relatively modest proton beam current of 1 μA.

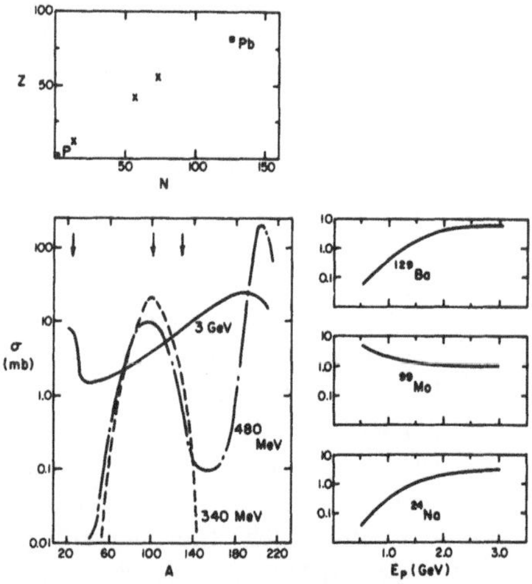

FIGURE 3 Measured cross sections[6] are plotted at the lower left as a function of product mass number for the interaction of lead with protons of various energies. The arrows mark the cuts displayed on the right. The simplified nuclide chart at the upper left illustrates the projectile, target, and the three specifically selected product masses.

By comparison, heavy ion beams have both advantages and disadvantages. At low bombarding energies ($E/A \leq 10$ MeV) heavy ions afford a highly specific production technique, with similar ≈10 mb cross sections[7,8], as shown in Figure 4. The nuclides in the region of peak production can also be chosen very far from stability, resulting in producton rates there in excess of

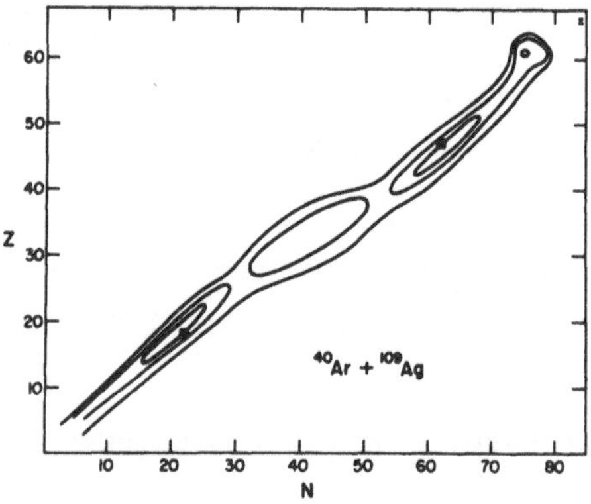

FIGURE 4 Cross section contours plotted as a function of residual nucleus for a reaction between 7 MeV/A ^{40}Ar and ^{109}Ag (each represented as a solid square). The contours are based partly on experiment[7-10] partly on calculations with ALICE[4], and partly on guess work. The lines correspond to 0.1, 1.0 and 10 mb, and the x in the upper right marks the location of the compound nucleus.

those from proton-induced reactions (see Fig. 2) despite a severe handicap: targets that cannot be thicker than a few mg/cm^2. Nearer stability, however, target thickness restricts the production rate from low-energy heavy ions to well below that from protons.

At higher energies, heavy ions behave in a similar way to protons, ultimately displaying a nearly constant 10 mb production cross section for all A values. The striking observation from Figure 5 is that they reach this level at

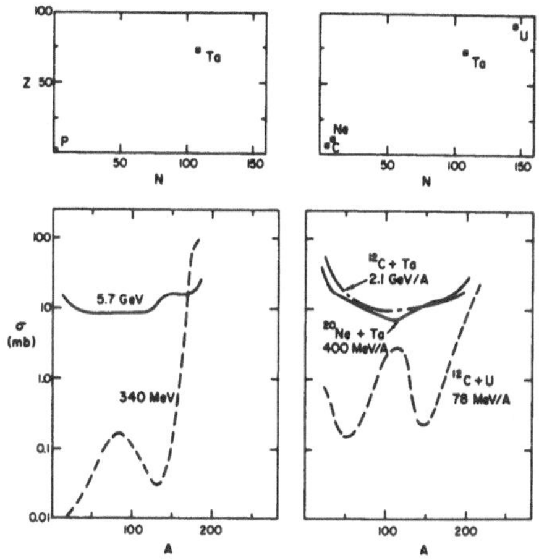

FIGURE 5 Measured cross sections[11,12] plotted for protons on Ta at the left and for various combinations of projectiles and targets as indicated on the right.

much lower specific energy (E/A) than protons do. Although it is not illustrated, there is also evidence[2,13] that even at these energies heavy ions have an advantage far from stability by producing a broader range of N and Z for each A. Most important, though, is the fact that thick-target techniques now become feasible for heavy ions as well. The consequence, shown in Figure 6, is that at 50 MeV/A, ^{12}C projectiles should be the equivalent of 600 MeV protons for isotope production and are possibly better for the most exotic products. Although no ISOL has yet exploited this observation, there seems little doubt that the best

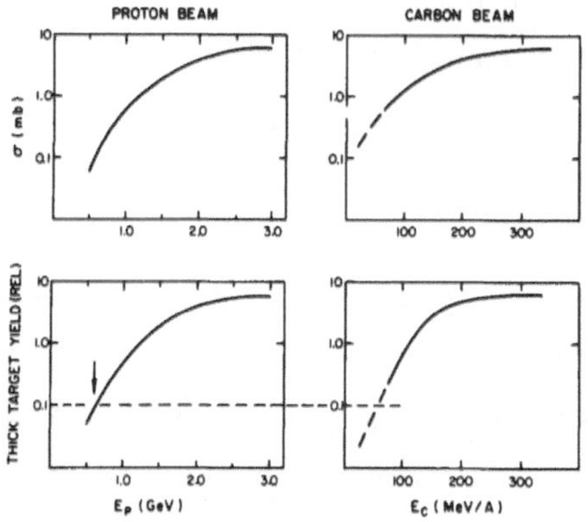

FIGURE 6 Cross sections and thick-target yields plotted as a function of beam energy for proton and carbon beams. The results approximate the yield of A ≈ 130 products from a heavy (Pb) target.

production results in Figure 2 could be matched and exceeded at a medium energy heavy-ion facility.

3. ION SOURCES

A comprehensive review of ISOL ion sources is beyond the scope of this work. Only a few examples are presented to give the flavour of a broad range of possibilities. An excellent detailed review was published[14] by Kirchner in 1981.

The FEBIAD[15] ion source is the workhorse of many ISOLs. It is shown in Figure 7 in the version used with

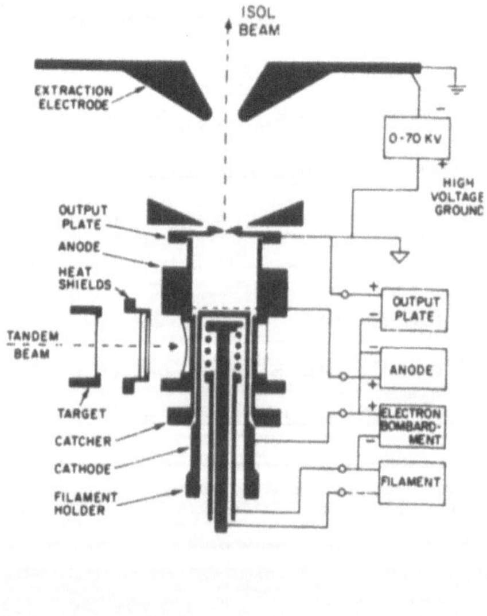

FIGURE 7 Schematic diagram of the Chalk River FEBIAD ion source[16].

the Chalk River ISOL[16], but other versions are used at GSI[17], where it was developed, and at ISOLDE, LISOL[18] and TRISTAN[19]. In the one shown, reaction products ejected from a thin target stop in the hot catcher that surrounds the cathode, from which they subsequently diffuse into the discharge region of the source. Special features of the FEBIAD include electron-bombardment heating of the covered cathode, and a grid to extract fast electrons into the discharge. It is characterized by a long cathode lifetime, high temperature, good efficiency (up to 15%), low pressure and low current density. It is also versatile.

While versatility is often a benefit it can, on occasion, be a curse. Most separators cannot resolve neighbouring isobars, so to produce a mono-isobaric beam it is necessary to have elemental selectivity in the ionization. One way to do so is to employ a source that depends on surface ionization - either positive or negative. One example is the OASIS[20] ion source shown in Figure 8. Once again, reaction products recoil into the source but here they are ionized on the hot (>2500°C) walls of the tungsten cavity. This type of source, like the FEBIAD, can easily be adapted to thick target use, in which case

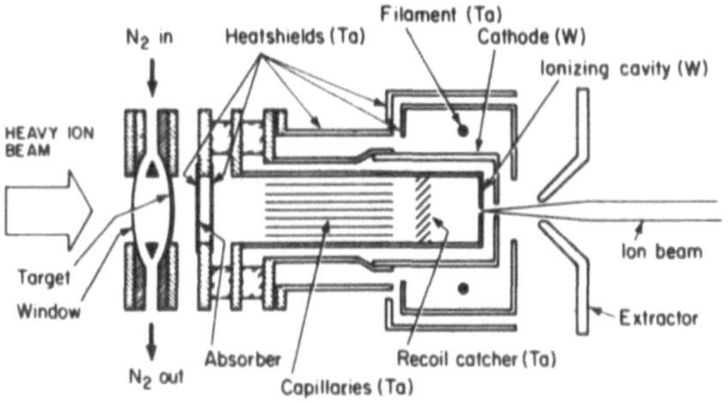

FIGURE 8 Schematic diagram of the OASIS surface ion source with a gas-cooled target[20].

product nuclei are allowed to diffuse out of a hot target and thence to the ion source itself. ISOLDE sources all operate this way[2,14].

Diffusion, whether from a catcher or a thick target, introduces unwanted inefficiency and time delay. In an attempt to avoid these difficulties, several separator groups have coupled a He-jet system, in which recoil nuclei are thermalized in helium gas, to an ion source through a skimmer. The only one to have achieved high efficiency (6±2% for ^{25}Aℓ in preliminary tests) by this technique[21] is the Chalk River group who use a modified Bernas-type source[22]. Their source, shown in Figure 9, utilizes slit geometry, which creates a favourable ratio of exit to entrance area; it also has small volume, a covered cathode and a high operating temperature.

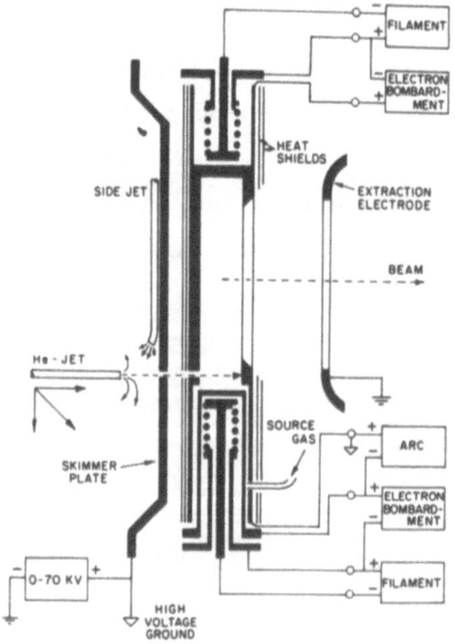

FIGURE 9 The Chalk River helium-jet source[21]. The source has an extraction slit of 1.0 mm x 30 mm; its body is cylindrical.

The most novel helium-gas ion source is IGISOL[23], illustrated in Figure 10. Here the recoils not only are thermalized in helium but also reach a 1^+ charge state through charge transfer with the helium and impurity molecules. These charged recoils are introduced directly into the separator. Very short delay times, and efficiencies up to ≈1%, have been obtained by this technique, though so far the resolution is rather poor (M/ΔM = 375 FWHM)[23].

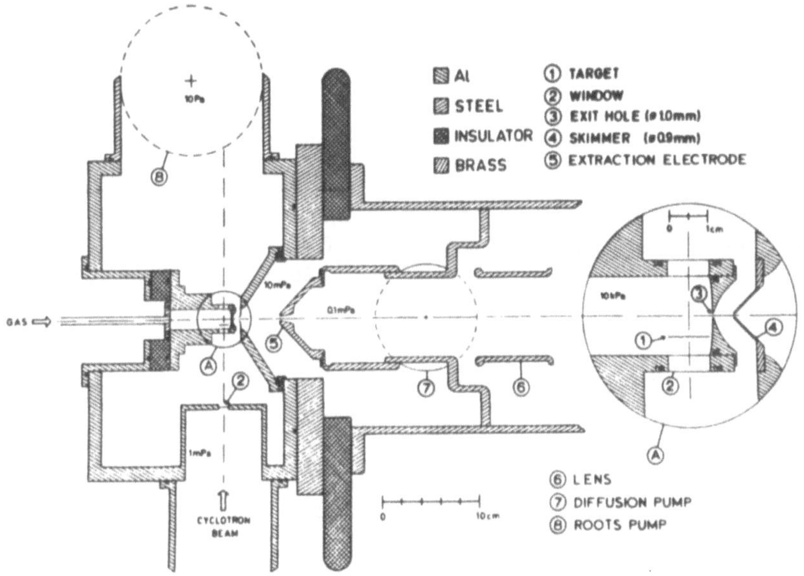

FIGURE 10 The IGISOL ion guide connected to the acceleration chamber of the isotope separator[23].

4. THE CHALK RIVER ISOL

Isotope separators come in a variety of shapes and sizes, each with its own characteristics. Rather than attempting to scan the field superficially, we have chosen to describe the Chalk River instrument[16,21] in some detail. Its outline and location in the new tandem-superconducting cyclotron (TASCC) facility is shown in Figure 11.

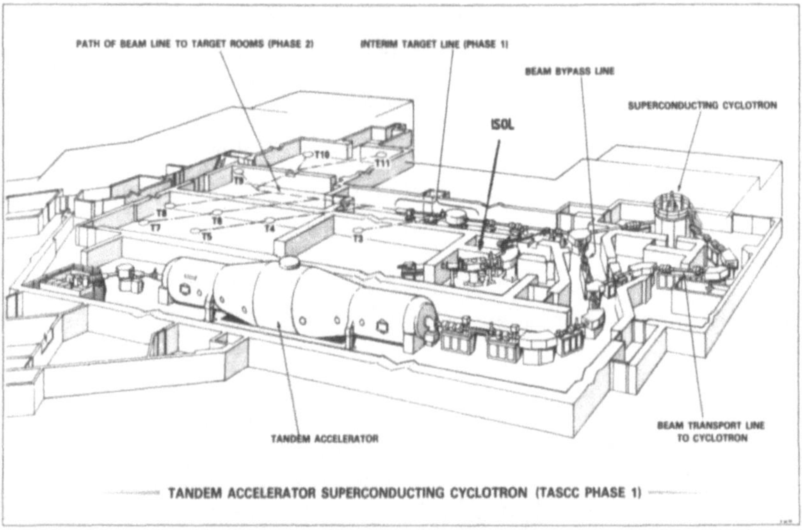

FIGURE 11 Layout of the TASCC facility at Chalk River. The ISOL is located at the middle right.

The ISOL beam formation is a one-step process, taking place within a few millimeters between the ion source orifice and the extraction electrode, whose position is adjustable through four mechanical degrees of freedom provided by remotely controlled precision drives. The beam shape is determined entirely by this geometry. The accelerating voltage is usually 40 kV (stability 5×10^{-6}), although it can be operated anywhere between 30 and 70 kV. The separator beam then passes through the 135° inhomogeneous magnet and into the collection chamber where an analyzing slit, mounted on a moveable carriage, defines the entrance to the beam transport line. There the selected beam is focussed by two electrostatic quadrupole doublets to a 2-mm diameter spot in a well-shielded counting chamber. Along the entire beam path the separator is equipped with appropriate beam diagnostic devices.

Probably the single most remarkable item employed in the separator is the pair of correction coils with which the magnetic sector is equipped. The radial field distribution in any magnet is given by

$$B = B_o\left[1 - \alpha \frac{r-r_o}{r_o} + \beta\left(\frac{r-r_o}{r_o}\right)^2 + \ldots\right]$$

where B_o is the field at the mean radius r_o. For the Chalk River magnet, α is nominally 0.5 and $\beta \approx \alpha^2$, but the correction coils (see Figure 12), called α- and β-coils, produce field contributions that independently adjust the

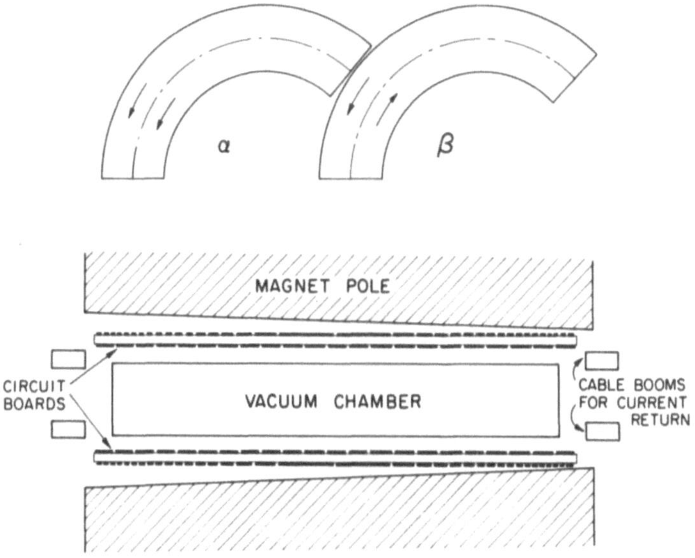

FIGURE 12 Design and wiring of the corrective coils for the Chalk River ISOL[16], only the horizontal dimension being to scale. The current rails (indicated by the heavy horizontal lines) of the α-coils are of nearly equal width; the current rails of the β-coils vary considerably in width. The arrows in the upper part of the figure indicate the current direction; both power supplies are reversible.

respective indices. As shown in Figure 13, α is largely responsible for the position of the image along the beam axis, and β can be used to minimize second order aberrations. With two simple adjustments the optimum beam profile can readily be obtained.

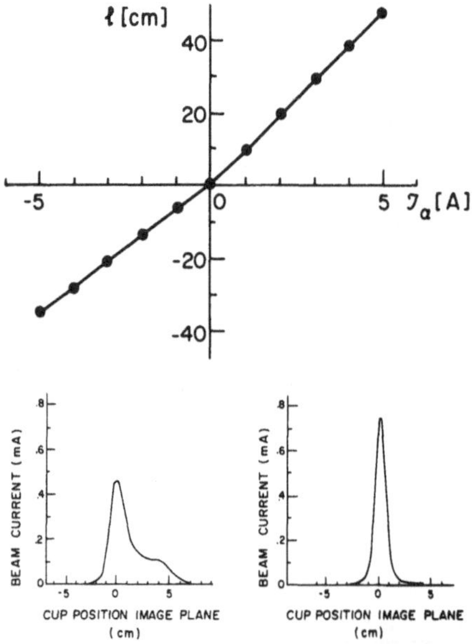

FIGURE 13 The effect of the α- and β-coils. The α-coils shift the image plane by as much as ±40 cm (top). The β-coils can minimize second-order aberrations (bottom, left) to produce the optimum beam profile (bottom, right).

By careful and continuous attention to component quality, and with the advantage given by the correction coils, the Chalk River group has achieved the highest resolution (M/ΔM = 20 000 FWHM) ever reported for an ISOL. The results are illustrated in Figure 14. With such high resolution it is possible to separate isobars under favourable conditions or even to measure directly the atomic masses of exotic nuclei[24].

Of equal importance to high resolution is high enhancement of a selected beam. Contamination, even at the percent level, from adjacent isotopes could be catastrophic to a precise experiment.

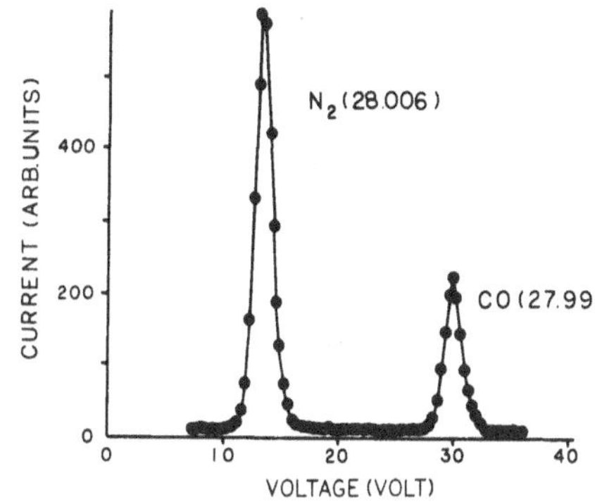

FIGURE 14 The CO-N_2 doublet at mass 28 measured with the Chalk River ISOL. The resolution of 20 000 is demonstrated in this scan obtained with the FEBIAD source and an orifice of 100 μm diameter. The separation between the two peaks is 10.5 MeV.

Figure 15 shows that, for the Chalk River ISOL, cross contamination is measured to be $<10^{-6}$ at A=20 and, by scaling, it must be $<10^{-5}$ at A=200.

5. ISOL-FED EXPERIMENTS

To many, an ISOL is seen as simply providing mass-identified samples of exotic nuclides. That is undoubtedly one of its important uses but it is by no means its only one: an ISOL capable of producing well-focussed, stable beams with high purity can support a wide variety of experiments

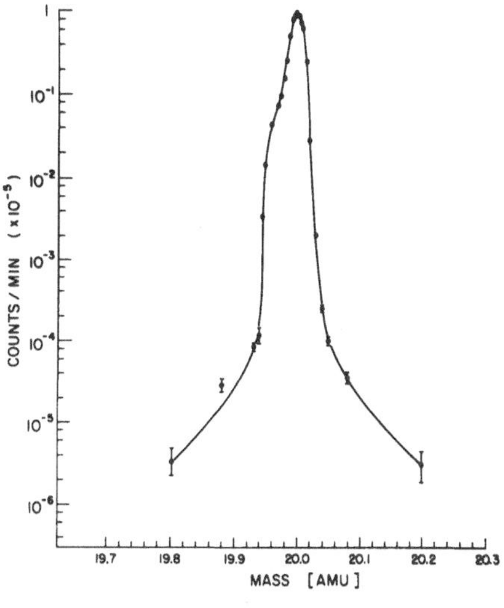

FIGURE 15 Beam profile for ^{20}Na. The bump at the 5% level on the low-mass side indicates that the β-coil is not perfectly adjusted[16].

that otherwise would be impossible even for nuclides quite near the valley of stability. Some examples are given here to illustrate the richness of the field.

5.1 Decays of Exotic Nuclei

The identification and spectroscopy of exotic nuclei employs techniques familiar to all. No specific examples are presented here. Instead, the nuclear chart in Figure 16 gives an overview, highlighting the extent to which exotic nuclei are being explored: there are now 2200 nuclides about which at least one property has been measured, and of these, many - indeed most at the outer fringes - have required the use of an ISOL. Should anyone imagine that the task is done, let them note that 5000 nuclides still remain unknown.

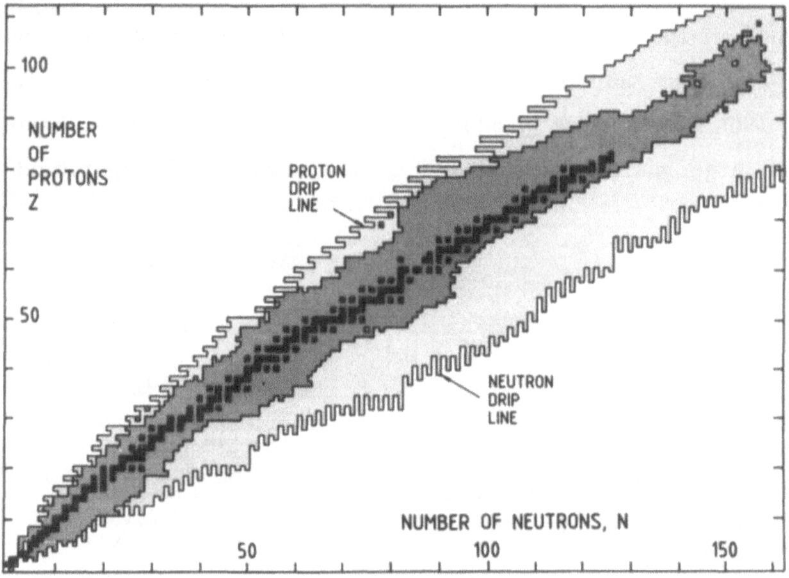

FIGURE 16 Chart of the nuclides. The stable nuclides are given by black squares. The dark stippled area represents nuclides identified[25] by 1981; the lighter area, nuclides predicted[26] to exist for an observable length of time.

5.2 Masses of Exotic Nuclei

Of particular interest to nuclear and astro-physicists alike are the masses of nuclei far from stability. Quite extensive isotopic chains of masses for alkali elements have been measured with a mass spectrometer connected on-line to ISOLDE[27]. More recently, mass measurements have been reported[24] for which the Chalk River ISOL was used directly as a mass-measurement device.

5.3 Weak Interaction Studies

Precision measurements of ft values for superallowed $0^+ \rightarrow 0^+$ β-decay transitions have yielded important confirmation[28,29] of the Conserved Vector Current Hypothesis. One significant aspect of such experiments is the determination of half-lives, for which radioactive sources of assured high purity are essential. Figure 17 shows the measured decay[30] of ISOL-produced $^{26m}A\ell$; the only "contaminant" is the constant room background at 1 to 2 counts/s.

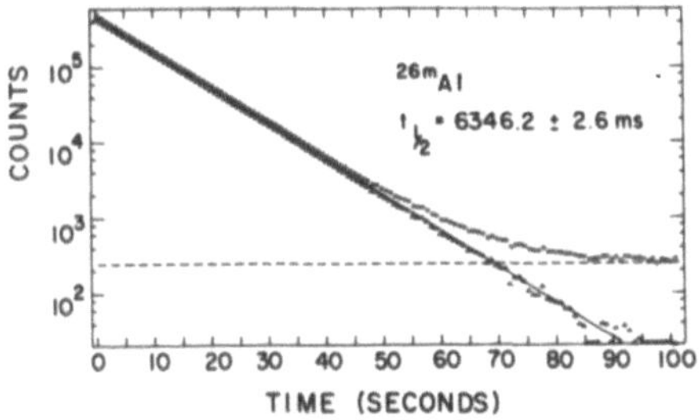

FIGURE 17 Decay curve for $^{26m}A\ell$, measured[30] for samples collected with the Chalk River ISOL. The dashed line indicates the contribution of the constant-rate room background.

The vector coupling constant, determined from $0^+ \rightarrow 0^+$ decays, can also be measured in non-0^+ nuclear decays providing that experimental information in addition to the ft-value is acquired. In one such case - the $2^+ \rightarrow 2^+$

superallowed decay of ^{20}Na (decay scheme in Figure 18) - the additional results were β-ν-α triple correlation coefficients determined from the energy shifts in delayed-α-decay groups measured as a function of the energy and angle of coincident β particles. The measurement[31] required 5 x 10^4 atoms/s of ^{20}Na ($t_{1/2}$ = 448 ms) to be deposited in a 2-mm-diameter spot under highly stable

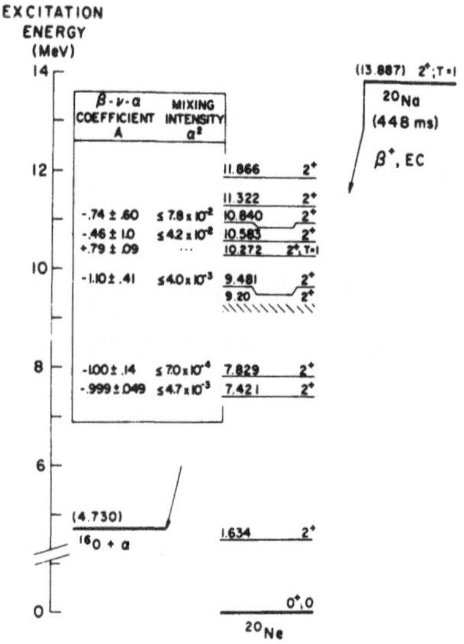

FIGURE 18 Partial decay scheme of ^{20}Na. All 2^+ states in ^{20}Ne are T=0 except the T=1 analogue state, which is indicated. The triple correlation coefficients and isospin mixing amplitudes are shown next to the daughter states to which they apply[31].

conditions for 7 days; the data appear in Figure 19. The resultant value of the vector coupling constant[31] - the first ever obtained from a $2^+ \rightarrow 2^+$ decay - was found to agree with the conventionally obtained value.

5.4 Laser Spectroscopy

On-line isotope separators provide the only available means for studying the atomic spectroscopy of short-lived isotopes. In the past 10 years, ISOL-based laser facilities have determined numerous nuclear ground (and isomeric) state spins, as well as magnetic dipole and electric quadrupole moments from measured hyperfine structure. They

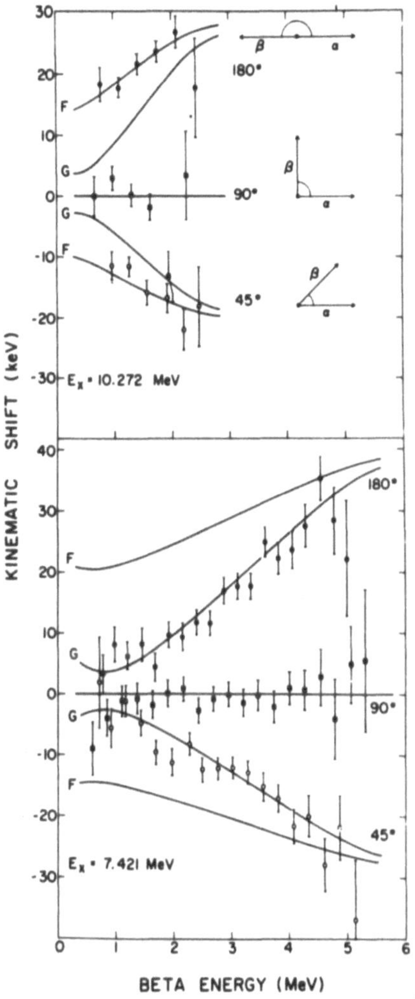

FIGURE 19 Coincidence data[31] for two $^{20}Na \rightarrow {}^{20}Ne$ transitions (labelled by excitation energy in the daughter) showing the energy shift of α groups as a function of β energy and relative angle. The curves show the expected kinematic shift for a pure Fermi transition (F) and a pure Gamow-Teller (G) one.

have also established, from observed isotope shifts, the variation in the mean-square nuclear charge radius along several isotopic chains[32]. As an example of the equipment used, Figure 20 shows the collinear-beam laser apparatus at ISOLDE. The method is particularly powerful since it produces high resolution through suppressed Doppler broadening, and is also very sensitive. Just how sensitive

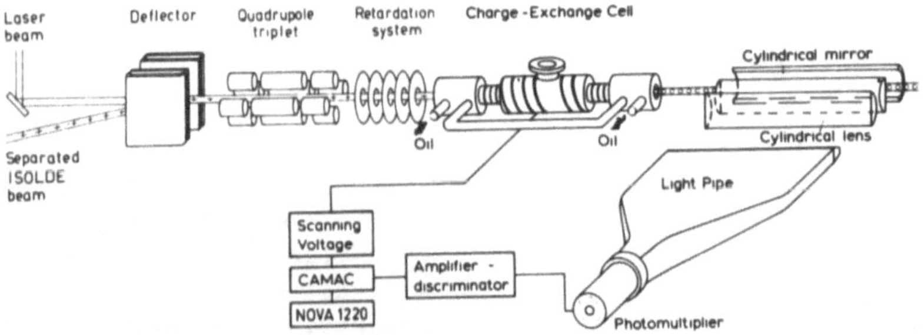

FIGURE 20 Set up for on-line collinear-beam spectroscopy at ISOLDE[32]. The mass-separated ion beam is bent into the laser beam by a pair of deflection plates, focussed slightly and neutralized in the charge-exchange cell; excitation of the atoms is detected via their fluorescence.

it is can be seen from Figure 21, which shows recent results from the UNISOR laser group[33]. The $^{193}T\ell$ spectrum at the top of the figure was obtained in a little over 3 hours fron an ISOL beam of 3.5×10^5 atoms/s. The central element of the UNISOR system is a cylindrically-symmetric fibre-optic detector, which collects light over 10 cm of the beam path and directs it onto a photomultiplier. The overall system efficiency is 1 photon per 1000 atoms in the strongest line of $^{203}T\ell$.

5.5 Nuclear Orientation

The separators at Leuven[18] and at Daresbury[34] each are coupled directly to a dilution refrigerator operating at temperatures of the order of 10 mK. Nuclei in the ISOL beams can thus be oriented by the local field they experience when implanted in a cooled ferromagnetic host.

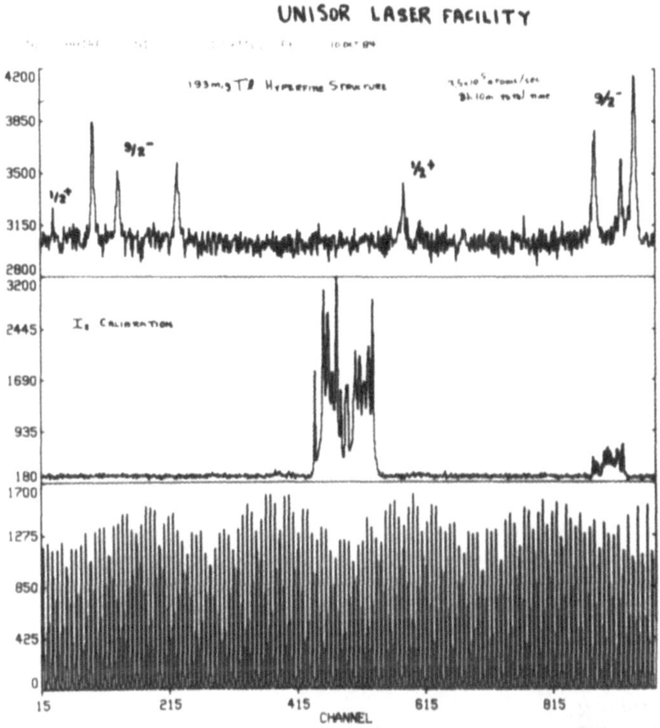

FIGURE 21 Fluorescence spectrum (top) from a $^{193m,g}Tl$ fast-atom beam, compared with calibration lines from an I_2 saturated absorption system (middle) and a calibrated etalon (bottom)[33].

The beauty in doing so is that nuclear spins and mixing ratios in the daughter can be obtained from <u>singles</u> measurements, since these properties produce characteristic anisotropies in the angular distributions of decay radiation. In addition, the magnetic moment of the implanted parent nucleus can be obtained from the temperature variation of the anisotropy, as shown in Figure 22. By any other method, only nuclei with half lives greater than ≈1 hr could be oriented, yet with these ISOL-based systems, nuclides with half-lives as short as 14 s (^{118}Cs) have now become accessible.

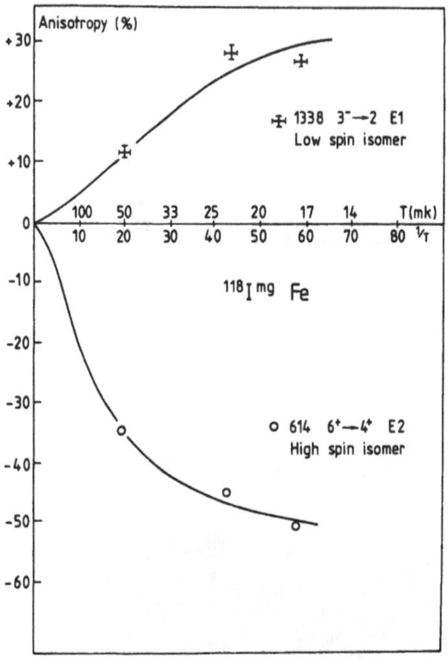

FIGURE 22 Anisotropies measured for two γ-ray transitions following the β-decay of ^{118}I, plotted as a function of temperature. The data at the top yield μ = 3.7±0.5 nm for the 7⁻ isomer in ^{118}I, while those at the bottom correspond to μ = 1.4±0.2 nm for the 2⁻ state.

6. CONCLUSION

With this necessarily brief overview, it is hoped to convey the spirit of a diverse and exciting field of study, not merely of a single class of experimental device. By now, an ISOL has become a means to many ends. It is not simply a tool for prospecting at the fringes of nuclear stability; it is a refined instrument, uniquely valuable in the study of nuclei throughout the nuclear chart.

REFERENCES

1. Electromagnetic Isotope Separators and Techniques Related to their Applications, edited by H.L. Ravn, E. Kugler and S. Sundell, Nucl. Instr. and Meth. 186 (1981).
2. T. BJORNSTAD et al, reference 1, p.391.
3. G. RUDSTAM, Nucl. Instr. and Meth. 139, 290 (1976).

4. M. BLANN, University of Rochester report UR-NSRL-181 (1978).
5. C. BRUSKE et al, reference 1, p.61.
6. R. WOLFGANG et al, Phys. Rev. 103, 394 (1956).
7. A.G. ARTUKH et al, JINR preprint E7-81-355, Dubna 1981.
8. H.C. BRITT et al, Phys. Rev. C13, 1483 (1976).
9. M. HILLE et al, Nucl. Phys. A252, 496 (1975).
10. J. GALIN et al, Nucl. Phys. A255, 472 (1975).
11. D.J. MORRISSEY et al, Phys. Rev. 21C, 1783 (1980).
12. W. LOVELAND et al, Phys. Rev. 23C, 253 (1981).
13. H. RAVN, private communication.
14. R. KIRCHNER, reference 1, p.275.
15. R. KIRCHNER and E. ROECKL, Nucl. Instr. and Meth. 133, 187 (1976).
16. H. SCHMEING et al, reference 1, p.47.
17. R. KIRCHNER and E. ROECKL, Nucl. Instr. and Meth. 139, 291 (1976).
18. J. VERPLANCKE et al, reference 1, p.99.
19. R.L. GILL et al, reference 1, p.243.
20. J.M. NITSCHKE, Nucl. Instr. and Meth. 206, 341 (1983).
21. H. SCHMEING, E. HAGBERG and J.C. HARDY, to be published.
22. J. CHAVET and R. BERNAS, Nucl. Instr. and Meth. 51, 77 (1967).
23. J. ÄRJE et al in JYFL Annual Report 1983-84 (University of Jyväskylä) p.16.
24. K. SHARMA et al; in 7th International Conference on Atomic Masses and Fundamental Constants, to be published.
25. W. SEELMANN-EGGEBERT et al, Chart of the Nuclides (Kernforschungszentrum Karlsruhe GmbH, Germany; ed.5, 1981).
26. S. MARIPUU, At. Data Nucl. Data Tables 17, 411 (1976).
27. C. THIBAULT, in 4th International Conference on Nuclei far from Stability, edited by P.G. Hansen and O.B. Nielsen, CERN report 81-09 (1981) p.47.
28. J.C. HARDY and I.S. TOWNER, Nucl. Phys. A254, 221 (1975).
29. I.S. TOWNER and J.C. HARDY, in 7th International Conference on Atomic Masses and Fundamental Constants, to be published.
30. V.T. KOSLOWSKY et al, Nucl. Phys. A405, 29 (1983).
31. E.T.H. CLIFFORD et al, Phys. Rev. Lett. 50, 23 (1983).
32. R. NEUGART, reference 1, p.165.

33. H.K. CARTER *et al*, to be published.
34. I.J.D. MACGREGOR, I.S. GRANT and P.M. WALKER, Vacuum 34, 3 (1984).

TESSA: ITS PRESENT STATUS AND FUTURE DEVELOPMENT

P.J. TWIN
Science and Engineering Research Council
Daresbury Laboratory
Daresbury
Warrington WA4 4AD
U.K.

The Total Energy Suppression Shield Array (TESSA) has been developed to study discrete line and continuum γ-ray spectroscopy. Its properties are detailed and its performance illustrated. Future developments including POLYTESSA and compact BGO shields are discussed.

1. INTRODUCTION

The Total Energy Suppression Shield Array (TESSA) has been operating on the 20 MV Tandem at the Daresbury Laboratory since the beginning of 1983. The instrument was developed by the University of Liverpool and the Daresbury Laboratory from a simpler suppression shield array[1] which the University of Liverpool and the Niels Bohr Institute set up and operated on the Risø tandem between July 1980 and July 1982. In this talk I shall discuss the properties of TESSA, illustrate its power in the field of γ-ray spectroscopy with some recent data, and finally indicate the status of its successor, currently known as POLYTESSA.

2. TESSA

The study of the gamma decay from very high spin states demands an instrument capable of efficient selective detection of the weak γ-rays de-exciting these states whose population is usually achieved by a heavy-ion fusion evaporation reaction. An example of such a reaction is the bombardment of ^{36}S on ^{98}Mo at 150 MeV. The compound system ^{134}Ce decays by evaporating neutrons and protons and the main final product nuclei are ^{131}Ce (3n), ^{130}Ce (4n), ^{129}Ce (5n), ^{129}La (p4n). Entry contours in excitation energy versus spin space are shown in Figure 1 for some of these channels, together with some representative γ-decay paths down to the known yrast line on ^{130}Ce. The detail of these decay paths can only be established by a high resolution γ-γ coincidence experiment. Such measurements are made extremely difficult by the large continuum γ-ray background, the many reaction channels and hence final nuclei whose discrete γ-ray lines are present, and also the poor photopeak efficiency relative to total detection efficiency of germanium detectors. TESSA is designed to overcome these problems and thus enable discrete line and continuum γ-γ spectroscopy to probe into the higher regions of spin[2]. It combines high resolution Compton suppressed germanium detectors with a compact bismuth germanate (BGO) ball to measure the total energy amd multiplicity of the γ-radiation, as shown in Figure 2.

2.1. Suppression Shields

The suppression shields used in TESSA are 25 cm diameter by 20 cm long cylindrical NaI crystals with the germanium detector inserted parallel to the crystal axis such that γ-rays strike the curved surface of the germanium. This geometry provides the best suppression at low γ-ray energies through efficient detection in the shield of the high energy forward scattered γ-rays. The effect of such suppression is shown in Figure 3 for a ^{60}Co source. The dominant remaining features are the backscatter peaks close to 250 keV below the photo peaks. Figure 3 shows that these backscatter peaks are removed by the NaI crystal backscatter detectors positioned as shown in Figure 2. A cross-over timing circuit is also included to veto slow rise time pulses. The collimation of the γ-rays striking the germanium detector is achieved by tungsten alloy (heavy metal) hexagons in the outer ring of the BGO ball. They are designed so that γ-rays from the target are restricted to pass through a 4 cm diameter circle at the centre of the germanium detectors. This ensures good photo peak to total fraction before suppression, as shown in Table 1 and prevents scattering into the germanium from the cans and reflectors of the NaI detectors. The resulting suppressed and unsuppressed peak to total fractions are given in Table 1. The highest fractions are always achieved for

TABLE I Peak to total fractions for 25% germanium detectors in TESSA

Type of detector	^{137}Cs source unsuppressed	^{137}Cs source suppressed	^{60}Co source unsuppressed	^{60}Co source suppressed
GeLi	0.30	0.66	0.18	0.53
Ge n-type	0.32	0.75	0.20	0.60
Average	0.31	0.71	0.19	0.57
γ-γ fraction	0.09	0.50	0.04	0.32

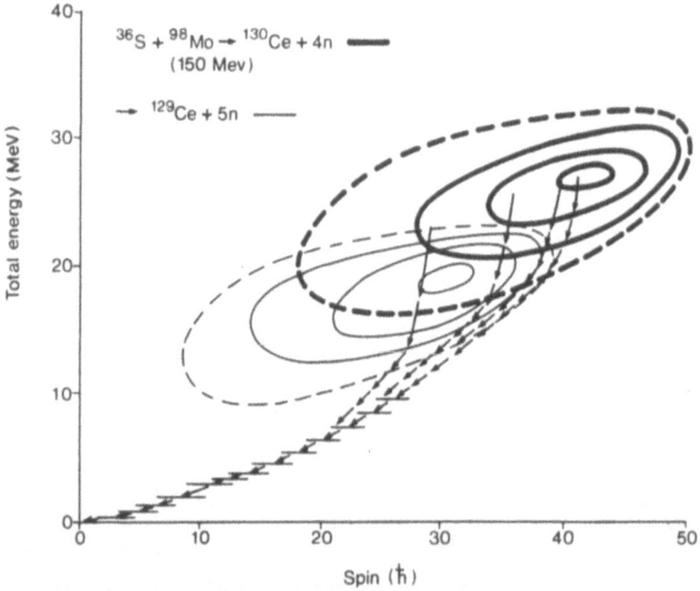

FIGURE 1 The entry contours in ^{129}Ce (5n channel) and ^{130}Ce (4n channel) following the bombardment of 150 MeV ^{36}S ions on ^{98} Mo. Examples of possible decay paths to the yrast line in ^{130}Ce are shown.

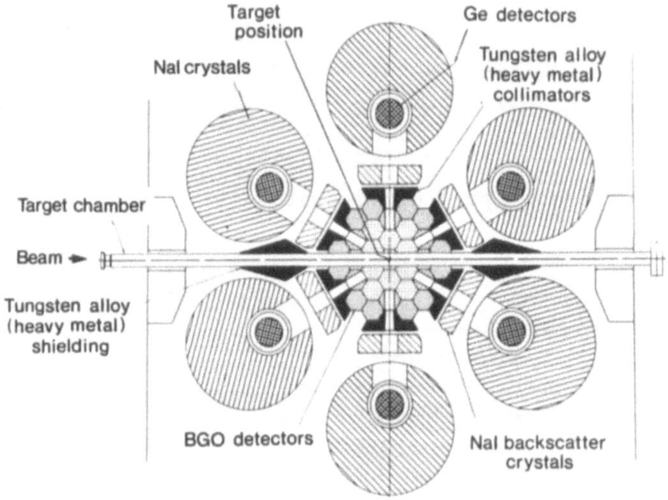

FIGURE 2 A schematic of the Total Energy Suppression Shield Array (TESSA) at Daresbury. It is a vertical centre line section and shows the detector systems are cylindrically symmetric about a horizontal axis through the target positions.

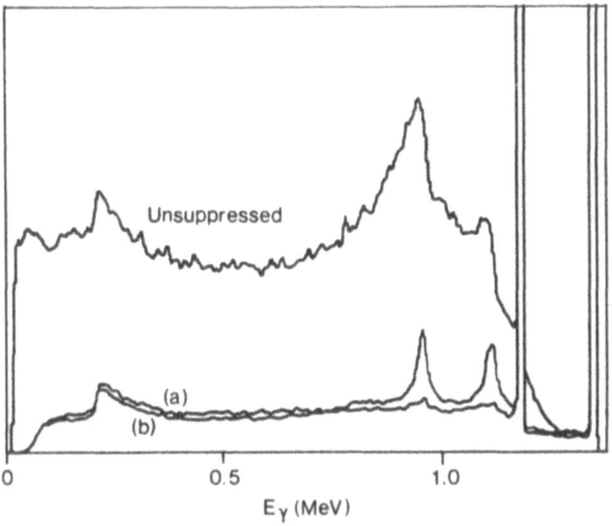

FIGURE 3 Effect of suppression on the 1.17 MeV and 1.33 MeV γ-rays from a ^{60}Co source as recorded by TESSA; (a) veto by the main NaI shield; (b) veto by NaI shield plus the backscatter NaI detector and poor crossover timing.

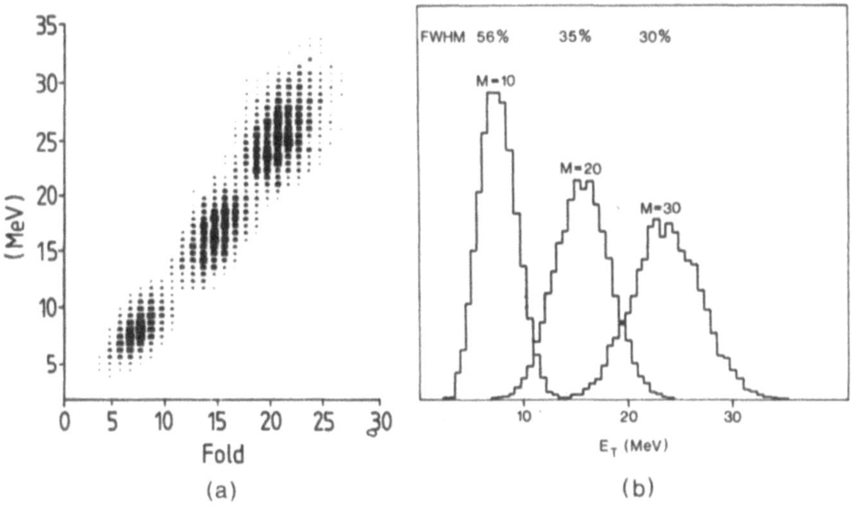

FIGURE 4 Response of TESSA; (a) Density plots of fold versus sum energy constructed from γ-ray source data for the detection of 10, 20 and 30 1.17 MeV γ-rays; (b) The multiplicity response for TESSA measured experimentally for the detection of 10, 20 and 30 1.17 MeV γ-rays.

n-type intrinsic detectors due to their very thin dead layers both on the outside curved surfaces and in the core.

The centres of the germanium detectors are 27 cm from the target and their collimated solid angles are 0.16% each. This small solid angle reduces Doppler smearing and pile up. The former is caused by the change in energy across the detector of γ-rays emitted from fast moving recoils ($\beta = 0.03$ in a typical H.I. reaction) and the latter from the detection of two γ-rays in the same germanium detector due to the high multiplicity. The Doppler smearing effect is smaller for the detectors at 30° and 150° to the beam direction and it is zero if the recoiling nucleus has stopped when the γ-rays are emitted.

The 6 germanium detectors constitute 15 pairs thus increasing the γ-γ efficiency. In a typical H.I. reaction experiment with TESSA the germanium singles rate is around 5K s^{-1} and the suppressed γ-γ rate is 100 s^{-1}. In γ-γ coincidence experiments the important quantity is the fraction of total recorded γ-γ events that are photopeak-photopeak events as the remainder just increase the Compton continuum. Taking average figures from Table 1 the suppression improves this fraction from 0.03 to 0.29 for 1.2 MeV γ-rays and 0.10 to 0.42 for 0.66 MeV γ-rays. It is this big increase in the fraction of useful γ-γ data that greatly assists carrying out detailed spectroscopy on low intensity γ-ray branches and makes useful E_γ-E_γ correlation experiments possible with germanium detectors.

2.2. Total Energy and Multiplicity Ball

Two multi-detector arrays of NaI detectors have been built.

These are the 'spin spectrometer' situated at Oak Ridge[3] and the 'crystal ball' at Heidelberg[4]. Both are spherically symmetric systems with 162 detectors (crystal ball) and 72 detectors (spin spectrometer) and their construction prevents suppressed germanium detectors being placed close to the target. The essential feature required for the ball for TESSA is thus compactness and hence a denser scintillator than NaI is needed. Bismuth Germanate (BGO) satisfies this requirement as it has a radiation length two and a half times shorter than NaI and an increased photoelectric efficiency for lower energy γ-rays. In order to place the suppressed germanium detectors close to the target the BGO ball has been designed as a cylinder. Each BGO crystal has a hexagonal cross section with dimensions of 38 mm face to face and 50 mm or 75 mm in length. The 62 detectors form a honeycomb arrangement such that in the 3 inner rings individual detectors subtend approximately equal solid angles at the target and in the outer ring pairs of detectors have this same solid angle. The resulting configuration is an effective 50 detector device with 46 detectors having solid angles within ± 10% for 1.17 MeV γ-rays and the remaining 4 detectors within 25%. The ball is designed to operate at high rates in excess of 300K s^{-1}. At present this is achieved by the integration and clipping of the signal from each BGO detector as the scintillator has a decay time of 0.3 μs. The analogue signals are added to provide the total energy information and a threshold at around 120 keV is set on each effective detector system to provide multiplicity data.

The ball actually measures the number of responding detectors (termed the fold) which is related to the multiplicity of the γ-ray shower, and the total energy detected (called the sum energy) which is related to the total energy emitted in the γ-ray shower. The pattern of the responding detectors is also recorded.

The detection efficiency and response of the ball has been measured using a source emitting two γ-rays in cascade with a known branching ratio. The total efficiency or probability of detection of a single γ-ray is an important quantity and it varies from 0.95 at 500 keV to 0.85 at 2 MeV. In some of the detected events the γ-ray is Compton scattered and the ball does not record the full γ-ray energy. This effect is taken account in the average fraction of energy recorded by the ball for a single source γ-ray, values of which vary between 0.83 at 500 keV to 0.60 at 2 MeV. The response functions of fold and sum energy are illustrated in Figure 4. The density plots indicate the probabilities of TESSA recording a particular sum energy and fold when sets of 10, 20 and 30 1.17 MeV γ-rays are emitted at the centre of the ball.

The projection of these density plots onto the sum energy gives the sum energy response function which has a full width half maximum of 30% for 30 1.17 MeV γ-rays.

The multiplicity information depends on the average number of responding detectors. The larger photo-electric contribution in BGO reduces the scattering between the detectors and the ratio of multiple to single hits in TESSA at 1.17 MeV is 0.17 compared with 0.43 for the spin spectrometer and 0.56 for the crystal ball. This big improvement for TESSA compensates in the multiplicity response for its lower total efficiency and the smaller number of detectors. The average number of responding detectors for a 1.17 MeV γ-ray is 0.85 for TESSA, 1.18 for the spin spectrometer and 1.32 for the crystal ball. The resulting multiplicity response is shown in Figure 4b and the FWHM at M = 30 is 30%.

The neutron response of the BGO ball has not been measured. However, it can be estimated from the dimensions and measured neutron efficiencies[5] and we conclude that on average 5 cm of BGO has 15% of the efficiency of 20 cm of NaI. A fraction of the neutrons come from scattering on support structures and for these the relative detection efficiency in the BGO ball compared to NaI balls is further reduced due to its much smaller volume. As the NaI balls detect on average 6-7 neutrons per event in a (H.I., 4n) reaction we conclude that the BGO ball detects around 1 neutron for the same case.

3. EXAMPLES OF DATA FROM TESSA

3.1. Discrete Line γ-ray Spectroscopy

The large peak/total values of the suppressed germanium detectors and reasonable efficiency for γ-γ coincidences enable 30 M events to be collected in a 3 day accelerator run thus forming the basis of the great effectiveness of TESSA in the field of discrete line γ-ray spectroscopy. The BGO ball also plays a role in providing some channel selection.

TESSA has been used extensively at Daresbury in the past eighteen months to study discrete γ-ray lines in nuclei around mass 80, around mass 130 and in many nuclei in the rare earth region. Examples of the quality of the spectra are shown in Figures 5 and 6. The nucleus ^{158}Er had previously been seen up to 38^+ [6] and various sugges-

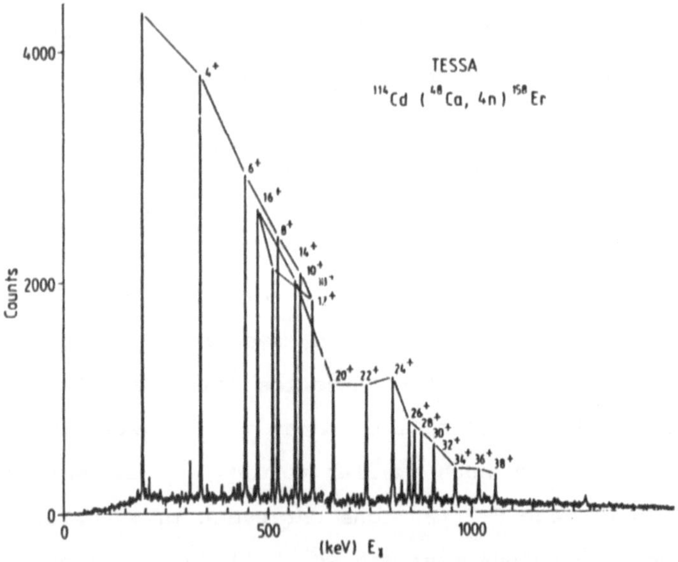

FIGURE 5 TESSA: a γ-γ coincidence spectrum of the yrast band in ^{158}Er produced by summation of gates on 20^+, 22^+, 26^+, 28^+, 32^+, 34^+ and 36^+ members of yrast band. The reaction was ^{114}Cd(^{48}Ca, 4n)^{158}Er at 205 MeV.

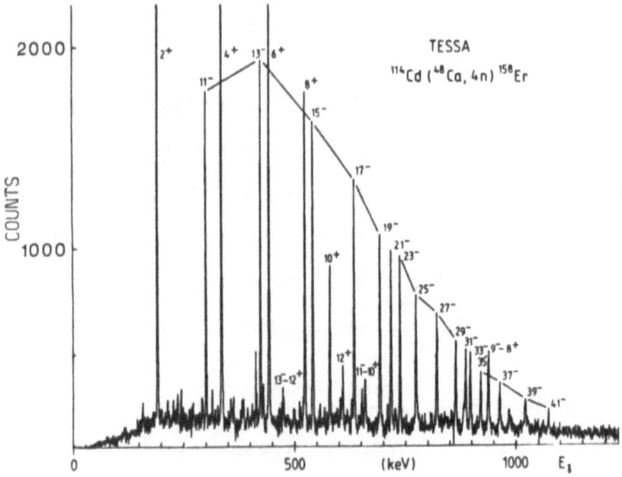

FIGURE 6 TESSA: a γ-γ coincidence spectrum of a negative parity side band in ^{158}Er produced by summation of gates on the 11^-, 13^-, 15^-, 17^-, 19^-, 21^-, 27^-, 29^-, 31^-, 33^- and 35^- members of the band.

tions put forward for the 40^+ decay. The TESSA spectrum for the yrast band in ^{158}Er is shown in Figure 5. It is the γ-γ coincidence spectrum formed by summing gates on the 20^+, 22^+, 26^+, 28^+, 32^+, 34^+ and 36^+ decays following the $^{114}Cd(^{48}Ca,4n)$ reaction. The spectrum clearly shows that there is a shape change at 38^+ and that the previous tentative proposals for the 40^+ decays were wrong[7]. The power of TESSA in the field of discrete line spectroscopy is best demonstrated by the spectra obtained for the lower intensity side bands. Figure 6 shows a γ-γ spectrum for the negative parity side band in ^{158}Er. The band is observed up to a spin of 41 $\hbar$ at which point the structure changes as in the yrast band. The low intensity of the background and its almost constant intensity down to 200 keV is characteristic of suppression shield data.

3.2. Unresolved (Continuum) γ-ray Spectroscopy

It has been proposed that at high spin there are many parallel collective bands in nuclei and that these constitute an E2 bump which forms part of the continuum. There has been considerable effort to identify the structures parallel to the diagonal in E_γ-E_γ correlation spectra[8] that would be characteristic of these bands. Data with NaI arrays have required various background correction procedures before a central diagonal valley could be seen. TESSA has the attributes required for continuum γ-ray spectroscopy; high resolution, excellent peak to total fractions, good channel selection and reasonable statistics.

The relatively simple response function of the suppressed germanium detectors, essentially a photo-peak containing over half the total intensity plus a flat background at lower energies, makes it straightforward to correct the spectrum for this background. This procedure has been carried out for the coincidence spectrum of the yrast band in ^{130}Ce which is shown in Figure 7. There is a rapid reduction in intensity of the discrete lines upto a spin of 24 $\hbar$ and a strong bump (the E2 bump) of unresolved continuum γ-rays centred around 1.2 MeV.

The continuum structures can be studied in a two dimensional E_γ-E_γ spectrum but it is difficult to identify the relevant structures in a grey scale intensity map due to variations in both γ-ray efficiency and the general background level with respect to γ-ray energy. These effects have been corrected for in the diagram of Figure 8. The horizontal and vertical stripes are coincidences of discrete and continuum γ-radiation with strong yrast γ-rays and the ridges parallel to the diagonal are characteristic of collective structures in a γ-ray energy region above the known discrete line regime. Cuts perpendicular to the central valley and integrating parallel to the valley are shown in Figure 9. These spectra are of raw data, with no correction for efficiency or background, and they constitute summations over $\bar{E}_\gamma$ in 100 keV widths with means of 1100 keV to 1500 keV. The ridges are clearly seen in these raw data spectra and their widths are approximately 20 keV, indicating the necessity of using high resolution germanium detectors. The instrumental response can be removed, as described above, and the intensity in the ridge compared with those of discrete line coincidences. These results indicate that in the spin 30-50 $\hbar$ region only 10-20% of the intensity decays down rotational bands. The remaining intensity is uncorrelated thus indicating the

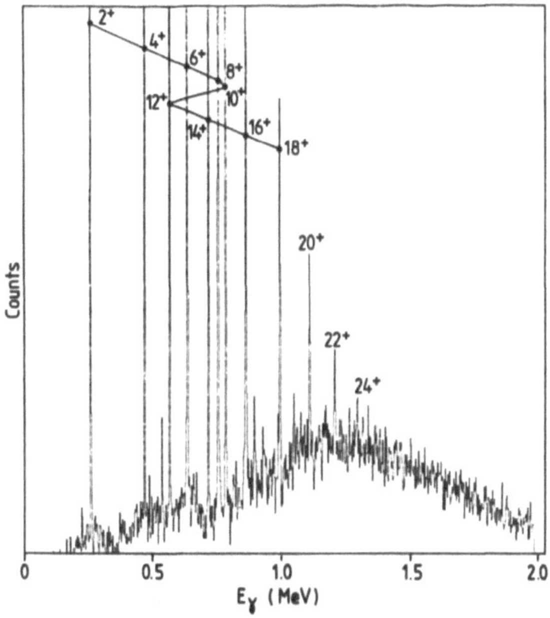

FIGURE 7 A γ-ray spectrum of the yrast band in ^{130}Ce selected by a γ-γ coincidence requirement and corrected for the detector response.

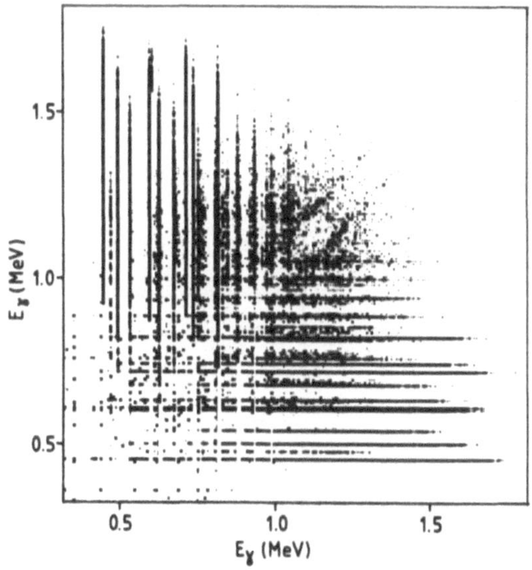

FIGURE 8 An E_{γ}-E_{γ} correlation grey-scale plot for ^{130}Ce as selected by the BGO ball. The ridges parallel to the diagonal indicate collective structures at γ-ray energies higher than those seen in discrete line studies.

high probability of out of band decay and yielding the conclusion that bands in the continuum are strongly mixed.

A second example of the power of TESSA in γ-ray continuum spectroscpy is the observation of superdeformation in ^{152}Dy. The nucleus ^{152}Dy is known to be spherical and the high spin states up to nearly 40 ħ to be irregularly spaced and single particle in origin[9]. It was studied with TESSA using the $^{108}Pd(^{48}Ca,4n)$ reaction at 205 MeV bombarding energy[10]. The E_γ-E_γ correlation spectrum, Figure 10, clearly shows a collective ridge from 800 keV to 1400 keV indicating that the nucleus becomes collective and the spherical shape changes to prolate at high spin. In the spectrum obtained from a cut perpendicular to the valley diagonal and summed over $\bar{E}_\gamma$ from 780 keV to 1330 keV the ridge is observed to have a width of only 7.5 keV close to the 6 keV experimental resolution again demonstrating the need for high resolution detectors. The small separation of the ridge from the valley diagonal is characteristic of a very large superdeformed prolate shape with an axis ratio of close to 2:1 and it is in excellent agreement with recent predictions[11-13].

A third example of the power of TESSA in studying continuum γ-rays is its ability to separate out the decay paths from different entry regions in spin and total energy for one particular final nucleus. When ^{116}Cd is bombarded by a 200 MeV ^{48}Ca beam the main exit channels are 4n (^{160}Er) and 5n (^{159}Er) with weaker channels of 3n (^{161}Er) and 6n (^{158}Er). The information from each event recorded in TESSA contains 2 coincident suppressed germanium data plus the sum energy and fold. In these studies one germanium signal is used to identify the ^{160}Er channel by requir-

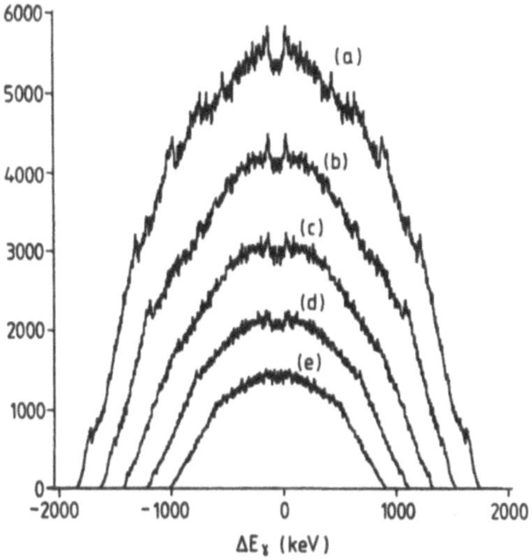

FIGURE 9 TESSA: cuts perpendicular to the valley diagonal with integration parallel to the valley of the E_γ-E_γ correlation shown in Figure 8. The γ-ray intensity as a function of the energy difference ΔE_γ is plotted for $\bar{E}_\gamma$ in 100 keV steps around a) 1100 keV, b) 1200 keV, c) 1300 keV, d) 1400 keV and e) 1500 keV.

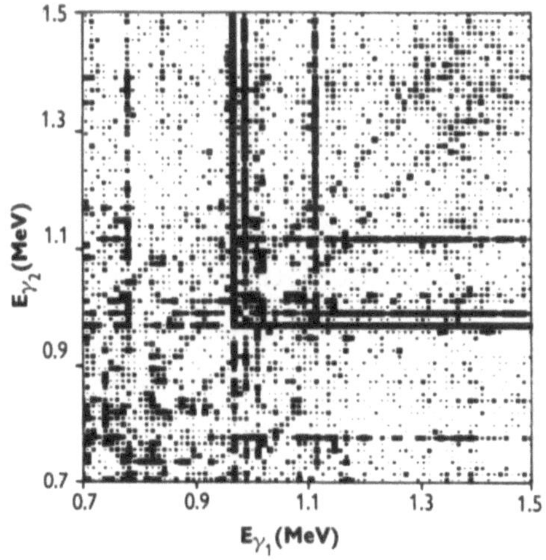

FIGURE 10 An E_γ-E_γ correlation spectrum for ^{152}Dy measured in the Ge escape-suppressed spectrometers. The channel has been selected using the BGO crystal ball. Vertical and horizontal stripes are due to discrete lines from yrast γ-rays below spin 40. The superdeformed prolate structure is identified by the ridges parallel and close to the $E(\gamma_1)=E(\gamma_2)$ diagonal for energies between 0.80 and 1.35 MeV.

ing that its energy corresponds to the γ-ray decay of the $2^+ \rightarrow 0^+$ or $4^+ \rightarrow 2^+$ transitions in ^{160}Er. The resulting sum energy and fold data are a measure of the entry point for this channel and contain no contamination from other exit channels, a fact confirmed by the spectrum generated from the second γ-ray signal. Various fold and sum energy conditions were then placed and the associated γ-ray spectrum generated. These were then normalised to have the same intensity in the low-lying yrast γ-rays in ^{160}Er and finally corrected for the response function and efficiency of the suppressed germanium detector. Figure 11 shows four of these spectra, corresponding to low fold and low sum energy, low fold and high sum energy, high fold and low sum energy, and high fold and high sum energy. For the same sum energy and an increase in fold the edge of the collective E2 bump is seen to move to higher γ-ray energies, an effect which was expected. However the edge of the bump also increases when the fold is constant and the sum energy is increased. This is completely in disagreement with the original concept of the nucleus cooling to the yrast line by statistical transitions and then following collective bands close to and parallel with the yrast line. The results show that instead the higher energy entry regions decay by higher energy E2 transitions and they are probably following lower moment of inertia collective bands far above the yrast line.

3.3 Sub-barrier Fusion

It has been established for some time that the energy dependence of fusion evaporation cross-sections cannot be reproduced by statistical model decay calculations[14,15].

Experiments have measured near barrier fusion cross sections[16,17] where enhancements of up to three orders of magnitude are observed relative to single dimensional barrier penetration calculations. This enhancement has recently been attributed to coupling to inelastic and transfer reaction channels[18]. This coupling leads to enhanced cross sections for evaporation residues for a few emitted neutrons and to greatly enhanced partial wave cross sections for large angular momenta[19].

TESSA has recently been used to study the angular momentum distributions which are present in the fusion of two heavy ions close to the Coulomb barrier. In these measurements the spectrometer is operated with only one suppressed germanium detector in coincidence with the BGO ball. The reaction used was ^{80}Se bombarding ^{80}Se and as the beam energy is lowered to the Coulomb barrier the γ-ray transitions associated with the 2n (^{158}Er), 3n (^{157}Er) and 4n (^{156}Er) channels become weaker compared with Coulex and other background. The suppressed germanium detectors are essential to cleanly set windows on these transitions so that the correct sum energy and fold spectra can be obtained. The Coulomb barrier is calculated to be close to 275 MeV and the sum energy versus fold plots are shown in Figure 12 for the 2n, 3n and 4n channels at a bombarding energy of 280 MeV. The statistics are low but the data clearly show the presence of a high fold component. The data were then projected onto the fold axis and following normalisation the spectrum were added to give the fold distribution for the total fusion cross-section. This is presented in Figure 13 for bombarding energies of 310 MeV, 280 MeV and 265 MeV. A multiplicity scale has been added to the fold scale and assuming that the decays are predom-

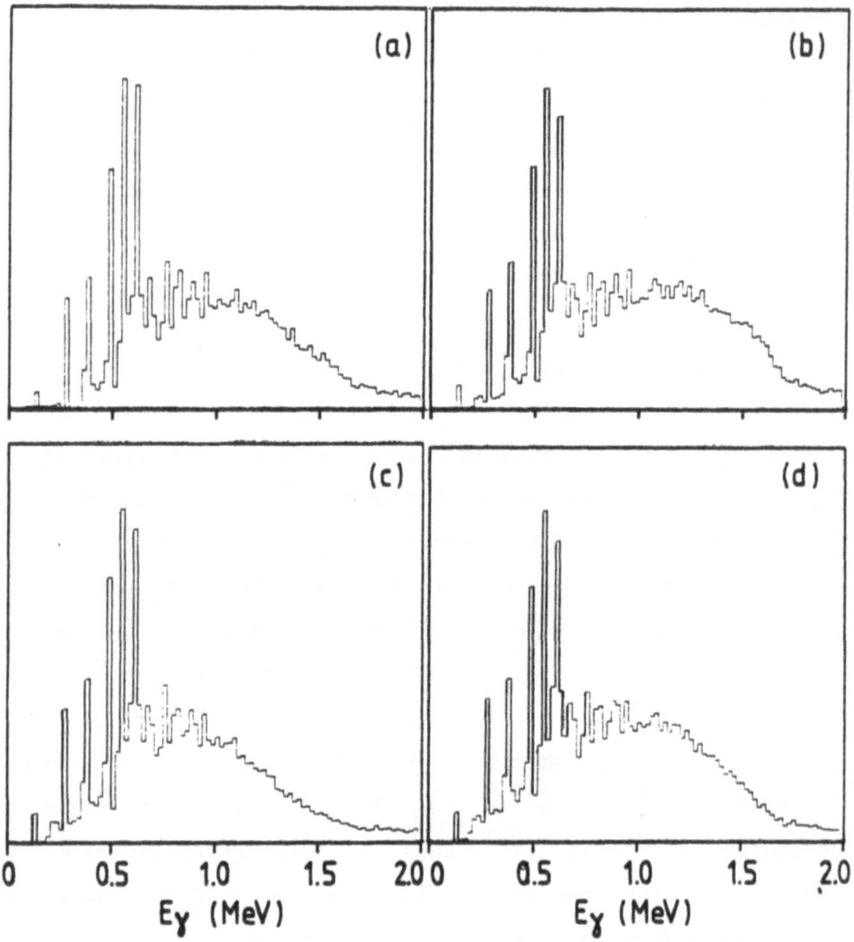

FIGURE 11 Gamma-ray spectra, corrected for the germanium detector response function and efficiency, of the decay from different entry regions of sum energy and multiplicity in ^{160}Er; a) low fold and high sum energy, b) high fold and high sum energy, c) low fold and low sum energy, and d) high fold and low sum energy.

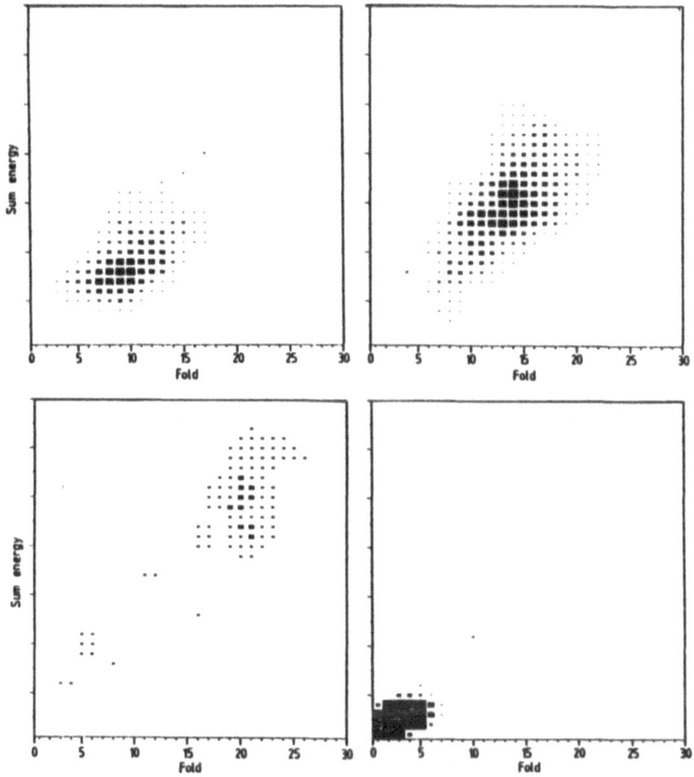

FIGURE 12 The sum energy versus fold plots for exit channels; a) ^{156}Er (4n), b) ^{152}Er (3n), c) ^{158}Er (2n) and d) coulex following the bombarding of ^{80}Se on ^{80}Se at 280 MeV.

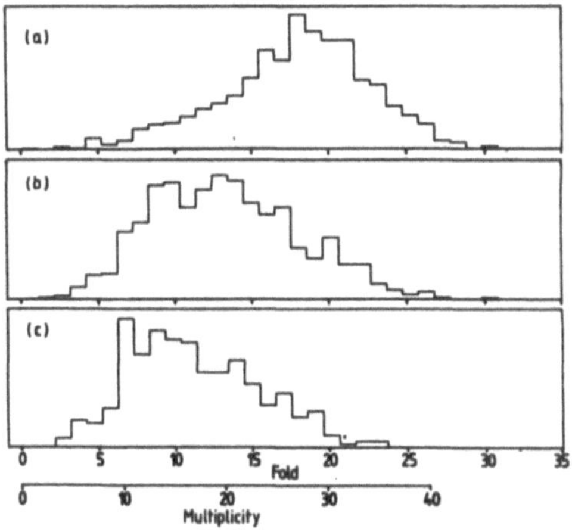

FIGURE 13 The fold distribution for the total fusion cross-section of ^{80}Se + ^{80}Se at bombarding energies of a) 310 MeV, b) 280 MeV and c) 265 MeV.

inantly quadrupole the high angular momentum tail extends upto a spin of 50 $\hbar$ at energies below the Coulomb barrier.

4. FUTURE DEVELOPMENTS

4.1. POLYTESSA

The TESSA array has proved itself a powerful versatile γ-ray spectrometer utilising both its high resolution suppressed germanium detectors and its BGO ball with sum energy and fold information. Its major limitation is its relatively low efficiency for γ-γ events and its negligible efficiency for γ-γ-γ events. There are two ways of improving the efficiency, increasing either the solid angle of each detector or the number of systems. Both approaches can be used but only if the suppression shields are made more compact and this can be achieved by using BGO instead of NaI. The solid angle cannot be greatly increased in experiments involving large numbers of coincidence γ-rays as the probability for detection of two γ-rays in one suppressed germanium detector becomes too large. Also the Doppler smearing from γ-rays emitted by moving recoils following H.I. reactions becomes larger than the inherent resolution of the germanium detectors.

These considerations have resulted in the specification of a new array (POLYTESSA) in which upto 30 suppression spectrometers can be placed in a 32 faced polyhedron structure which has 12 pentagon faces and 20 hexagon faces. The schematic in Figure 14a shows compact BGO shields in adjacent pentagon and hexagon faces with their associated germanium detectors subtending solid angles of 0.31% and 0.42% respectively. Thus with only 20 systems installed the γ-γ-γ events rate already becomes 6 times faster than the present γ-γ rate in TESSA. This factor increases to 20 with 30 systems and the γ-γ-γ-γ rate is then twice the present γ-γ rate in TESSA.

4.2. New Compact BGO-NaI Suppression Shield

A prototype of the new compact suppression shield has been fully tested. It consists of 3 elements, the BGO/NaI shield, the BGO catcher and the n-type HpGe detector, arranged as shown in Figure 14b.

The shield has an axial hole to accommodate the germanium detector and the system should give better suppression than the transverse design[20] except for the lowest energies in the γ-ray spectrum (i.e. below about 300 keV for γ-rays from a ^{60}Co source). The shield comprises two parts. The main cylindrical body is made from eight optically separated pieces of BGO each 15 cm long with a minimum thickness of 35 mm and viewed by a separate photomultiplier. The photomultiplier tubes are set at an angle of 25° to allow more space for mounting the BGO catcher and the germanium detector. Bismuth germanate has a low light output relative to sodium iodide (10-15%). For this reason, the front of the shield is made from sodium iodide where the γ-rays scattered back from the germanium will have low energies ($\lesssim$ 300 keV) and the greater density of the BGO is not needed to stop γ-rays of these energies. The NaI is made in one piece and is optically coupled to all the BGO pieces, thus any scintillation in the NaI is viewed through the BGO. The detector is assembled in an aluminium can and provision is made for a heavy metal collimator to shield the scintillator from direct

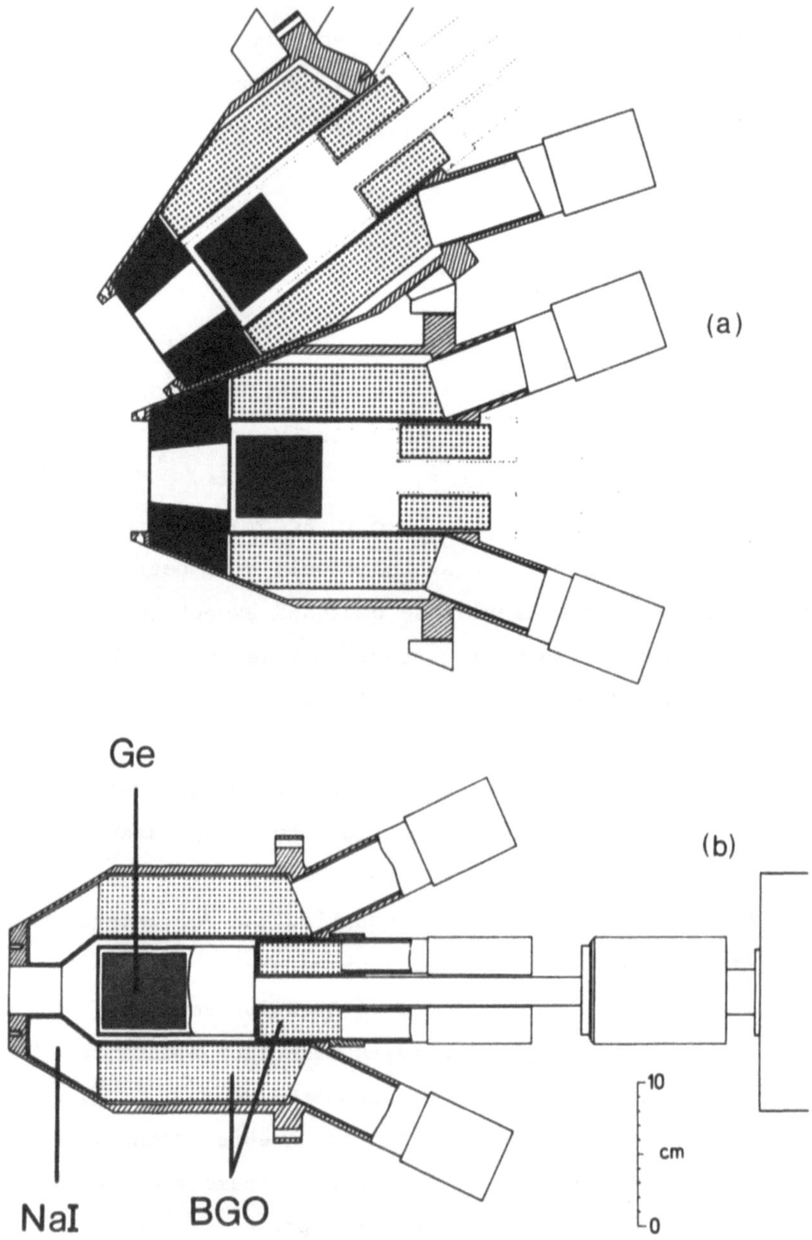

FIGURE 14 a) A schematic of compact BGO/NaI shields in adjacent pentagon and hexagon faces of POLYTESSA; b) The prototype compact BGO/NaI shield; in addition a 40 mm thick heavy metal collimator shielded the scintillator from direct radiation from the γ-ray source.

γ-radiation. The assembly was supplied by Harshaw Chemie BV, Holland.

It is important to use n-type HpGe detectors in escape suppression shields as the lack of outer dead layers on the Ge crystal is vital for good performance. One limitation of the cylindrical shield design chosen is the lack of scintillator to detect γ-rays that are forward Compton scattered. This difficulty has been resolved by using a new type of mounting for the Ge detector (a "Duet" supplied by EG & G Instruments Ltd.) as shown in Figure 14b. It allows the inclusion of a BGO catcher to detect the γ-rays which are forward Compton scattered. The catcher is an annular BGO detector whose inner and outer radii are fixed by the Ge detector and BGO/NaI shield respectively. The length is determined by the range of γ-ray energies of interest. The present design uses 6 cm as the maximum γ-ray energies of interest are 1.5-2 MeV. The detector is made in two halves each viewed by two photomultiplier tubes. The assembly was supplied by Harshaw Chemie BV, Holland.

An important property of the BGO based detectors is their performance with low energy γ-rays which determines the lowest threshold that can be used thus influencing the quality of supression. Figure 15 shows the spectra of 60 keV γ-rays from an ^{241}Am source indicating that thresholds as low as 10-15 keV can be set. A further indication of the quality of both the design and the crystals used in the shield is that the NaI resolution for 662 keV γ-rays is 11% in spite of the light having to cross the BGO before striking the photomultiplier.

The critical test of a system is clearly its performance as a suppression shield spectrometer and this is illustrated in Figures 16 and 17. Figure 16 shows the spectrum for ^{60}Co and it is very similar to that given in Figure 3 for the transverse NaI shields used in TESSA. The suppression factors (unsuppressed spectrum divided by suppressed spectrum) are shown in Figure 17 such that the effect of the catcher can be seen. Figure 17a is the suppression factor with full suppression including the BGO/NaI shield and the BGO catcher while in Figure 17b only the BGO/NaI shield is included. It is clear that the catcher improves the suppression factor over the whole spectrum but particularly below about 400 keV. The value obtained with full suppression of about 10 between 500 and 1000 keV is an excellent result. With a lower limit of 100 keV, the peak to total ratios obtained are 60% and 71% for ^{60}Co and ^{137}Cs respectively. These results are very similar to the best values found for the NaI suppression shields used successfully in TESSA2 (Table 1) with the advantage that the new spectrometer is compact and many more can be packed around the target. It can also be used in many flexible geometries as well as in POLYTESSA. For example the shields can be utilised in experiments involving recoil detectors and a wall of neutron detectors.

5. SUMMARY

The past eighteen months have seen the great success of TESSA in the field of γ-ray spectroscopy. It is planned for it to remain in operation mainly concentrating on experiments where the BGO ball sum energy and fold data are required. Next year, 1985, will herald the operation of POLYTESSA at Daresbury when high statistics γ-γ-γ and other

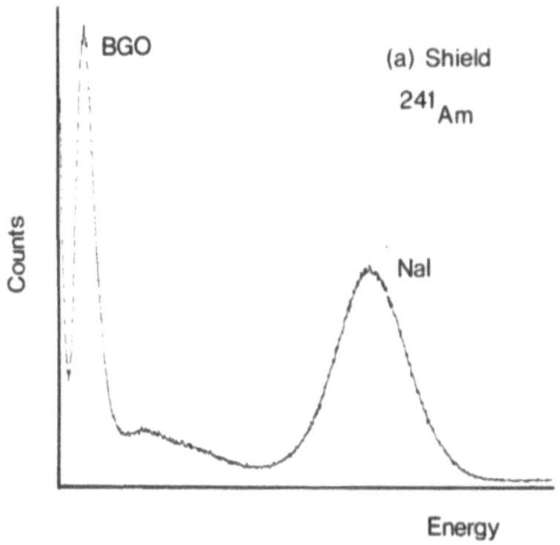

FIGURE 15 Compact BGO/NaI shield: spectrum taken using an ^{241}Am source with the signals from all photomultiplier tubes added. The signal was amplified and shaped using a standard spectroscopy amplifier (0.5 μs shaping time). The two separate components from the BGO and NaI reflect their different light outputs for a given γ-ray energy.

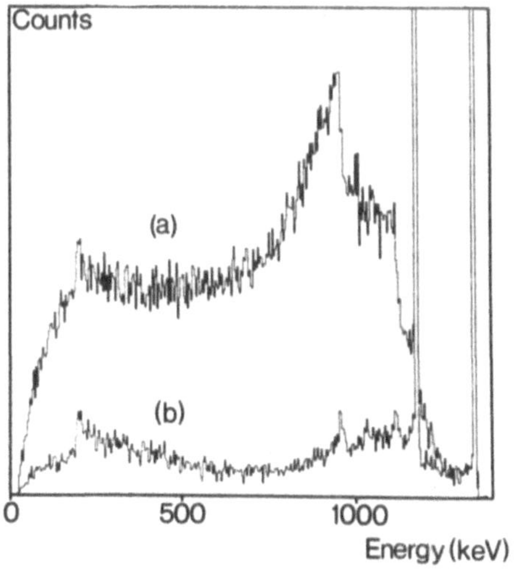

FIGURE 16 Compact BGO/NaI shield: spectra taken using a ^{60}Co source to show the performance of the system. The source was about 15 cm from the germanium detector. a) all γ-rays in the germanium detector, b) the fully suppressed spectrum, i.e. γ-rays in the germanium detector in anti-coincidence with both the shield and the catcher.

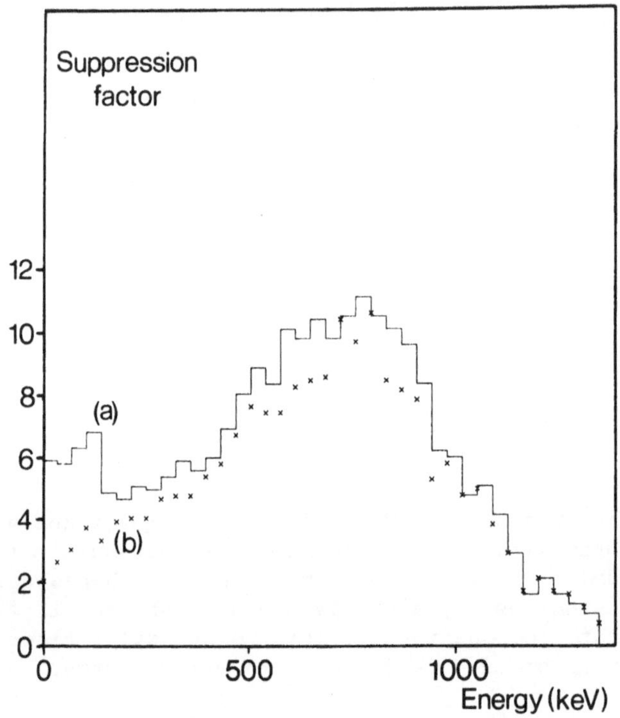

FIGURE 17 Suppression factors measured for a ^{60}Co source. a) The fully suppressed system (histogram) including sheild and catcher, and b) only the BGO/NaI shield used as a suppressor (crosses).

triple coincidence experiments will become feasible.

The work reported here is the result of much labour by many people. In particular P.J. Nolan, D.J.G. Love and A. Nelson from Liverpool University have played key roles. I thank C. Kalfas and B. Nyako who contributed much during their visits to Daresbury and Liverpool; D. Howe, A. Kirwan, M. Riley, J. Simpson, D.W. Gifford, P.D. Forsyth and J.F. Sharpey-Schafer from Liverpool; and B. Herskind and J. Garrett from N.B.I. Copenhagen. This work was supported by the Science and Engineering Research Council of the United Kingdom.

REFERENCES

1. P.J. TWIN, Workshop on nuclear structure at high spin, Risø, 1981.
2. P.J. TWIN et al, Nucl. Phys., A409, 343c (1983).
3. M. JÄÄSKELÄEINEN et al, Nucl. Instr., 204, 385 (1983).
4. V. METAG et al, Nucl. Phys., A409 331c (1983).
5. M.A. LONE et al. Proceedings of Int. Workshop on BGO, Princeton University, p237 (1982).
6. J. BURDE et al, Phys. Rev. Lett., 48, 530 (1982).
7. J. SIMPSON et al, Phys. Rev. Lett., 53, 648 (1984).
8. O. ANDERSEN et al, Phys. Rev. Lett., 43, 687 (1979).
9. T.L. KHOO et al, Phys. Rev. Lett., 41, 1027 (1978).
10. B. NYAKO et al, Phys. Rev. Lett., 52, 507 (1984).
11. I. RAGNARSSON et al, Nucl. Phys., A347, 287 (1980).
12. S. ÅBERG, Physica Scripta, 25, 23 (1982).
13. Y. SCHUTZ et al, Phys. Rev. Lett., 48, 1534 (1982).
14. J. GILAT et al, Phys. Rev., C7, 1973 (1973).
15. S. DELLA NEGRA et al, Z. Physic., A282, 75 (1977).
16. M. BECKERMAN et al, Phys. Rev. Lett., 45, 1472 (1980).
17. R. STOCKSTAD et al, Phys. Rev., C23, 281 (1981).
18. R. BROGLIA et al, Phys. Lett., 133B, 34 (1983).
19. S. LANDOWNE et al, Phys. Lett., 138B, 32 (1984).
20. R.M. LIEDER et al, Nucl. Instr., 220, 363 (1984).

THE BERKELEY HIGH-RESOLUTION BALL*

R.M. DIAMOND
Nuclear Science Division
Lawrence Berkeley Laboratory
University of California
Berkeley, California 94720

Criteria for a high-resolution γ-ray system are discussed. Desirable properties are high resolution, good response function, and moderate solid angle so as to achieve not only double- but triple-coincidences with good statistics. The Berkeley High-Resolution Ball involved the first use of bismuth germanate (BGO) for anti-Compton shields for Ge detectors. The resulting compact shield permitted rather close packing of 21 detectors around a target. In addition, a small central BGO ball gives the total γ-ray energy and multiplicity, as well as the angular pattern of the γ rays. The 21-detector array is nearly complete, and the central ball has been designed, but not yet constructed. First results taken with 9 detector modules are shown for the nucleus ^{156}Er. The complex decay scheme indicates a transition from collective rotation (prolate shape) to single- particle states (possibly oblate) near spin 30 $\hbar$, and has other interesting features.

*This work was supported by the Director, Office of Energy Research, Division of Nuclear Physics of the Office of High Energy and Nuclear Physics of the U.S. Department of Energy under Contract DE-AC03-76SF00098.

Four or more years ago Frank Stephens and I became interested in developing a high-resolution γ-ray system, both to extend discrete-line spectroscopy and to study continuum γ-rays. If it were possible to push discrete transition work 10-15 spin units higher, all the detailed studies possible at lower spins could be carried out and much more could be learned about how nuclei carry angular momentum and how they change with spin. Questions to be answered are how rapidly do the pairing correlations quench, do alignments and backbends continue to high spin, and what are the changes in shape and in collective motion with increase in spin? To achieve this goal appeared to require both the "looking down" technique of the NaI balls then being constructed, namely, to select a limited population of initial states by cuts on the fold distribution and on the total-energy response of the ball, and a "looking up" scheme in which a gate is set on a discrete line of moderately high spin and the transitions above it are looked for in coincidence. But the highest spin lines observed might represent at most a few percent of the total population at that spin, and to learn about the majority of the states and their properties, one would also have to study the continuum spectrum. We had already observed, by comparing NaI and Ge γ-γ correlation spectra, that there were structures in the continuum region that required 5-10 KeV resolution to be seen. This is better than can be done with NaI detectors, but, as explained below, a good response function is also very important in coincidence studies and Ge detectors are poor in that regard.

So at that time we were able to convince C. Gruhn to undertake a feasibility study of liquid Xe detectors as modules for a 4π high-resolution ball. In theory such a system would have an intrinsic resolution only a factor of two worse than solid-state Ge detectors, and it has a number of attractive properties including a much better response function. But this undertaking proved to be an ongoing developmental problem, principally because of very low-level (ppb) impurities, and appeared that it might take years to bring to fruition. So after a year we decided to base our multidetector system on already existing Ge detectors. However, you have already heard at this Conference from Dr. Lindblad[1] on the progress they have made in Stockholm during the last three years with liquid Ar and Xe counters, including the operation of a liquid Xe detector with about 8 KeV resolution for ~1 MeV γ rays. So such counters still may live up to their theoretical promise.

TABLE I Criteria for optimum γ-ray system.

1.	High energy resolution
2.	Good response function
3.	Good efficiency
4.	Total-energy spectrometer
5.	Multiplicity filter
6.	Prompt initial timing signal

In Table I are listed three primary criteria important for a high-resolution array, and three secondary features we wanted. As mentioned, Ge detectors provide the highest resolution possible today with reasonable efficiency for γ-ray energies in the range of 50 KeV to a few MeV (~2 KeV resolution at 1 MeV). Can we do better? With a single detector, no. In this case the average resolution can be defined as the reciprocal of the number of resolvable points in the energy range of interest. Consider a range

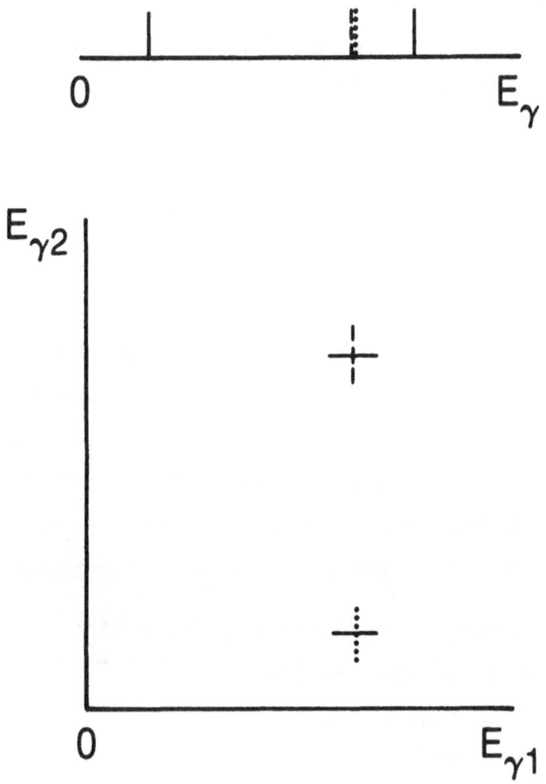

FIGURE 1 The overlapping lines in the singles spectrum, |⁝, are resolvable in the doubles spectrum by their different cascade partners.

of 0-1 MeV and a constant resolution of 2 KeV; this gives 500 resolvable points. If two γ rays fall on the same point, they cannot be resolved. But the γ rays we are interested in are usually not single events but members of a cascade and are correlated with the other members. If two (or more) detectors are used, the coincidence relationships between the two overlapping γ rays and the other members of their individual cascades may serve to resolve them, as illustrated in Fig. 1. The 500 resolvable points of the single detector have become 500^2 points in the two-dimensional array of the two coincident detectors, and it is much less probable that the two pairs of γ rays coincide than the original single lines. Still higher order coincidences give still higher effective resolution for coincident cascades of γ rays by enormously increasing the number of possible resolvable points; a triple coincidence would have 500^3 possible resolvable points. Thus, a desirable feature for our array is to have as many detectors as close as is reasonably possible (see below) to the target, in order to favor high-order coincidences.

A second criterion is to have a good response function, i.e., every γ ray that strikes the detector should result in a voltage pulse proportional to the original γ-ray energy. A Ge counter does not fulfill this requirement very well. It is a relatively low Z, moderately dense material that interacts with γ rays in the energy range of interest principally by Compton scattering. With a 5 x 5 cm Ge detector (approximately 20% the efficiency of a 7.6 x 7.6 cm NaI detector for 1.33 MeV γ rays) a 1.33 MeV γ ray has an absorption probability of ~3/4 and a peak to total (P/T) ratio of 15-20%. This means that ~1/4 of such γ rays do not interact at all with the detector, and that of those that do, only 15-20% give useful full-energy peaks. In a doubles coincidence measurement, only 2-4% of the events obtained are good peak-peak values; the remaining 96-98% are unwanted garbage. This is a very undesirable situation, and the solution has been known for some time; put Compton-suppression shields around the Ge detectors. The preceding talk by Dr. Twin on TESSA illustrated how this was done at Daresbury[2]; large 20 x 25 cm NaI shields were placed on six Ge detectors arranged around the target, giving a markedly improved peak/total ratio and resulting in the beautiful spectra you have just seen. Not having such NaI shields on hand, we decided to try for bismuth germanate (BGO) shields, for, if successful, they offered a great advantage. The material is considerably denser than NaI (7.13 to 3.67 g/cm^3) and of higher averge Z, so that it has a γ-ray absorption length 2-1/2 times smaller than NaI. This means more compact shields, so that more Compton-suppressed detectors can be placed near the target to give the higher order coincidences discussed in the previous paragraph. But bismuth germanate also has a serious handicap; the light output is only 10-15% that of NaI. In addition, at that time (1981) the companies that marketed it had not made BGO crystals large enough to serve this purpose, as shields of the order of 13 cm or more in diameter and length were needed.

It took a year to obtain a prototype version of the BGO shield. The unit was not quite large enough, and it had broken in two during machining and had been cemented back together. Indeed it looked as if it might be a long, hard period before single crystals of BGO of sufficient size could be processed into suitable shields. But upon testing the broken prototype, the cemented fracture did not seem to cause any problem, and the taper at the front of the shield appeared to reflect light so well that the phototubes at the back gave a bigger signal when the source was placed on the front piece (whose light had to pass through the fracture) than when it was put on the back piece. So the problem was solved by making the shield in six longitudinal pieces that were put together to form the detector shell of suitable size. I believe that all of the Compton suppression shields that have been made from BGO

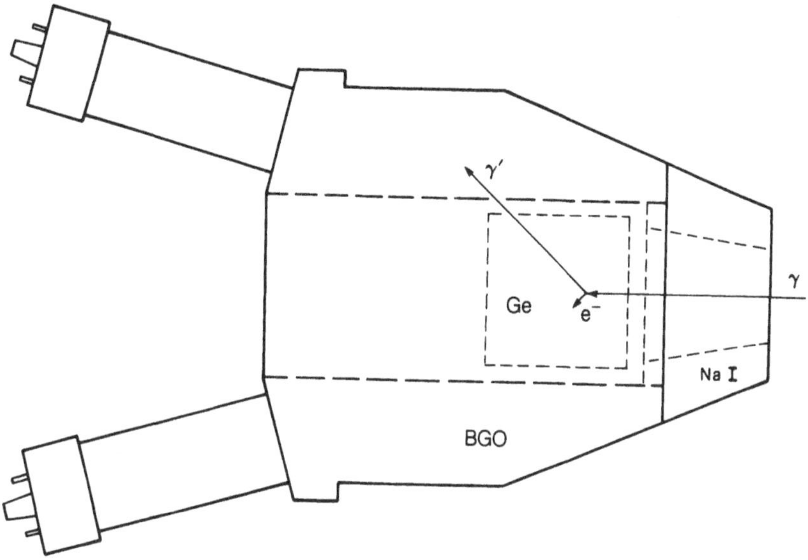

FIGURE 2 Outline of BGO Compton-suppression shield showing 5 x 5 cm Ge detector inside, two of six photomultiplier tubes on back surface, and NaI "cap" on front taper.

have been produced in this way. In our design a 1-1/2" photomultiplier tube was placed on the back of each of the six sections to gather the light as efficiently as possible. A schematic drawing of our BGO shield showing two of the tubes is given in Fig. 2. It is a coaxial design, so that there is no shield material at the front opening where the γ rays come in to the Ge detector, nor at the rear where the snout of the Ge detector enters. Lack of coverage at the latter position means that Compton-scattered γ rays in the angular range of 0°-25° (multiple Compton scattering in the Ge crystal and primary interactions throughout the whole volume of the Ge smear these angles out) will not be caught by the shield, leaving the complementary non-full-energy transition in the resulting spectrum. We do not believe these low-energy transitions pose a serious problem, although there are attempts being made to reduce its magnitude by reducing the snout behind the Ge detector to a minimum diameter and then surrounding it with additional shield material. (Of course the transverse shields do not have this hole, but they have one at the side instead.) But the opening at the front is more disturbing, as the (low energy) γ rays Compton-scattered through about 135°-180° correspond to leaving in the spectrum the Compton edges just below the full energy peaks. The edges appear just where we are interested in having a clean, low-background region. Figure 3 shows two spectra of ^{60}Co taken with one of our Ge detectors, a Compton-suppressed one taken with the BGO shield and an unsuppressed one normalized to the same full-energy peak heights. (In the former, the Compton-edge peaks just mentioned can be clearly seen.) The peak/total ratio for ^{60}Co above a 300 KeV threshold improves from ~20% to 50%. But for coincidence experiments there is the even more marked improvement shown in Table II. Note, for example, that with bare Ge detectors full-energy triple

TABLE II Ge detector 20% (counts above 300 KeV).

	Ge^1	Ge^2	Ge^3	Ge^4
Peak/Total	0.2	0.04	0.008	0.0016
Compton-suppressed Peak/Total	0.5	0.25	0.13	0.06
Improvement Factor	2.5	6	16	39

coincidences are only 0.8% of the events, so that triple coincidences have not been used in in-beam spectroscopy. But with a 50% P/T ratio, 12.5% are good events making them

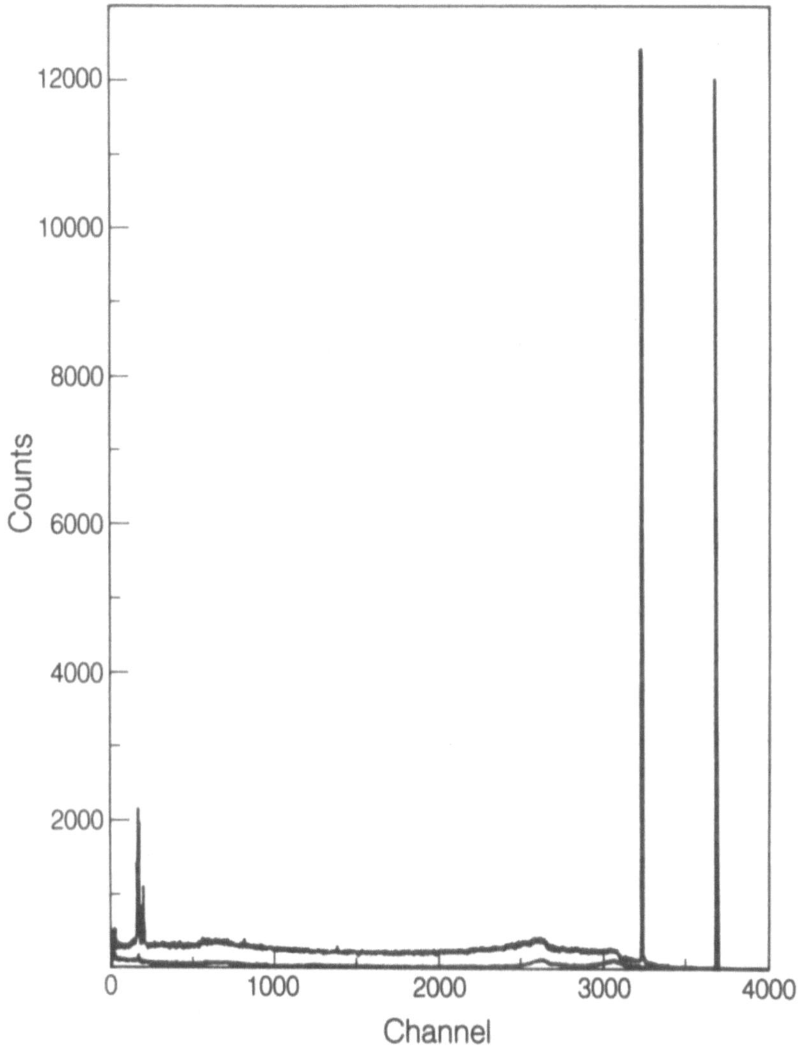

FIGURE 3 Spectra of ^{60}Co taken with bare 20% Ge detector and with Compton-suppressed detector (no NaI cap).

quite useable. And still higher ratios are possible. By the addition of a NaI "cap" to the front of the BGO shield, as shown schematically in Fig. 2, we have obtained a P/T ratio of 55% for the ^{60}Co peaks (300 KeV threshold) and, most importantly, have almost completely wiped out the Compton-edge peaks. Other designs using larger BGO shields can give P/T ratios of 60-65% for ^{60}Co, so tremendous improvement in the Ge response function is possible.

The system efficiency involves the questions of how many Ge detectors at what distance from the target. With the design shown in Fig. 2 involving 13.5 x 13.5 cm shields, some units will touch at their tapered cones when 13-14 cm from the target. If we take 15 cm as the distance from the target to the face of the 5 x 5 cm Ge detector inside the BGO shield, we find we can place 21 detectors at this distance and still leave the top and bottom above the target clear. This will be necessary to accommodate the photomultiplier tubes for a small, central BGO ball between the target and the Ge modules. Of course, the number of detectors is a matter of choice and depends somewhat on the geometrical arrangement selected. Too small a number hurts the higher order coincidences and the statistics; too large a number is costly, and after a moderately large number one does not gain very rapidly, as additional detectors require an expansion in the radial distance from the target. Another consideration on the distance from the target involves the amount of summing to be tolerated in an individual detector. At 15 cm from the target, the geometric solid angle of our detectors, diminished by the Ge transparency, is $\sim 5 \times 10^{-3}$. Sixty percent (or more) of such γ rays interacting with the Ge detector are rejected by the shield (plus NaI cap) so the overall efficiency is $\lesssim 2 \times 10^{-3}$. If the γ-ray cascades have an average multiplicity of 20, this gives ~4% summing, on average. This is probably O.K. for most purposes; if not the geometric efficiency must be reduced.

Lastly, but not least, Doppler effects must be considered. The usual method of producing nuclei in high-spin states is with (H.I.,xn) reactions. In a typical case, ~180 MeV ^{40}Ar on ^{124}Sn, the recoiling compound nucleus has a velocity ~2.5% that of light. The Doppler shift for a γ ray emitted by such a moving source is

$$\frac{E - E_0}{E_0} = \frac{v}{c} \cos \theta \tag{1}$$

where θ is the angle between the detector and the recoiling nucleus. So near 0° or 180° the Doppler shift is quite large, ~25 KeV for a 1 MeV γ ray. But the broadening due to the finite opening-half-angle ϕ of the detector

$$\frac{\Delta E}{E_0} \approx 2 \frac{v}{c} \sin \theta \sin \phi \tag{2}$$

is quite small, $\leq$ 1/2 KeV for θ near 0° or 180° and ϕ = 9.5°. By using a thin target or multiple thin targets, the full Doppler shift is obtained, but since this corresponds to a change in gain it can be corrected for in the amplifier or ADC or in the analysis program with no loss in resolution. On the other hand, for angles of θ near 90°, the Doppler shift becomes small, but the broadening becomes a maximum. For the 180 MeV ^{40}Ar + ^{124}Sn example, $\Delta E/E_0 \approx 0.8\%$ or 8 KeV broadening for a 1 MeV transition. Such a smearing of the energy ruins the resolution of the detector, but can be reduced or avoided altogether under certain conditions. For example, if such a product nucleus recoils out of the target into a lead or gold backing, it will stop in the order of picoseconds. If the transitions are emitted after stopping, there will be no Doppler shift or broadening. Thus, nuclei that do not have fast, col-

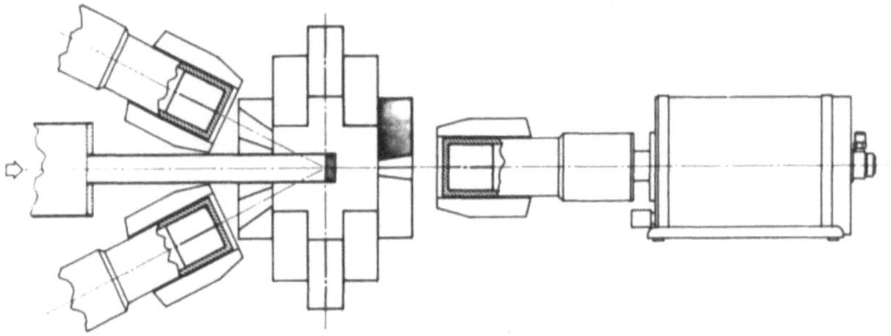

FIGURE 4 Vertical cut through the High-Resolution Ball system, showing one Ge detector in each ring. Not shown are photomultiplier tubes and NaI caps on BGO shields.

lective transitions in their de-excitation cascades can, and do, take picoseconds to de-excite and do not present a problem. But if the γ rays are fast, something has to be done to alleviate the Doppler broadening; the detectors must be pulled back further or collimated or both in order to decrease the detector opening angle, and thus the broadening.

With these considerations in mind we have designed[3] the High-Resolution Ball to consist of 21 BGO-shielded Ge detectors arranged (in three rings of seven detectors) around a small, central "ball" or "castle" of 40 BGO sectors. Some of the latter have holes through which the Ge detectors see the target, and the 40 together form a sum spectrometer and multiplicity filter around the target, as well as giving the angular pattern of the γ rays emitted. A sideview of the arrangement is given in Fig. 4, which shows one detector from each ring. Not shown are the NaI caps on the BGO shields, nor are the photomultiplier tubes drawn in on either the shields or the central ball sectors. For the BGO sectors in the upper half of the central

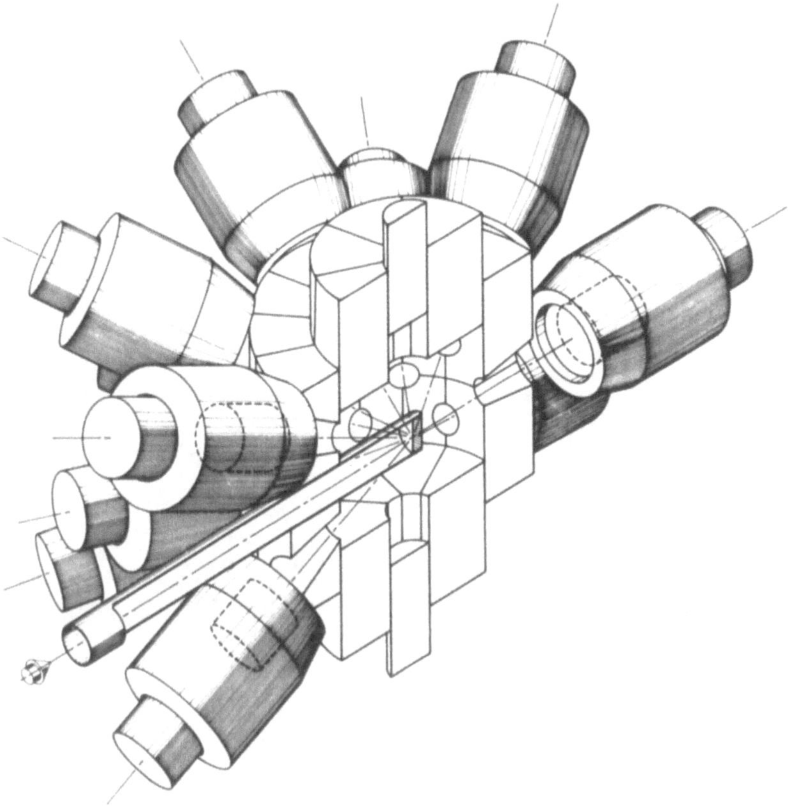

FIGURE 5 Perspective view of ~1/2 the system.

ball, the photomultiplier tubes go up from the top faces of the sectors, and for those in the lower half, the tubes go downwards from the bottom faces. A cut-away perspective drawing of the system is shown in Fig. 5.

With 21 of our detectors at 15 cm from the target, an event rate of 10^5/second (the order of our usual rate in a (^{40}Ar,xn) reaction), and an average γ-ray multiplicity of 20, we estimate the double-, triple-, and quadruple-coincidence rates to be 11 K/s, 2.2 K/s and 280/s, respectively. We have already performed experiments with 9, 12, and 15 Compton-suppressed Ge detectors (but without the

FIGURE 6 Photograph of 12 Compton-suppressed Ge modules in place around a small target chamber.

central BGO ball), and do get the corresponding calculated rates.

Where we were 3-4 months ago (Summer, 1984) is shown in Fig. 6, a photograph of 12 Compton-suppressed Ge modules in place around the target inside a small chamber. The beam-line to the chamber passes between the two middle modules in the horizontal plane. The Ge detectors are out of sight inside the BGO shields, but their small liquid-nitrogen dewars are visible, as are the pre-amps for the shields. We now have all 21 Ge detectors and 16 of the BGO shields; the last of the latter are due by February, 1985. We shall go out to bid for the NaI caps for the BGO shields before the end of this year. By the beginning of 1985 we should have a new system of fast 8-channel multiplexers and of ADC's and a fast CAMAC crate in operation for acquiring data. All that will remain to build is the central BGO ball, and its design has been completed.

One of the goals of the High-Resolution Ball is to push the limits of discrete γ-ray spectropscopy to higher spin into the bottom of the present continuum region, in order to gain the much more detailed information possible with discrete-line studies. This will be accomplished by cutting down the number of de-excitation pathways being observed both by cuts on the total energy and multiplicity measured with the central ball, and by simultaneously gating on single, high-spin discrete lines (and "looking up" in coincidence above the last known discrete lines). Since we do not have a central ball yet, we can only do the latter at present, and for our first experiments we have chosen to look at the region around $Z = 64$ and $N \geq 82$. We have taken data on some eleven nuclei. This choice was made for three reasons. First of all, most nuclei in this region are expected to have long enough feeding times (picoseconds) into the range of high-spin states which we want to study that they will be stopped in lead-backed targets before they emit their de-excitation γ rays. Thus, there will be no problem with Doppler broadening. Secondly, the level schemes are apt to be quite complex, as these are not good collective rotors having bands with approximately rotational spacings, but involve mainly single-particle states with possibly weakly collective interactions. For this reason many of them have not yet been studied to high spin and they represent a challenge, though a difficult one. Thirdly, they are in a region where the physics appears to be varied and interesting, involving

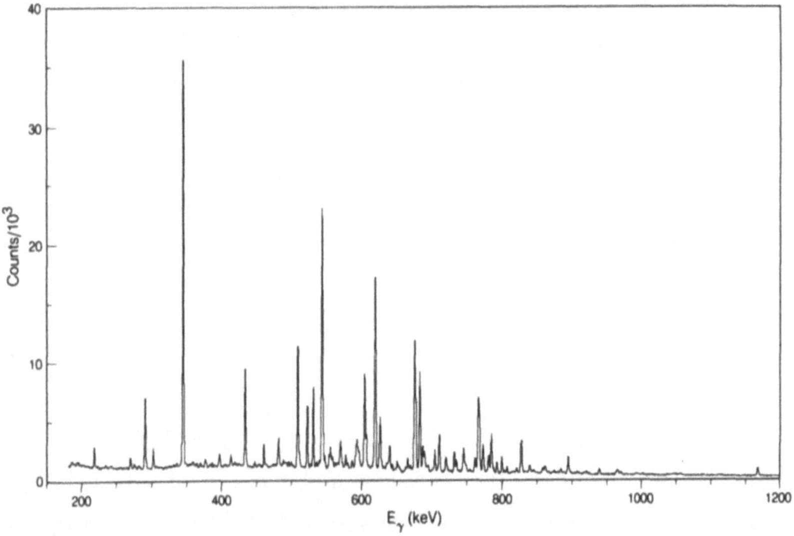

FIGURE 7 Spectrum of ^{156}Er in coincidence with the 452 KeV, $4^+ \rightarrow 2^+$ ground-band transition.

shape changes, and possibly the theoretically predicted transformation to superdeformed shapes (2:1 axis ratio) at relatively low spins and also band terminations.

I shall show you some data on ^{156}Er which are being analyzed by Frank Stephens. They were taken with 9 Compton-suppressed Ge detectors in two days running at the 88" cyclotron with a ~1 pna beam of 175 MeV $^{40}Ar^{8+}$ on a lead-backed ~1 mg/cm^2 ^{120}Sn target. The data consist of ~1.5 x 10^8 double- and ~10^7 triple-coincidence events. A spectrum in coincidence with the 452 KeV, $4^+ \rightarrow 2^+$ transition of the ground band shows the number of lines observed in ^{156}Er, essentially clean of $^{157,159}Er$ (Fig. 7). Actually, the total γ-γ projection spectrum is clean enough so that the small peak at 884 KeV, the $28^+ \rightarrow 26^+$ transition that has only ~3% the strength of the ground-band $4^+ \rightarrow 2^+$ transition, can be used as a gate to provide the very nice

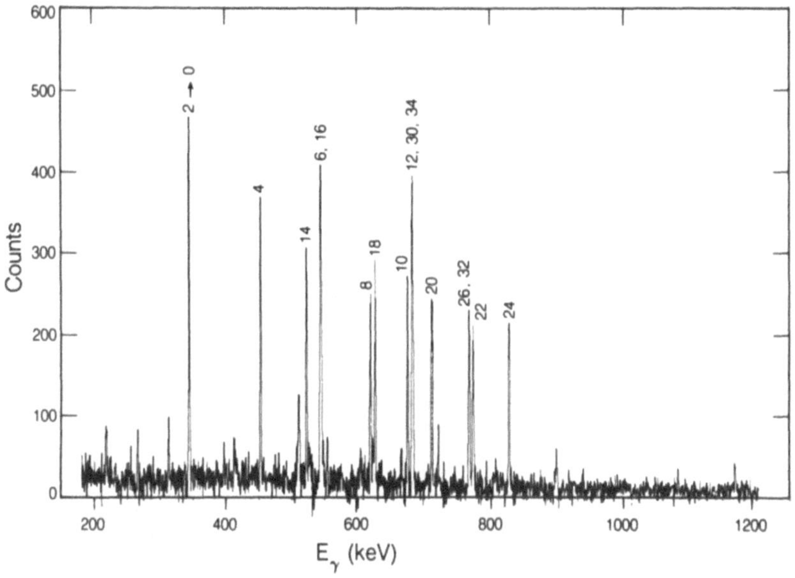

FIGURE 8 Spectrum of ^{156}Er in coincidence with the 884 KeV, $28^+ \rightarrow 26^+$ transition.

spectrum shown in Fig. 8. The ν $i_{13/2}$ backbend at spin 12 ħ shows clearly, as does the discontinuity at spin 26 ħ, but also one at 30 ħ. This part of this cascade is no longer rotational, but appears to be made of single-particle states. It would be interesting to determine the transition lifetimes here. Note that in this positive-parity yrast band there are two lines that are double and one that is triple. In fact, of the more than 100 lines Frank has assigned to the decay scheme, more than half are double, triple, or higher fold, and there are six transitions at 766 ± 1 KeV. A spectrum brought back as the result of a double gate on 344 KeV is shown in Fig. 9. That is, the $2^+ \rightarrow 0^+$ transition of the ground band is weakly coincident with another γ ray of the same energy, and the result-

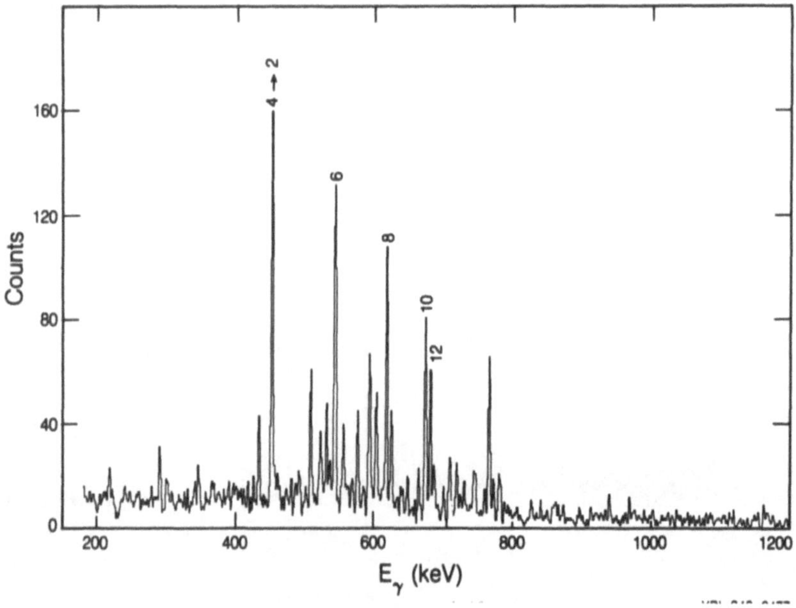

FIGURE 9 Triple coincidence spectrum of ^{156}Er with a double gate set on two 344 KeV transitions. The spectrum has been smoothed once using a simple three-channel algorithm.

ing triple-coincidence spectrum is shown. The ground band is observed through the first backbend, and then there are a number of transitions in the energy range of 550-600 KeV which must involve the other 344 KeV transition, but they have not all been placed in the scheme yet.

The preliminary level scheme for ^{156}Er is given in Fig. 10. Positive-parity levels above spin 26 ħ and negative-parity levels above ~23 ħ are mostly new; a few lower levels have been changed from previous work.[4,5] The number of transitions and levels have been doubled over those in earlier schemes. Spin assignments for the new states are tentative and based solely on the angular correlations of the cascade transitions. The new spin and parity assign-

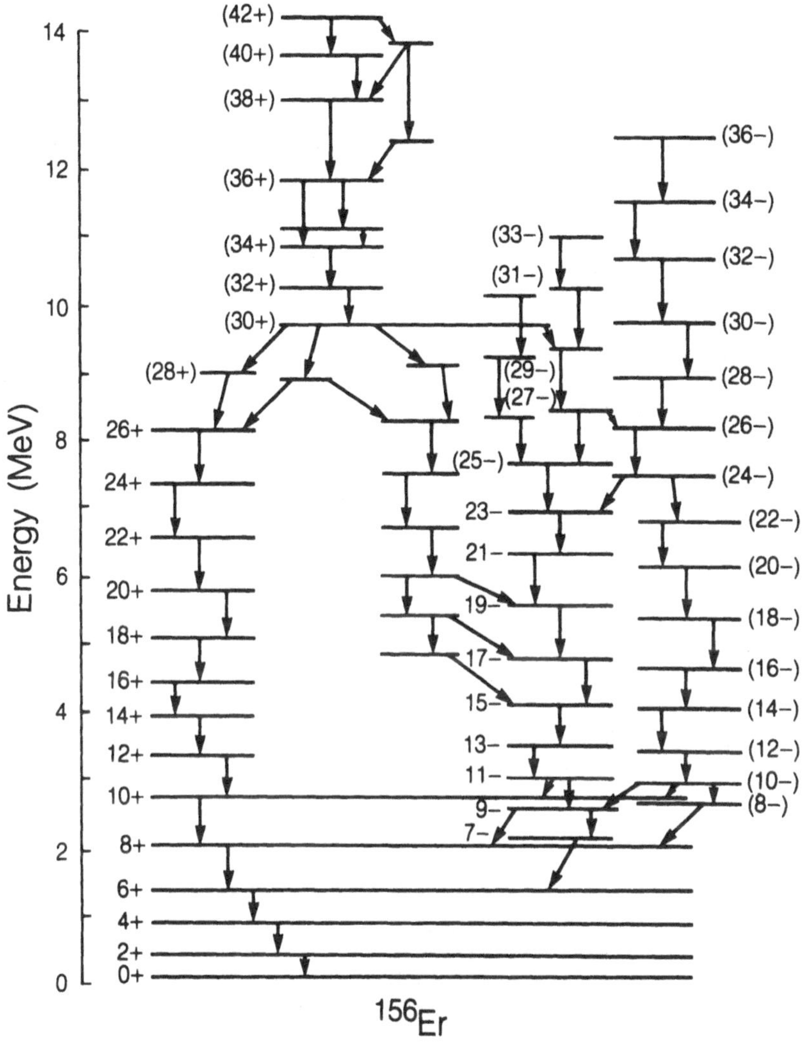

FIGURE 10 Preliminary level scheme for ^{156}Er.

ments for the right-hand band in the figure are also tentative and will have to be checked by conversion-electron measurements. A plot of energy vs. I(I + 1) for the levels is shown in Fig. 11. The lines are drawn only

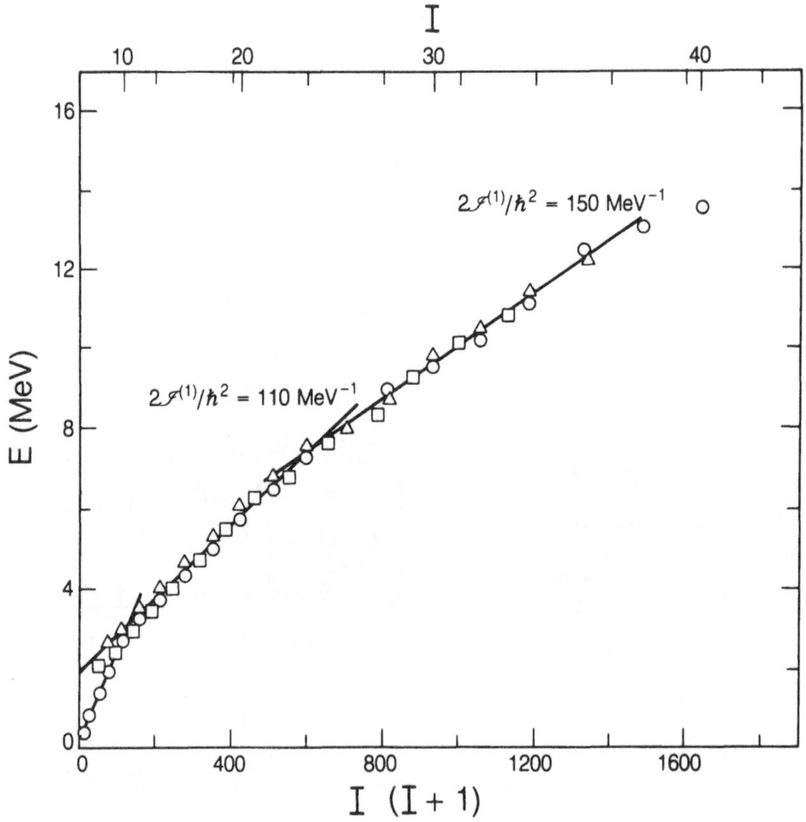

FIGURE 11 Plot of excitation energy E vs. I(I + 1) for the three main bands of ^{156}Er. Corresponding to Fig. 10, they are: the left-most (positive-parity) band, including the non-collective region at the highest spins, circles; the right-most (negative-parity, even-spin) sequence, triangles; and the center (negative-parity, odd-spin) band, squares.

to guide the eye, but the points do seem to group into three different regions with different slopes. First comes the ground band below spin 10 ħ (0 quasiparticles) then all bands between spins ~12 ħ and ~25 ħ (2 quasiparticles), and finally all levels above spin ~25 ħ (4 or more quasiparticles). The increasing moment of inertia given by the three slopes is probably the result of decreasing pairing and increasing particle alignment as the number of quasiparticles increases. Although $\mathcal{J}^{(1)}_{eff}/\hbar^2 = I/\hbar\omega$ is an integral quantity and so not very sensitive to local structural changes, it is somewhat surprising that all the points seem to lie so well on the three straight-line segments. The right-hand band in Fig. 10 has been reassigned as a negative-parity band, probably the unfavored even-spin partner of the already assigned negative-parity odd-spin band. Both of these start out with an aligned spin of 7–8 ħ relative to the ground band and show the "blocked" $\nu\ i_{13/2}$ backbend at $\hbar\omega \approx 0.35$ MeV with an additional alignment of 7–8 ħ, as expected for the lowest negative-parity bands in this region. Above the backbend the odd-spin band splits into two very similar branches (which was not expected) and the even-spin band continues, possibly as a still-collective one. This scheme can be compared with Cranked-Shell-Model calculations,[6] and the agreement is reasonably good, especially for the predicted onset of non-collective or single-particle behavior (oblate shape) taking over from collective rotation (prolate shape) in the neighborhood of spin 30 ħ. A low-lying calculated state of spin 42 ħ is interesting and probably corresponds to the highest spin state observed experimentally and tentatively given that spin. It is possible that the observation of the high-spin positive-parity states even depends upon this state, as it may be the source of the lifetime (>1 picosecond) necessary to hold up the following cascade long enough for the recoiling nucleus to stop first in the lead backing. Otherwise the lines would not be sharp, particularly in the detectors near 90° to the beam. It will be interesting to compare this decay scheme with those of the neighboring nuclei that we are also studying in order to observe the systematics of the changes in nuclear structure and shape in this region of the Periodic Table and see how well they agree with theoretical predictions.

It has become quite clear that the use of Compton-suppressed Ge detectors is an exciting step forward in nuclear γ-ray spectroscopy. The beautiful spectra taken by the Daresbury and Copenhagen groups during the past 1-2 years give eloquent testimony. Their work was done with six large NaI suppressors. The use of bismuth germanate permits much more compact shields, and so allows more Ge detectors to be grouped closer to the target or radioactive source than is possible with NaI. As a result, higher-order coincidences can be obtained which correspond in favorable cases to effectively higher resolution. In addition, adequate statistics can be obtained in a much shorter time. In this talk I have discussed our design of a high-resolution Ge array, where it is now in construction, and some first results from a partial array whose spectra do show the superior qualities we had hoped for. In closing I would like to point out that although I have only shown results of discrete γ-ray studies, we intend to use the array for continuum work also, as the shielded Ge detector response function is about as good as that of the large NaI detectors usually used in continuum studies. To learn about the properties of the highest spin states we shall have to study their average behavior as given by continuum γ-ray measurements; this will require significant improvements in our continuum techniques. But we can look forward to great increases in our knowledge of nuclear structure at high spin and at high temperature from a combination of the detailed measurements of discrete γ-ray studies and of the average behavior exemplified in continuum work.

Acknowledgements: Many people have helped on this project and deserve thanks. I can only mention a few: A. Dancosse for the design and construction of the holder, M.K. Lee for the LN filling system, M.K. Lee, D. Landis, and F. Gin for the electronics, and R. Belshe for software development. And the project owes most to the thought and care of Frank Stephens.

REFERENCES

1. TH. LINDBLAD, L.BAGGE, J. BIALKOWSKI, K. FRANSSON and S. LEVEN, On the Energy Resolution of LAr and LXe Ionization Chambers, this Proceedings (Harwood Academic Publishers, New York, 1985).
2. P. J. TWIN, P. J. NOLAN, R. ARYAEINEJAD, D. J. G. LOVE, A. H. NELSON and A. KIRWAN, Nuclear Physics, A409, 343e (1983); P. J. TWIN, TESSA: Its Present Status and Future Developments, this Proceedings (Harwood Academic Publishers, New York, 1985).
3. R. M. DIAMOND and F. S. STEPHENS, The High Resolution Ball (proposal) (unpublished, 1981).
4. A. W. SUNYAR, E. DER MATEOSIAN, O. C. KISTNER, A. JOHNSON, A. H. LUMPKIN and P. THIEBERGER, Physics Letters 62B, 283 (1976).
5. T. BYRSKI, F. A. BECK, C. GEHRINGER, J. C. MERDINGER, Y. SCHUTZ, J. P. VIVIEN, J. DUDEK, W. NAZAREWICZ AND Z. SZYMANSKI, Physics Letters 102B, 235 (1981).
6. J. DUDEK and W. NAZAREWICZ, private communication (1984).

THE GSI-LBL PLASTIC BALL/WALL SPECTROMETER*

H.-A. GUSTAFSSON, K.G.R. DOSS, H.H. GUTBROD, B. KOLB, H. LOHNER, B. LUDEWIGT, A.M. POSKANZER, T. RENNER, H. RIEDESEL, H.G. RITTER, A. WARWICK AND H. WIEMAN
Gesellschaft fur Schwerionenforschung
D-6100 Darmstadt, West Germany

and

Nuclear Science Division
Lawrence Berkeley Laboratory
University of California
Berkeley, California 94720

For the high multiplicity events occurring in relativistic nuclear collisions an electronic 4π detector with particle identification has been built.[1] It consists of 815 $\Delta E - E$ telescopes and 176 TOF telescopes covering 97 per cent of 4π. Very good particle identification has been obtained for hydrogen and helium isotopes and also π^+ have been detected with high efficiency. This spectrometer was used at the Berkeley Bevalac to study the reactions Ca + Ca, Nb + Nb and Au + Au at energies between 0.15 and 1.05 GeV/nucleon. Various exclusive observables like the degree of thermalization, the multiplicity dependence of composite particle formation, and freeze-out densities are discussed. From global analyses the collective flow of nuclear matter has been deduced. A method of separating efficiently participant and spectator particles has been

*This work was supported by the Director, Office of Energy Research, Division of Nuclear Physics of the Office of High Energy and Nuclear Physics of the U.S. Department of Energy under Contract DE-AC03-76SF00098.

applied in order to disentangle the two collective effects: The "side-splash" of the participants and the "bounce-off" in the fragmentation region.

1. INTRODUCTION

In relativistic nuclear collisions measurement of the charged particle multiplicity of an event has been found to be an important characteristic of the underlying reaction mechanism.[2] Further progress towards full event analysis is made very difficult because of the large dynamic range in fragment energy, fragment mass, and fragment multiplicity. In emulsion and streamer chamber studies, attempts have been made to fully reconstruct the events, but limits have had to be accepted in the quality of particle identification and total number of events analyzed. A detector system that can favorably complement 4π detectors like emulsions, AgCl detectors, or streamer chambers, has to have the capability of detecting and identifying as many charged particles as possible in high multiplicity events. Therefore, the concept of a 4π detector consisting of many individual $\Delta E - E$ and time-of-flight telescopes was chosen, that promises fast data analysis.

For the coverage of most 4π, the Plastic Ball was built based on the geometry of the Crystal Ball at Stanford (Fig. 1). It completely surrounds the target, except for the extreme forward and backward angles. The Plastic Ball consists of 815 $\Delta E - E$ particle identifying detector modules where the ΔE counter is a $CaF_2(Eu)$ crystal and the E counter is a plastic scintillator. They are optically coupled and read out by one

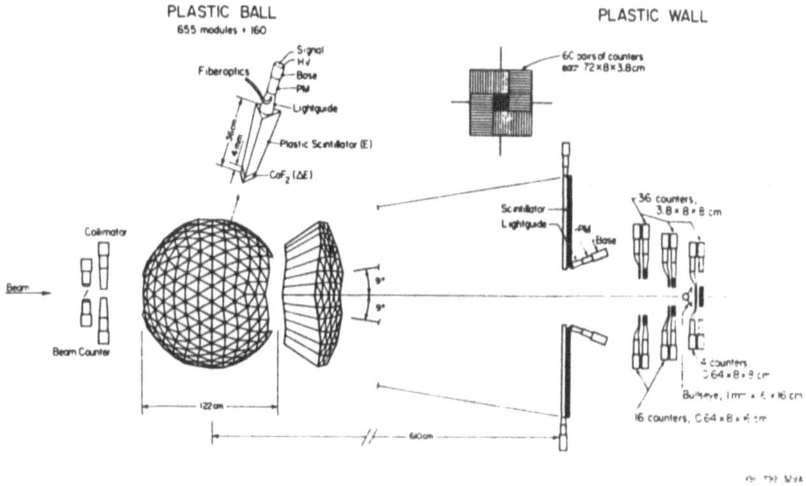

FIGURE 1 Schematic view of the Plastic Ball and the Plastic Wall

photomultiplier with subsequent separation of the signals by pulse shape analysis. This design allows one to cover a solid angle of up to 97 per cent of 4π.

The most forward angles (0 deg. - 10 deg.) are covered with a multi-element time-of-flight system (the Plastic Wall). The inner part of the Wall, covering the angular region from 0 deg. to 2.5 deg., and the outer part, from 2.5 deg. to approximately 10 deg. The Plastic Wall serves in addition as a trigger counter for the Ball.

During many years of studying mostly inclusive observables in heavy ion collisions at relativistic energies with light or medium heavy projectiles a wealth of information on nuclear fragmentation, particle production, and statistical properties of hot and dense nuclear matter has been accumulated and the complexity of the systems under investigation has been more and more uncovered. Various models based on rather different assumptions like the intranuclear cascade or the hydrodynamical model are able to reproduce the inclusive cross sections. Therefore the call for decisive experiments using exclusive observables became inevitable in order to draw a consistent picture of the behaviour of nuclear matter under extreme conditions and to gain information on the energy per nucleon as a function of density and temperature, i.e. the nuclear equation of state, from which the properties of nuclear systems created in energetic heavy ion collisions may consistently be derived. First attempts to relate the measured pion multiplicity to the equation of state have been made recently.[3] Since heavy beams up to uranium are now available at the Bevalac, the total mass in the reaction volume is as close to nuclear matter as possible in the laboratory. Data from 4π detectors like the Plastic Ball are ideally suited to study the emission patterns and event shapes of high energy nuclear collisions which might be able to distinguish between predictions of cascade and hydrodynamical models.

2. THE PLASTIC BALL

2.1. General Characteristics

Each of the 815 detector modules represents a particle identifying telescope with a ΔE and E detector using a slow and a fast scintillator read out via one photomultiplier (Fig. 2). The ΔE counter is a CaF_2(Eu)

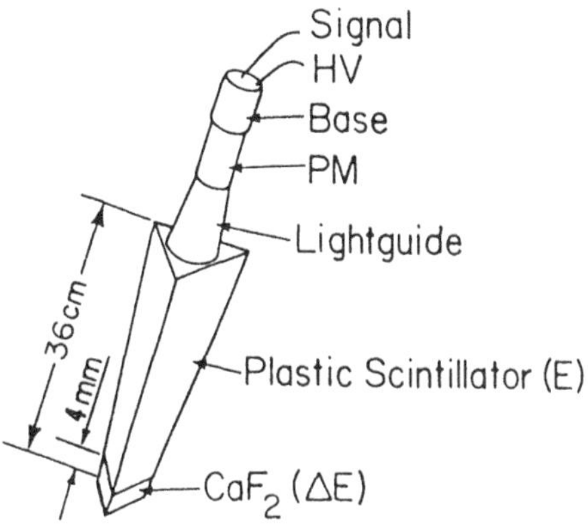

FIGURE 2 Schematic view of a single Plastic Ball module.

crystal with a characteristic decay time of 1 μs for the emission of the scintillation light. The ΔE light is read out through the E counter (i.e., the E counter serves as a light guide to the ΔE counter), which consists of a plastic scintillator. The E signal is collected within 10 ns.

The thickness of the ΔE CaF_2(Eu) crystal was chosen to be 4 mm with 35.6 cm as the length of the E plastic scintillator. This allowed good ΔE signals even for minimum ionizing particles and also a minimum low energy cutoff for particles stopping in the ΔE counter. In order to obtain clean proton spectra up to 200 MeV, the detector length was chosen to stop 240 MeV protons. This additional length assures that the punch through deuterons do not disturb the proton spectra below 200 MeV.

2.2. Pion Identification

In designing the Plastic Ball, special care was taken to detect the positive pions. As the yield of the π^+ is only about 10 per cent of the proton yield at beam energies well above the pion production threshold, in a pure ΔE - E identification scheme the pions would be overshadowed by the background produced by heavier particles. Therefore, the π^+ are additionally identified by their delayed decay. Stopped π^+ decay into a μ^+ and a neutrino with a mean life of 26 ns. The μ^+ subsequently decays into a positron and two neutrinos with a mean life of 2.2 μs.

3. THE PLASTIC WALL

The Plastic Wall covers an area of 192 cm x 192 cm and provides fine position resolution coverage of the angular region between θ = 0 deg. and the forwardmost sections of the Plastic Ball (θ ~ 10 deg.).

The outer Wall serves to extend the acceptance of the Ball-Wall system from θ = 10 deg. to θ = 2.5 deg. detecting particles in the same range of mass and charge as the Ball itself, but measuring velocity and Z instead of energy, Z and A as in the Ball. The inner Wall extends the angular measurement to 0 deg. In addition, the center region of the inner Wall makes up the event trigger in the first round of experiments.

3.1. The Outer Wall

The outer Wall region consists of 60 pairs of counters, each 72 x 8 x 3.8 cm^3, providing coincidence units sensitive to charged particles passing through both bars of a pair. Neutron and gamma background is rejected by this coincidence requirement.

The coordinates of the point of incidence of the charged particle on the outer Wall are determined to within the width of the module (8 cm) in one direction and measured to the same accuracy along the length of the pair by measuring the light travel time to opposite ends of each of the two scintillators. A time resolution of 350 ps (fwhm) was achieved for minimum ionizing protons. The mean time of the pair gives the flight time from the target and hence the particle velocity.

3.2 The Inner Wall

The inner Wall covered the region within 2 deg. of the beam and was finely divided because of the high multiplicity of fragments near the beam. Its main purpose was to form the fast trigger with the upstream beam counter and collimator to decide which events were to be recorded. Of course, it also gave nuclear charge and velocity information for the particles it detected and could be used for centering the beam on the Wall.

The inner Wall consisted of a 6 x 6 array of 36 thick scintillators 8 x 8 cm^2 each, 3.8 cm thick. In front of each was a 0.64 cm thin scintillator. For the center four counters the thin and thick scintillators were of the same area and in one to one correspondence, but for the remaining 32 thick scintillators, only 16 thin scintillators were used, each 8 x 16 cm^2 in area and covering a pair of thick detectors. In addition, in front of the center four pairs of scintillators was placed one 16 x 16 cm^2 "Bull's Eye" scintillator, only

1 mm thick. The purpose of the bull's eye was to reject beam particles, and it was made thin to minimize nuclear interactions in the detector itself.

4. PERFORMANCE

4.1 The Plastic Ball

The tests performed at LAMPF low energy pion beam line and at the Berkeley 184" cyclotron with 800 MeV α-particles showed that the energy resolution of a

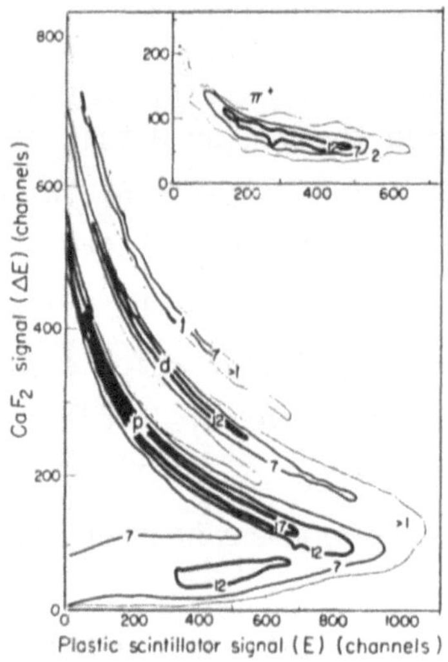

FIGURE 3 ΔE – E contour diagram of 655 modules after gain matching and scattering out reconstruction of approximately one million reaction products. The contour lines are labeled as to their relative height in arbitrary units. The upper contour diagram for π^+ is obtained by requiring a delayed decay signal measured in the 10 μs TDCs.

single module is sufficient to achieve the desired particle identification.

The first Bevalac experiment was performed in June 1981 with an 800 MeV/nucleon Ne beam on a Pb target and ^{40}Ca beams at 400 MeV/nucleon and 1.05 GeV/nucleon on a calcium target. The quality of the particle identification for the 655 modules between 30 deg. and 160 deg. is shown in Fig. 3. A cut in the ΔE - E plane perpendicular to the particle identification lines is selected with ΔE and E projected on this cut in a certain energy range (Fig. 4). The dashed curve

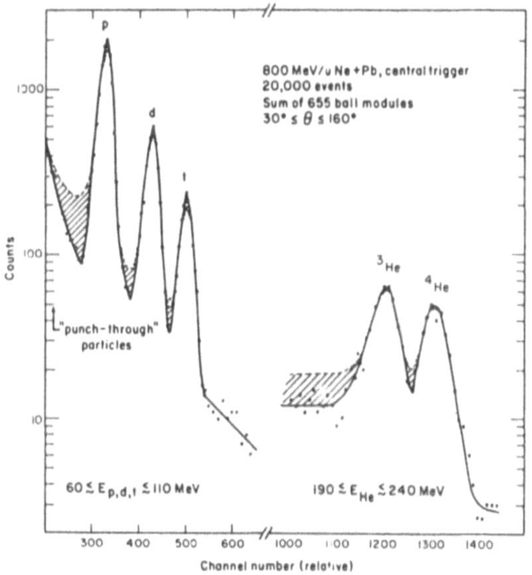

FIGURE 4 Particle identification spectrum for 655 modules after gain-matching and with and without scattering out reconstruction.

represents the raw data after gain matching, whereas the solid line shows the particle separation after all scattered out particles have been reconstructed (Fig. 4). The effectiveness of the reconstruction has been shown before.[4]

4.2 The Plastic Wall

Since the Plastic Wall can detect beam velocity particles the time calibration is taken from each individual experiment. In the 800 MeV/nucleon ^{20}Ne on Pb experiment a time resolution of ~ 350 ps was observed between the inner Wall detectors and the beam counter. For the outer Wall, where dominantly hydrogen and helium fragments are detected, a pulse height resolution of ~20-25 per cent was good for an excellent separation of hydrogen, helium, and lithium.

These performance figures of the Ball and Wall allow one to draw a figure for the system's response as a function of fragment momentum and rapidity. Fig. 5 shows the response of the Ball and the Wall for protons or α-particles.

5. RESULTS

5.1 Stopping and Thermalization

By measuring in each event the momentum distribution of the fragments in the center of mass system the degree of thermalization can be determined. For an isotropically expanding system one finds in the center-of-mass frame from simple phase space considerations:

$$\frac{2}{\pi} \sum_i |p_\perp^i| = \sum_i |p_\parallel^i| \quad . \tag{1}$$

The sums in eq. (1) contain the perpendicular, $p_\perp$, and

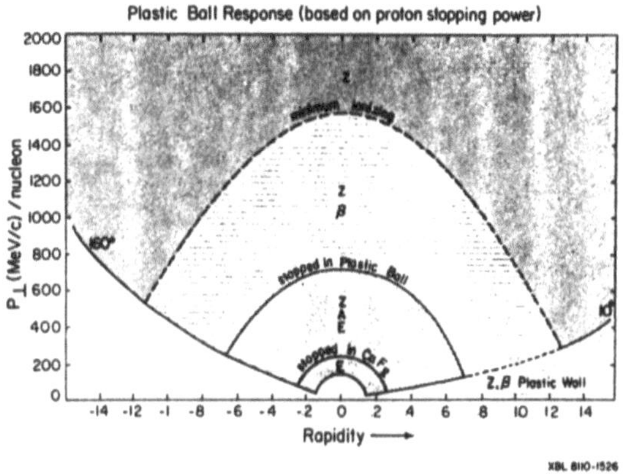

FIGURE 5 Plastic Ball and Plastic Wall response in the transverse momentum p versus rapidity y plane. The limits labeled "stopped in CaF_2" and "stopped in Plastic Ball" refer to protons and ^4He fragments.

longitudinal, $p_{\parallel}$, momentum components of all particles in an event. A global stopping of the two nuclei in the center of mass system would show up by the two sides of eq. (1) being equal, or $p_{\perp}$ even larger if flow into the transverse direction exists. In the presence of transparency $p_{\parallel}$ would always be larger. Isotropy is a necessary but not sufficient condition for thermalization. If in addition the energy distributions are of Maxwell-Boltzmann type the emitting system may be called thermalized.

Plastic Ball data[5] are shown in Fig. 6. In Fig. 6 (top) contour lines of the event yield accumulated with the minimimum bias trigger are shown in the $p_{\perp}$ versus $p_{\parallel}$ plane for the Ca + Ca system. The peak at

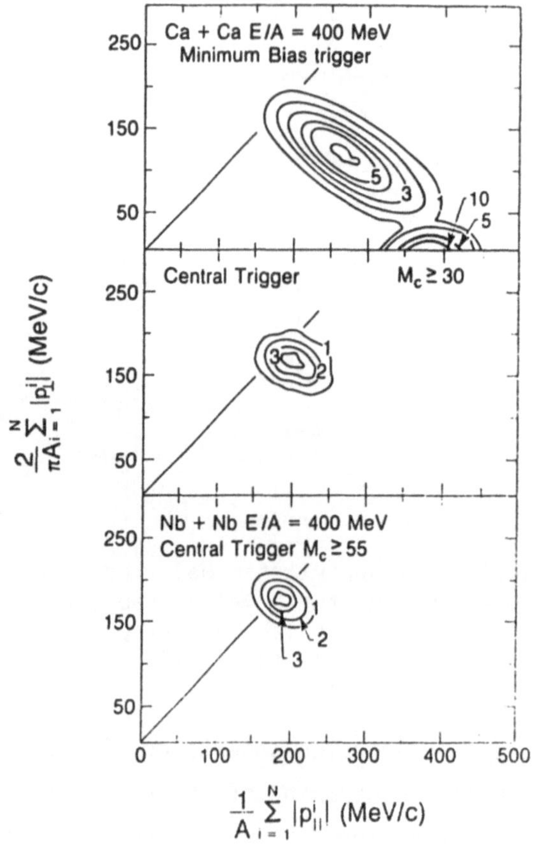

FIGURE 6 Contour plots of the average particle momentum components perpendicular and parallel to the beam axis for Ca + Ca (top and center) and Nb + Nb (bottom) at E/A = 400 MeV. The diagonal line corresponds to isotropic events.

small $p_\perp$ but large $p_\parallel$ corresponds to peripheral reactions and is dominated by projectile fragments. This contribution vanishes as the trigger is changed to a "central" one. Fig. 6 (center) shows central trigger events with a charged particle multiplicity larger than 30. The maximum of the yield is shifted towards the diagonal but only a few events reach it, which corresponds to full stopping of the nuclei. In the lower part of Fig. 6 for the reaction of 400 MeV/ nucleon Nb + Nb the central trigger events with charged particle multiplicities M_c larger than 55 almost fulfill the stopping and isotropy condition on the average.

5.2 Collective Flow

Direct evidence of the compression effects predicted by an equation of state would be collective flow of the nuclear matter upon re-expansion. Data from 4π detectors like the Plastic Ball are ideally suited for studying the emission patterns and event shapes which might be able to reveal this effect and distinguish between predictions of cascade and hydrodynamical models.

The sphericity tensor

$$F_{ij} = \sum_{v} p_i(v) p_j(v) w(v) \quad (3)$$

is calculated from the momenta of all measured particles for each event. It is appropriate to choose the weight factor w(v) in such a way that composite particles have the same weight per nucleon as the individual nucleons of the composite particle at the same velocity. For the Plastic Ball analyses the weight $w(v) = 1/2m(v)$ (kinetic energy flow) is used.

The sphericity tensor approximates the event shape by an ellipsoid whose orientation in space and whose aspect ratios can be calculated by diagonalization. The shapes predicted by hydrodynamical and intranuclear cascade calculations are quite different. The hydrodynamical model predicts prolate shapes along the beam axis only for grazing collisions but with decreasing impact parameter the flow angle increases, and reaches 90 deg. (with oblate shapes) for zero-impact-parameter events. Simple cascade calculations on the other hand predict zero flow angles at all impact parameters.

Approximately 50,000 Plastic Ball events accumulated with a minimum-bias trigger have been analyzed for the two systems Ca + Ca and Nb + Nb at 400 MeV/nucleon.[6] The distribution of the flow angle θ (angle between the major axis of the flow ellipsoid and the beam axis) is shown in Fig. 7 for different multiplicity selections. A striking difference between the Ca and Nb data can be observed. For all but the highest multiplicity bins the distribution of the flow angles for the Ca data is peaked at 0 deg. For Nb, however, there is a finite deflection angle increasing with increasing multiplicity. The same analysis has been performed with filtered events from a cascade code calculation (Fig. 7). For both systems studied the distributions are always peaked at 0 deg. It is not so evident that the Ca + Ca collision differs from its simulation with the cascade model, whereas a new collective phenomenon definitely appears in the larger-mass system which is not accounted for by the present cascade models.

In this analysis each event was parametrized by an ellipsoid, but it is of interest to study the shapes in more detail. The fact that finite flow angles are seen in the data indicates that in those events a reaction plane exists that is defined by the flow axis and the beam axis. All events can be rotated by the azimuthal angle ϕ determined by the flow analysis so that their

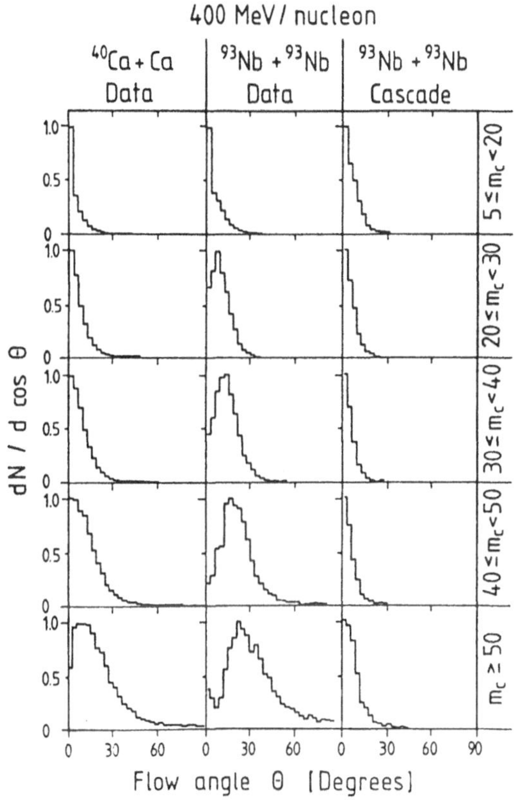

FIGURE 7 Frequency distributions of the flow angle θ for two sets of data and a cascade calculation for different multiplicity bins. For the case of Ca the multiplicities are half the indicated values.

individual reaction planes all fall into the x-z plane, with the z axis being the beam axis. For those rotated events the invariant cross section in the reaction plane $[d^2\sigma/dy\ d(p_x/m)]$ can be plotted, where p_x is the projection of the perpendicular momentum into the reaction plane and y is the center-of-mass rapidity. In addition, because of detector inefficiency near the target rapidity, the graphs in Fig. 8 have been

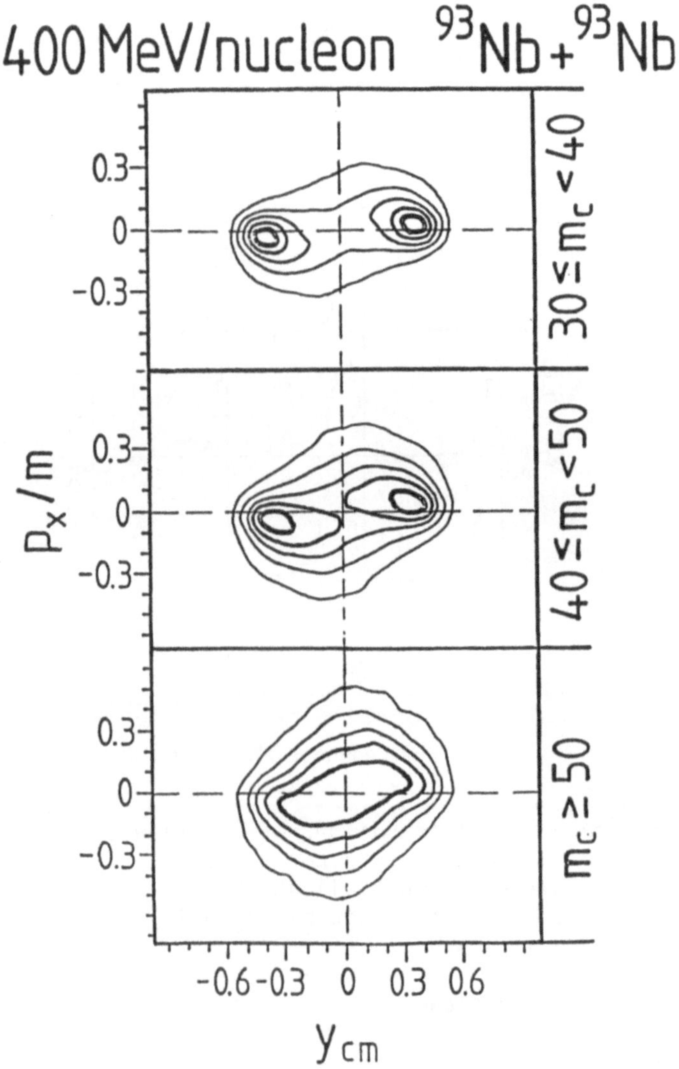

FIGURE 8 Contour plots (equally-spaced linear contours without offset) of P_x/m as a function of c.m. rapidity for multiplicities of 30-39, 40-49, and greater than 50.

reflected about 90 deg. in the center of mass for didactic reasons.

The highest level contour, near projectile rapidity and just above the horizontal axis, results largely from the projectile remnants and indicates a definite bounce-off effect. The bounce-off process is a slowing down of the projectile fragments and a sidewards deflection in the reaction plane. The data show that the multiplicity dependence of the orientation of the outer contour lines from the lower left to the upper right follows the trend indicated by the flow-angle distributions (Fig. 7). However, the position of the bounce-off peak from the projectile remnants changes only slightly with multiplicity. Thus one can conclude that the strong sideward peaking seen in Fig. 7, which we will call side splash, is mainly due to the midrapidity particles. It should be noted that the bounce-off and side-splash effects appear to be in the same plane.

The hydrodynamical prediction of the flow angle seems to be qualitatively in agreement with the measurement, but the present models do not predict two separate effects as seen in the data.

5.3 Chemical Freeze-out Density

One of the major goals in the study of composite particle production is to extract information about the size and density of the participant volume. The d/p ratio determines the volume of the system at freeze-out. To determine the density one needs the number of baryons in this volume. We define N_p as the participant baryon charge multiplicity. The participant baryon charge multiplicity, N_p, will be abbreviated to proton multiplicity.

The definition of d_{like} is given by

$$d_{like} = d + 3/2(t + {}^{3}He) + 3({}^{4}He) \qquad (4)$$

and the quantity p_{like} is defined as

$$p_{like} = p + d + t + 2({}^{3}He + {}^{4}He) \qquad (5)$$

The functional form of the observed d_{like}/p_{like} ratios can be understood in terms of the coalescence model.[7] We have used an improved version of the model which is a complete 6 dimensional phase space calculation relating the radii of the deuteron and the participant zone to the coordinate space, and the temperature of the interacting region to the momentum space. In this model the radius r_p and the temperature T of the interacting region as well as the deuteron radius r_d are related to d_{like} and p_{like} through

$$d_{like}/p_{like} = 6((A - Z)/Z)(1 + 2(r_p/r_d)^2)^{-3/2} N_p \times$$
$$(1 + 2mTr_d^2/3)^{-3/2} \qquad (6)$$

where the factor for (A - Z)/Z makes up for the difference between neutron and proton number and m is the nucleon mass. The radius r_p assuming a spherical source is parametrized as $r_p = r_o((A/Z)N_p)^{1/3}$, where A/Z is the factor converting the participant baryon charge multiplicity to participant baryon multiplicity.

In the fits to the observed ratios, r_o and r_d were free parameters. The temperature T was taken to be the mean temperature obtained from Boltzmann fits[5] to the proton spectra at 90 deg. in the center-of-mass system. The fits to the experimental ratios are presented as solid curves in Fig. 9. The r_0 values

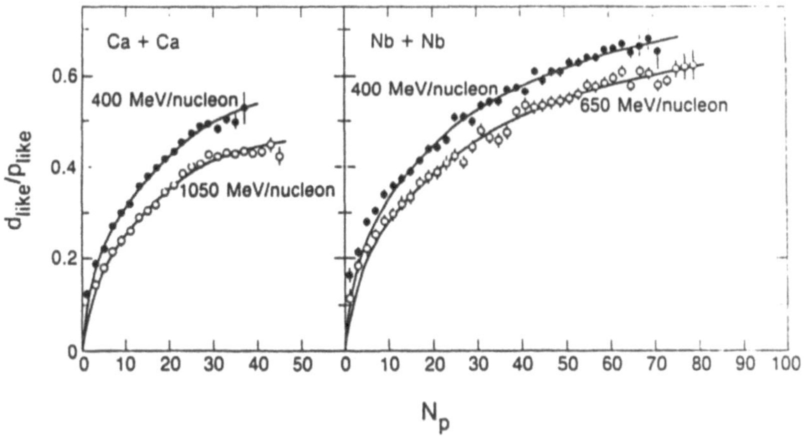

FIGURE 9 d_{like}/p_{like} as a function of proton multiplicity (N_p) for the two systems Ca + Ca at 400 and 1050 MeV/nucleon and Nb + Nb at 400 and 650 MeV/nucleon. The curves shown are from fits to eq. (6).

are the rms values for a Gaussian density distribution. To convert these values to equivalent sharp sphere radii the values have to be multiplied by $\sqrt{5/3}$. The mean sharp sphere r_0 values obtained from the four cases studied correspond to freeze-out densities of 50 to 100 per cent times normal nuclear density.

5.4 Thermal Freeze-out Density[8]

Proton-proton correlations were measured with the

Plastic Ball which has several features that are of significant value in this application: a large solid angle and rapid data-collection ability, and the ability to detect most of the charged particles in each event, thus obtaining a direct measure of the participant multiplicity which is necessary for the determination of the freeze-out density.

Proton-proton correlations came into use for source-size measurements followowng the correlation predictions of Koonin which, in addition to the second-order interference effect due to quantum statistics, also include nuclear and Coulomb-induced final-state interactions between the two outgoing protons. At $\Delta p = 0$ the correlation function is zero as a result of the Coulomb repulsion. A peak in the correlation function at $\Delta p = 20$ MeV/c is generated by the attractive nuclear force. The amplitude of this enhanced correlation is strongly dependent on the source-size parameter r_g and increases as the source gets smaller. In the general form of Koonin's calculation the proton source density is treated as isotropic with a Gaussian distribution in space and time, and the proton momenta are assumed to be independent of position. The Gaussian spatial density, $p(r) = \exp[(-r/r_g)^2]$, may be thought of as representing the proton positions at the point of their last scattering. For our analysis we assume that the time parameter is zero, i.e., all the protons are emitted at the same time.

Radii have been extracted by comparing measured correlation functions to theoretical predictions which depend on the source-size radius. In practice, the correlation function is obtained from the data as a function of Δp (either one of the proton momenta transformed into the center of mass of the pair) by summing all p-p pairs, event by event, in bins $[N_{true}(\Delta p)]$ according to Δp, and dividing by p-p pairs from different events $[N_{mixed}(\Delta p)]$ to give

$$F(\Delta p) = \text{norm} \times N_{true}(\Delta p)/N_{mixed}(\Delta p) \quad . \tag{7}$$

In forming the mixed-pair sums each proton was paired with each proton in the preceding five events. It should be pointed out that the Plastic Ball has a reduced efficiency for detecting pairs in the low-Δp region since these pairs will tend to go both into the same detector. However, the correlation function is well determined at Δp = 20 MeV and above and thus gives a good measure of the radius. In Fig. 10 we show an example of a measured correlation function along with the distorted theoretical correlation function.

The extracted radii are shown as a function of proton multiplicity in Fig. 11. The error bars represent statistical errors only. For both the Ca + Ca and Nb + Nb systems the radius increases with multiplicity. The radii are shown with a fit using the function $r_g = r_0(N_pA/Z)^{1/3}/(\frac{5}{2})^{1/2}$ where r_0 is a reduced radius (inversely proportional to the density). N_p is the participant baryon charge multiplicity and A/Z has been included to reflect the presence of neutrons. The extracted radii are the radius parameters for a Gaussian source distribution, and so by inclusion of $(\frac{5}{2})^{1/2}$ in the above function r_0 becomes the radius parameter of an rms-equivalent sharp sphere. The value of r_0 extracted from the fits is 1.9 fm for the two systems

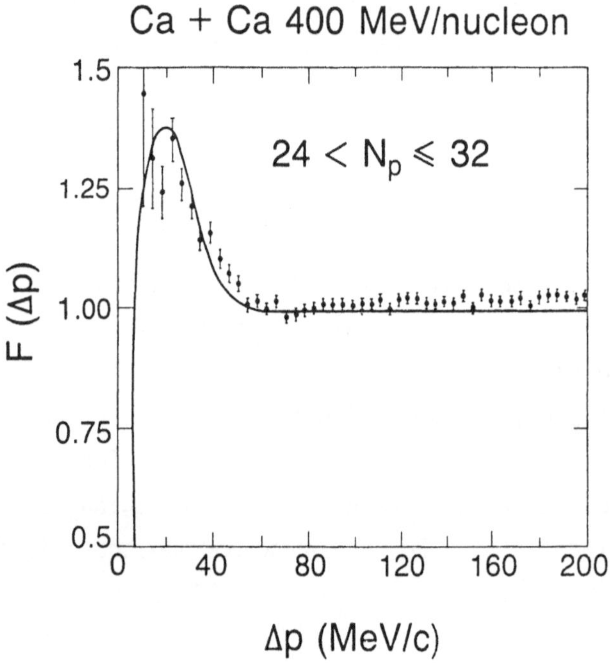

FIGURE 10 The measured proton-proton correlation function for Ca + Ca with a proton multiplicity from 25 to 32. The solid curve is a least-squares fit interpolation between distorted theoretical correlation functions for radii of 4 and 5 fm which differ from each other by 25 per cent at the peak.

studied. Comparing this directly with r_0 = 1.2 fm, the sharp-sphere r_0 for normal nuclei, yields a participant-region thermal freeze-out density about 25 per cent of normal nuclear density.

6. CONCLUSION

Results from the analyses of Plastic Ball data have been presented which show the various interrelations of the exclusive observables studied here.

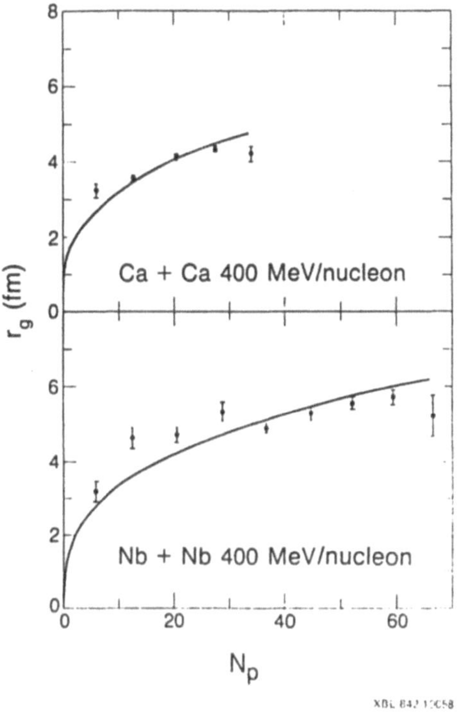

FIGURE 11 Extracted Gaussian source radii as a function of proton multiplicity (N_p) for the two systems Ca + Ca and Nb + Nb at 400 MeV/nucleon. The curves are fits to the results with the function $r_g = r_o(N_pA/Z)^{1/3}/(\frac{5}{2})^{1/2}$.

From deuteron to proton ratios we are able to extract freeze-out densities for the chemical freeze-out stage occurring at earlier times in the expansion process than the thermal freeze-out stage[9] which is accessible by two-particle correlation analysis. The two different collective effects - the bounce-off in the fragmentation region and the side-splash of the participants, which are prominent in the Nb + Nb data, offer another piece of information that has to be included in a consistent description of high energy heavy ion interactions. Up to now the hydrodynamical model including a full quantum statistical calculation for composite particle formation seems to be the most promising tool to describe the above mentioned features of the data and to finally deduce the nuclear equation of state.

ACKNOWLEDGEMENTS

Drs. A. M. Poskanzer and H. Lohner are thanked for their help in preparing this paper.

This work was supported by the Director, Office of Energy Research, Division of Nuclear Physics of the Office of High Energy and Nuclear Physics of the U.S. Department of Energy under Contract DE-AC03-76SF00098.

7. REFERENCES

1. A. BADEN, H. H. GUTBROD, H. LÖHNER, M. R. MAIER, A. M. POSKANZER, H. RIEDESEL, H. G. RITTER, H. SPIELER, A. WARWICK, F. WEIK and H. WIEMAN, Nucl. Instr. Meth. 203, 189 (1982)

2. A. SANDOVAL, H. H. GUTBROD, W. G. MEYER, A. M. POSKANZER, R. STOCK, J. GOSSET, HJ.-C. JOURDAIN, C. H. KING, G. KING, CH. LUKNER, NGUYER VAN SEN, G. D. WESTFALL and K. L. WOLF, Phys Rev. C21, 1321 (1980); W. G. MEYER, H. H. GUTBROD, CH. LUKNER and A. SANDOVAL, Phys. Rev. C22 179 (1980); A. I. WARWICK, A. BADEN, H. H. GUTBROD, M. R. MAIER, H. G. RITTER, H. STELZER, F. WEIK, H. WIEMAN, S. B. KAUFMAN, B. D. WILKINS, E. P. STEINBERG and J. PETER, Phys Rev. C27, 1083(1983)

3. R. STOCK, R. BOCK, R. BROCKMANN, J. W. HARRIS, A. SANDOVAL, H. STRÖBELE, K. L. WOLF, H. G. PUGH, L.

S. SCHROEDER, R. RENFORDT, A. DACAL and M. E. ORTIS, Phys. Rev. Lett. 49, 1236 (1982)

4. H. H. GUTBROD, M. R. MAIER, H. G. RITTER, A. I. WARWICK, F. WEIK, H. WIEMAN, and K. L. WOLF, IEEE Transactions on Nuclear Science, Vol. NS-28, No. 1 (February 1981)

5. H.-A. GUSTAFSSON, H. H. GUTBROD, B. KOLB, H. LÖHNER, B. LUDEWIGT, A. M. POSKANZER, T. RENNER, H. RIEDESEL, H. RITTER, A. WARWICK and H. WIEMAN, Phys. Lett. 142B, 141 (1984)

6. H.-A. GUSTAFSSON, H. H. GUTBROD, B. KOLB, H. LÖHNER, B. LUDEWIGT, A. M. POSKANZER, T. RENNER, H. RIEDESEL, H. RITTER, A. WARWICK and H. WIEMAN, Phys. Rev. Lett. 53, 1590 (1984)

7. H. H. GUTBROD, H. LÖHNER, A. M. POSKANZER, T. RENNER, H. RIEDESEL, H. RITTER, A. WARWICK, F. WEIK and H. WIEMAN, Phys. Lett. 127B, 317 (1983)

8. H.-A. GUSTAFSSON, H. H. GUTBROD, B. KOLB, H. LÖHNER, B. LUDEWIGT, A. M. POSKANZER, T. RENNER, H. RIEDESEL, H. RITTER, A. WARWICK, F. WEIK and H. WIEMAN, Phys. Rev. Lett. 53, 544 (1984)

9. J. W. HARRIS and R. STOCK, LBL Report 17054 (1984)

THE HISS SPECTROMETER*

DOUGLAS E. GREINER
Nuclear Science Division
Lawrence Berkeley Laboratory
University of California
Berkeley, CA 94720

This talk describes the Heavy Ion Spectrometer System (HISS) facility at the Lawrence Berkeley Laboratory's Bevalac. Three completed experiments and their results are illustrated. The second half of the talk is a detailed discussion of the response of drift chambers to heavy ions. The limitations of trajectory measurement over a large range in incident particle charge are presented.

1. THE HISS FACILITY

The HISS spectrometer is designed to allow multiparticle measurements with good resolution over a large solid angle. The main ingredient is the HISS superconducting dipole that provides up to 6 tesla-meters of bending over a 1 meter gap. Figure 1 shows that the vertical acceptance of the magnet can be as large as 30 degrees for targets placed in the

*This work was supported by the Director, Office of Energy Research, Division of Nuclear Physics of the Office of High Energy and Nuclear Physics of the U.S. Department of Energy under Contract DE-AC03-76SF00098.

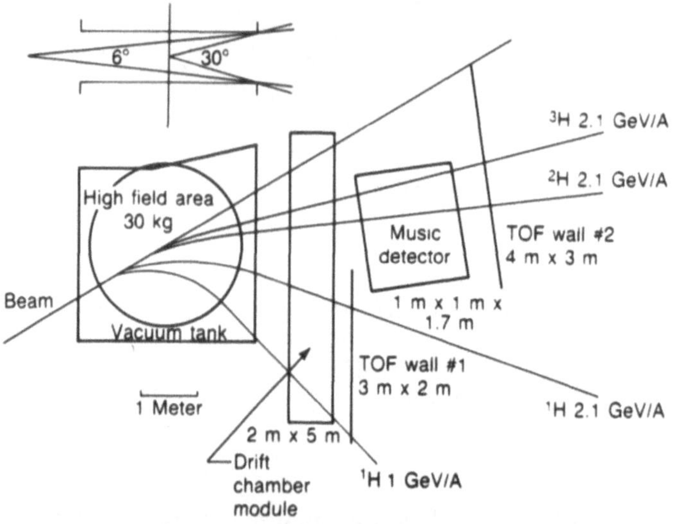

FIGURE 1 HISS Phase II detector system.

center of the dipole. The detector positions are arbitrary. Figure 1 shows a setup that accepts reaction products from infinite rigidity down to 1 GV.

The detectors shown in Fig. 1, with the exception of the drift chamber module, have been built and tested. Hans Sann spoke about the Multiple Sampling Ionization Chamber earlier in the conference, so I will not discuss it further in this talk. There is a high resolution velocity detector, which will be added to the facility, that is not shown. It is based on total reflection of Cerenkov light and will be used for the higher charged particles. Mat Baumgartner will speak about it later in this conference. Experiments to date have used a pair of 1 m x 2 m drift chambers as the tracking device. These chambers were prototypes and will no longer be used. We have done extensive evaluations of a test model of the proposed new, large tracking chamber, and these results will comprise the second half of my talk.

2. EXPERIMENTS AT HISS

A list of HISS experiments is shown in Table I. It is clear that there is a large backlog of experiments, including the proposed measurements, which are now approved. In order to show how HISS is used and to give you an understanding of the versatility of the facility, I will show the detector configurations and some of the results of a few of the experiments that have completed data taking.

The first experiment to complete data taking at HISS was experiment 512H, an INS/LBL collaboration led by Shoji Nagamiya. The goal of this experiment was to utilize the large solid angle capability of HISS to measure the high momentum particle spectra at 90 degrees in the center of mass. This experimental configuration is shown in Fig. 2. The A-arm uses the HISS 1 m x 2 m drift chambers and TOF walls along with INS constructed multiwire chambers inside the gap of the magnet. The B-arm was a test of in-field tracking and particle identification for a later experiment. With the field at 1.8 tesla only very energetic particles beyond the nucleon-nucleon kinematic limit would be detected in the A-arm. The resultant proton center-of-mass energy spectrum is shown in Fig. 3. The large solid angle coverage extends the measurements to over three orders of magnitude lower cross sections. It is clear that temperature is a local concept; also, the very large momentum region is not predictable with the conventional cascade calculation. The deuteron and triton spectra are shown in Fig. 4, along with the coalescence-type scaling from the proton spectrum. It is remarkable that this scaling works over 8 orders of magnitude.

TABLE I List of HISS Experiments.

No.	Title	Institution	Spokesperson	Hours
Approved HISS Experiments				
512H	Large Pt+multiplicity	INS/LBL	Nagamiya	150 complete (June 1982)
513H	^{12}C multiparticle	LBL/SSL	Greiner	138 complete (June 1983)
516H	^{56}Fe multiparticle	LBL/SSL	Lindstrom	108
517H	Coulomb decay	LLL/LBL	Berman	93 complete
521H	p-p corrrelations	UCLA/LBL	Carroll	97 (dropped Sept. 1983)
593H	0 degree pi+z flow	INS/LBL	Hashimoto	249 complete (Dec. 1983)
614H	Limits of stability	LBL/SSL	Symons	100
655H	Missing Gamow-Teller	UCD/LBL	Brady	48
662H	Neutron spectra	KSU/LBL	Madey	262 (May 1984)
683H	^{16}O and ^{56}Fe excitation function	LSU/LBL/SSL	Wefel	96
684H	pi and k correlations	UCR/LBL	Beavis	212
690H	Nuclear radii	INS/LBL	Tanihata	160 (July 1984)
734H	Exotic nuclei	UCD/LBL	Brady	42 (18 used)
			Total	648 run hours (1026 pending)
Proposed Experiments				
758H	New nuclear isotopes	UCD/INS	Brady	120
761H	Entropy production	MSU/LBL	Westfall	120
763H	Velocity Detector Debug	LBL	Olson	48
			Total	288

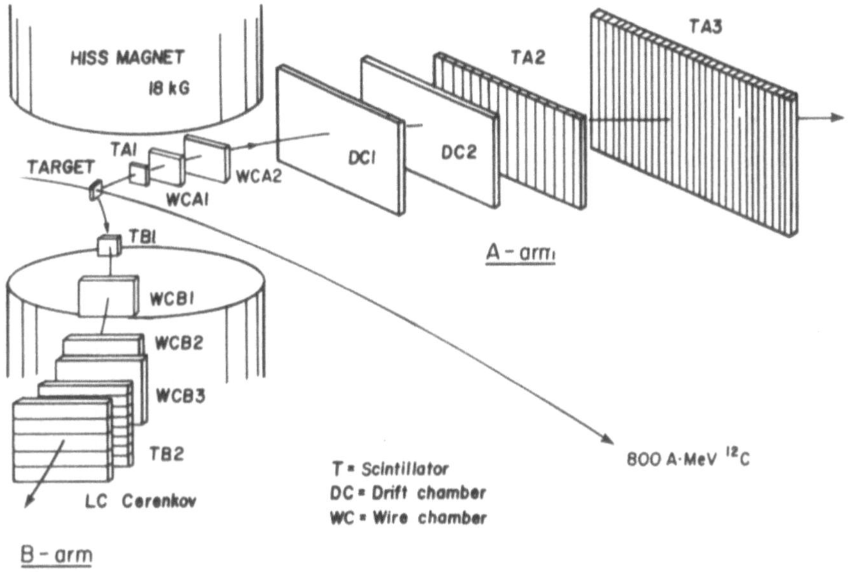

FIGURE 2 Experimental configuration for experiment 512H.

The second experiment involved looking at projectile fragments accompanied by pion production. This INS/LBL experiment 593H used the HISS detectors to observe the projectile fragments and triggered on pions detected by chambers and Cerenkov detectors placed to catch pions bent through 180 degrees by the HISS field (see Fig. 5). It is interesting to speculate, before looking at Fig. 6, as to how much of the original charge of the La still continues in the forward direction. Figure 6 shows the sum of the fragment charges detected in the A-arm when the reaction produces a pi-minus between 50 and 300 MeV/c momentum near 0 degrees. The spectrum peaks at about half the projectile charge, a rather high price to pay to produce a pion. This experiment was also performed on an argon beam. In Fig. 7 we see that the TOF wall has moderate charge resolution through Z=18.

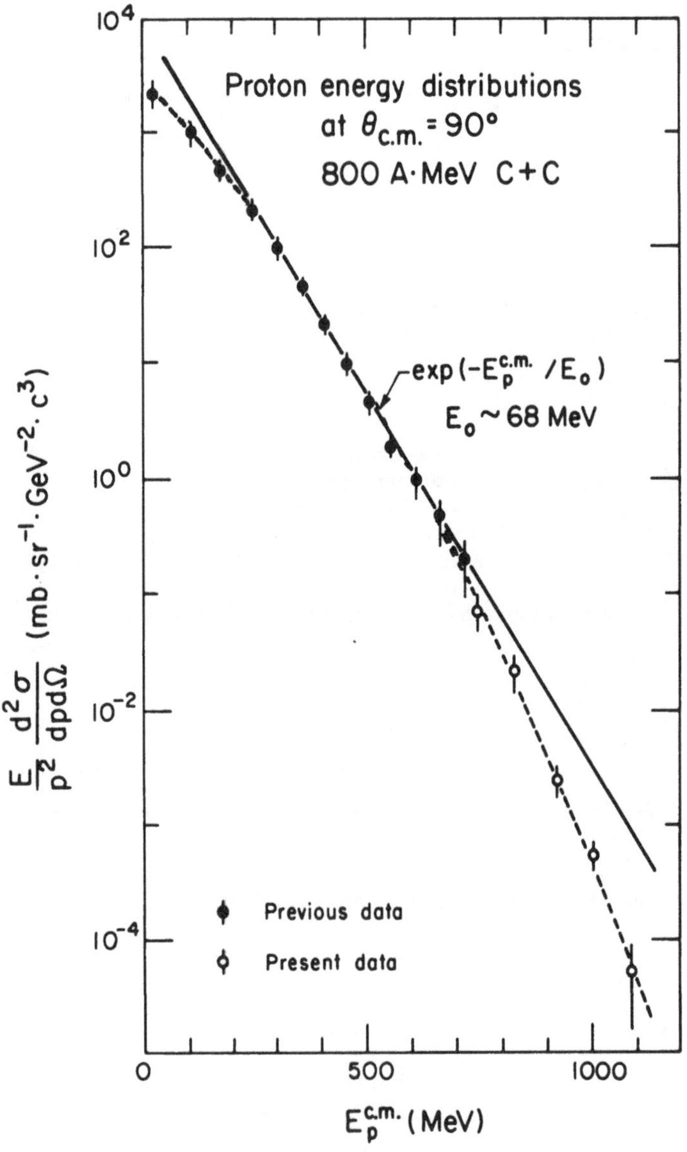

FIGURE 3 Proton center-of-mass spectrum from experiment 512H.

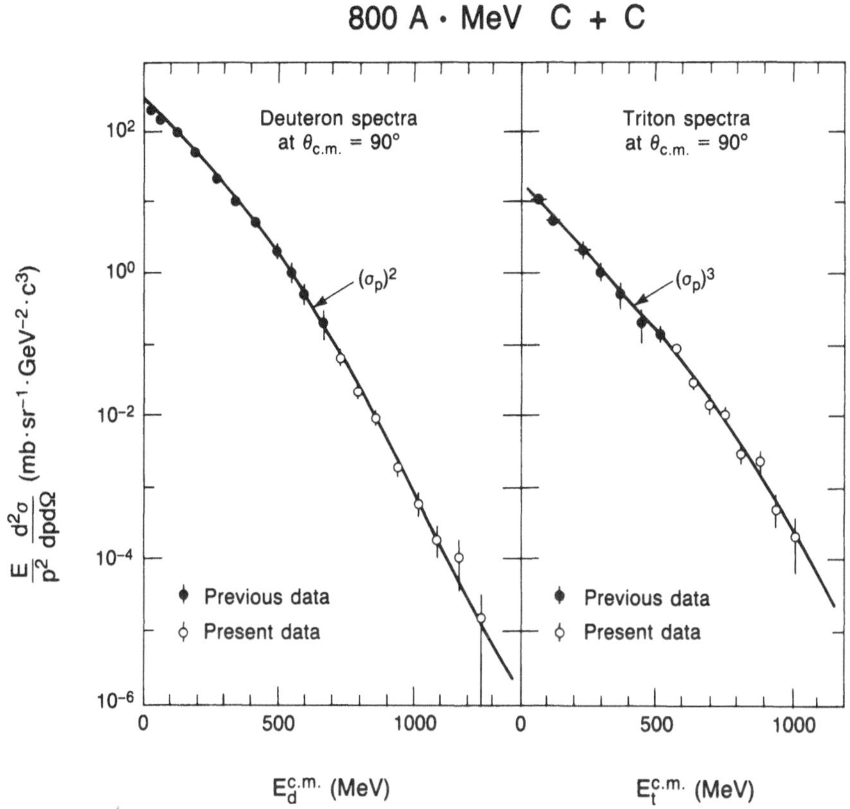

FIGURE 4 Deuteron and triton center-of-mass spectrum from experiment 512H.

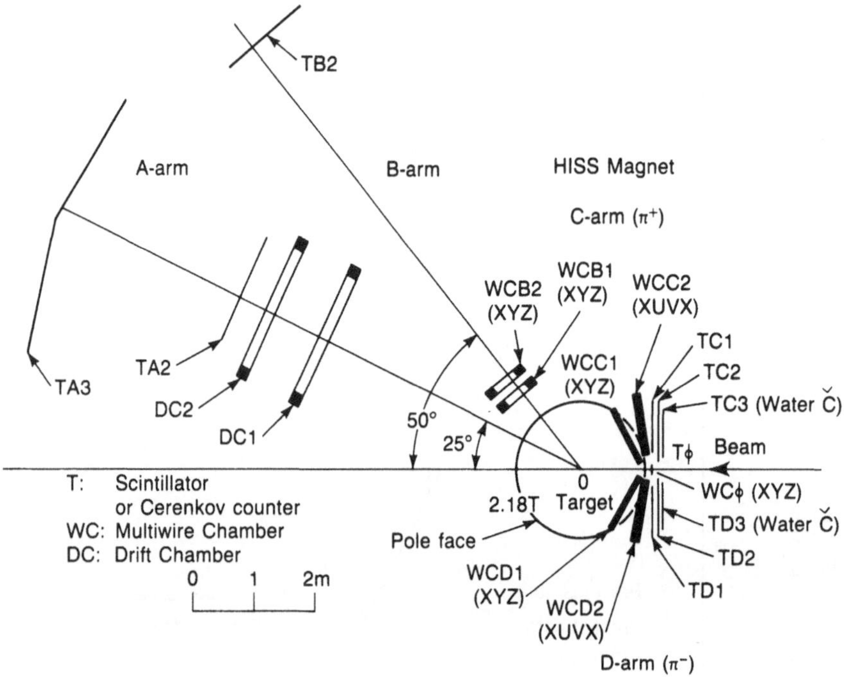

FIGURE 5 Experimental configuration of experiment 593H.

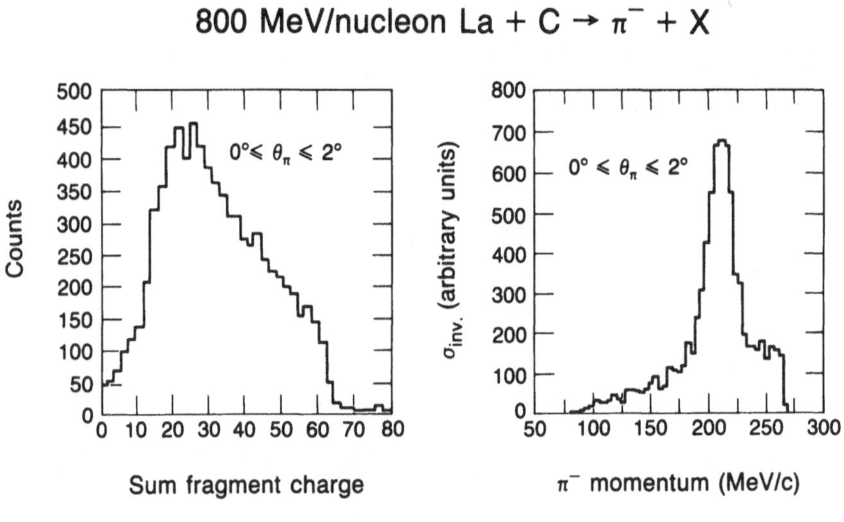

FIGURE 6 Fragment charge sum and pion momentum distributions from experiment 593H.

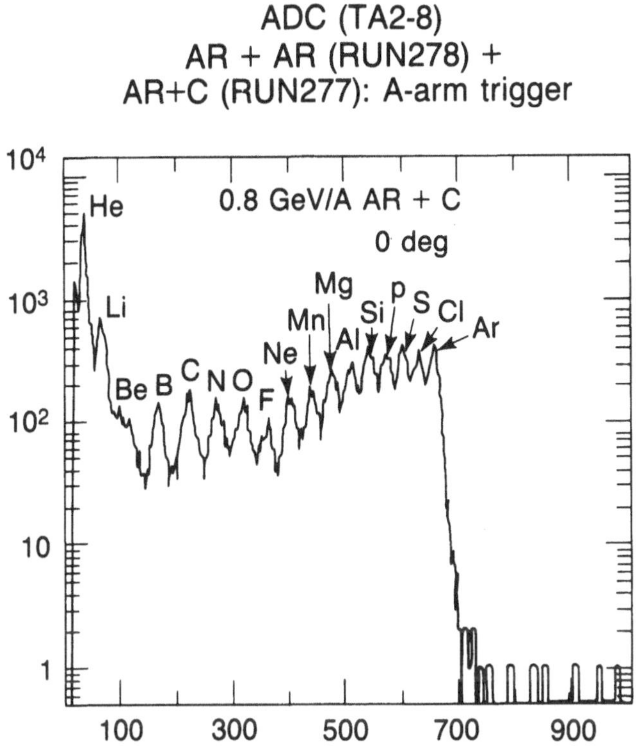

FIGURE 7 Pulse height distribution from TOF wall.

The third experiment (513H) is a HISS group effort designed to shed some light on the reaction mechanism. The basic idea, shown in Fig. 8, is to directly measure the energy transfer in the reaction by detecting all of the projectile fragments allowing reconstruction of the excitation energy these particles have received. The people involved in the experiment are listed in Table II. In Figs. 9 and 10 we see the detector positions, angular acceptance, and rigidity coverage. The drift chambers before the dipole determine the incoming trajectory that allows us to correct for the emittance of the beam. The target is in the center of the magnet; this gives us transverse momentum coverage up to ± 400 MeV/c for protons. Fragments are detected by the

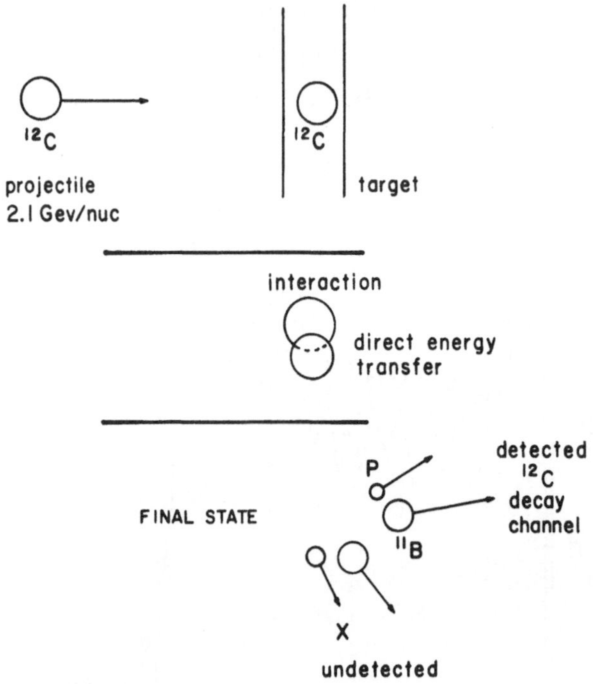

FIGURE 8 Reaction mechanism.

TABLE II Heavy ion spectrometer system group.

M. Baumgartner	LBL	J. Engelage	UCLA	D. Olson	LBL
E. Beleal	LBL	I. Flores	UC/SSL	J. Porter	LBL
F. Bieser	LBL	D. Greiner	LBL	H. Sann	GSI
M. Bronson	LBL	P. Lindstrom	LBL	T.J.M. Symons	LBL
H. Crawford	UC/SSL	C. McParland	LBL	R. Wada	GSI

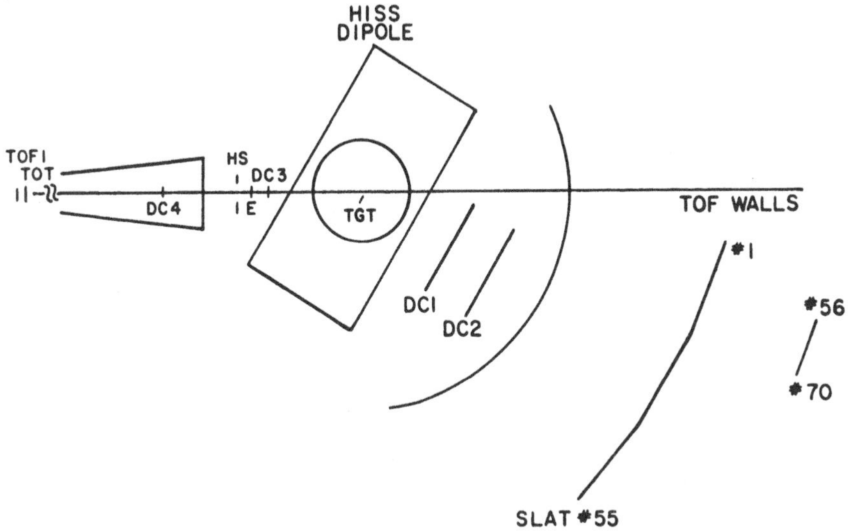

FIGURE 9 Experimental configuration of experiment 513H.

drift chambers and TOF walls, with a second wall for redundancy in the high flux region near Z/A = 0.5. Triggering requires a good beam trajectory, Z=6 incoming and no Z=6 coming out, determined by a small scintillator placed near the TOF wall in the position of the beam focus. The relative positions of the various projectile fragment peaks in rigidity space are shown in Fig. 11. The desire to see ^{1}H and ^{3}H in the same event determines the lateral extent of the desired coverage. Performance of the TOF wall as a charge sensitive detector is shown in Fig. 12; charge resolution is about 0.2e. The single plane resolution of the drift chambers is shown in Fig. 13.

The proton spectrum is quite broad. In Fig. 14a,b we plot the hit positions of ^{1}H and ^{3}H for events containing at least one proton and one triton. It is clear that the proton distribution is quite broad, and we could even use more coverage. It is interesting to look at the invariant

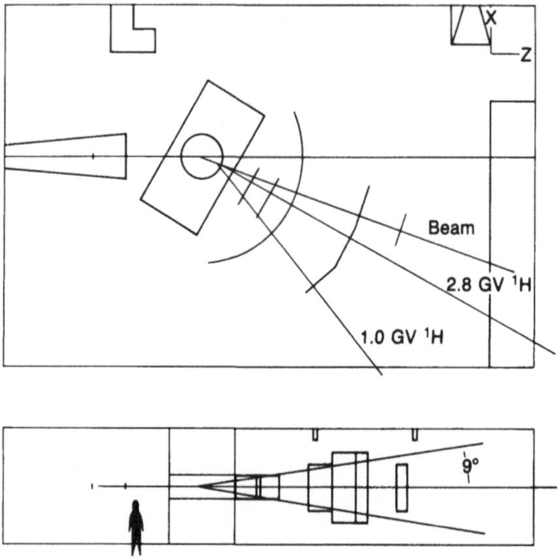

FIGURE 10 Rigidity limits for experiment 513H.

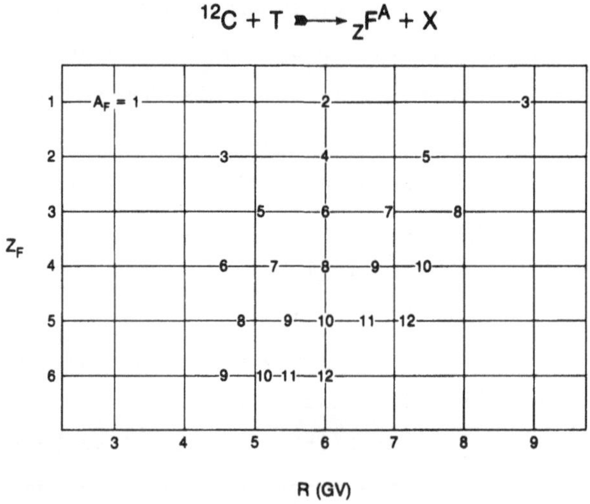

FIGURE 11 Central rigidities of secondary fragments.

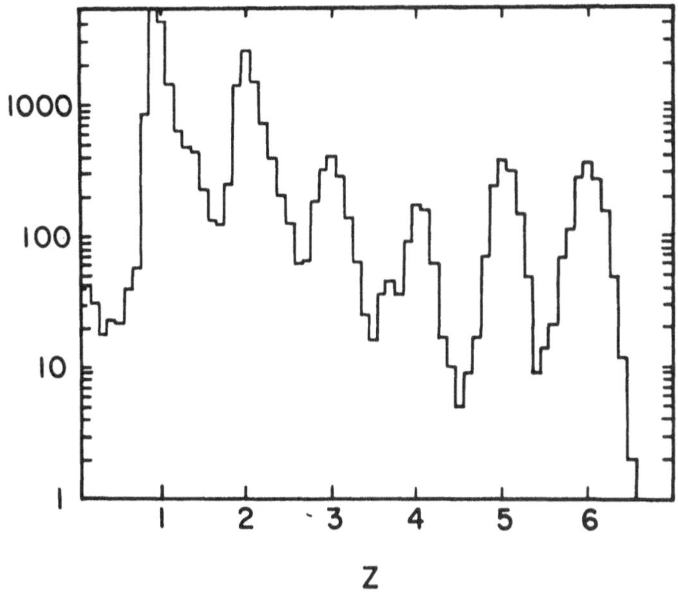

FIGURE 12 Charge distribution from TOF wall.

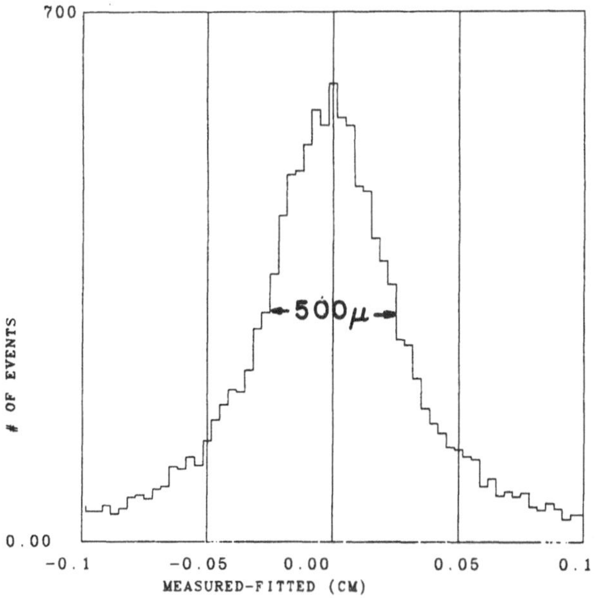

FIGURE 13 Single plane position resolution (Z=6).

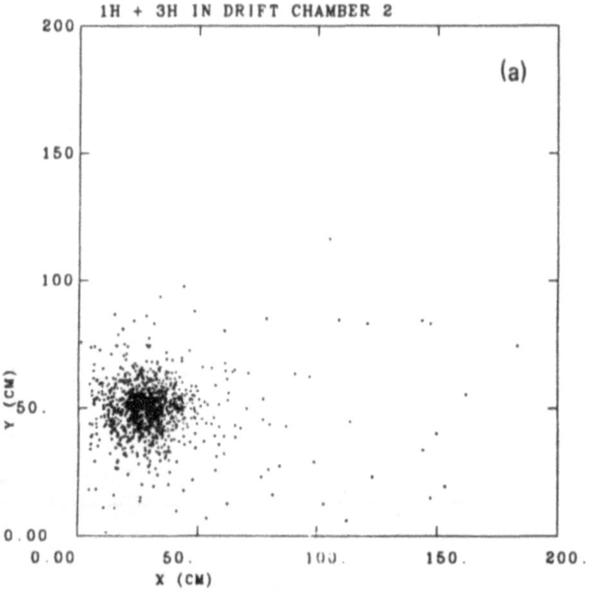

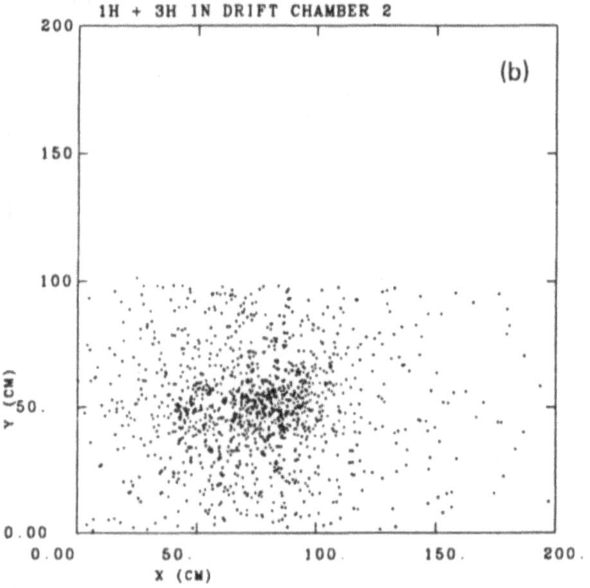

FIGURE 14 Hit pattern: (a) tritons and (b) protons.

mass of this particle pair in Fig. 15. One sees the expected peak at the mass sum due to the many events where the proton and triton come out at low relative momentum along with a long tail of several hundred MeV where possible bound states can show up. We would expect to see bound state effects in the corresponding plot for events containing two helium products (Fig. 16). Here we have a very much finer energy scale and hints of structure presumably produced by ^{8}Be bound states. These examples represent a first analysis of 1% of the available data.

3. TRACKING CHAMBERS FOR HEAVY IONS

We turn now to the technical problem of determining the trajectories of heavy ions at relativistic energies. To demonstrate the effects of the copious delta ray production

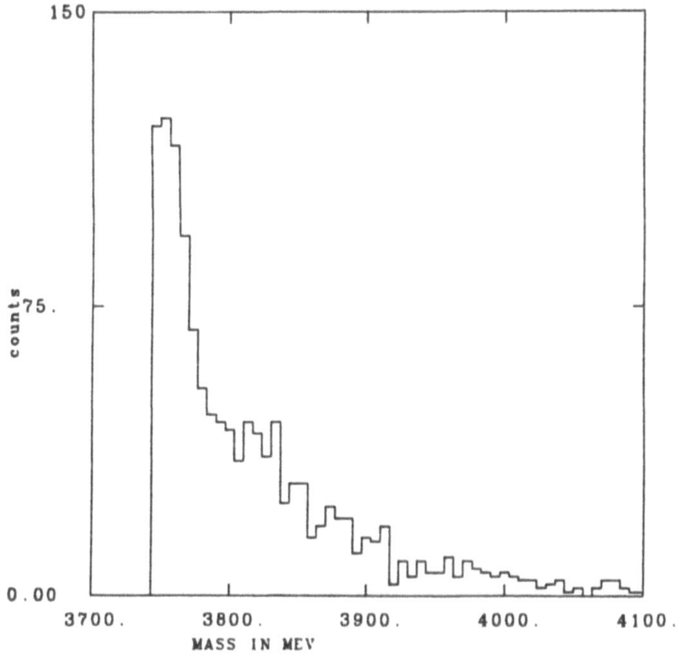

FIGURE 15 Proton plus triton invariant mass.

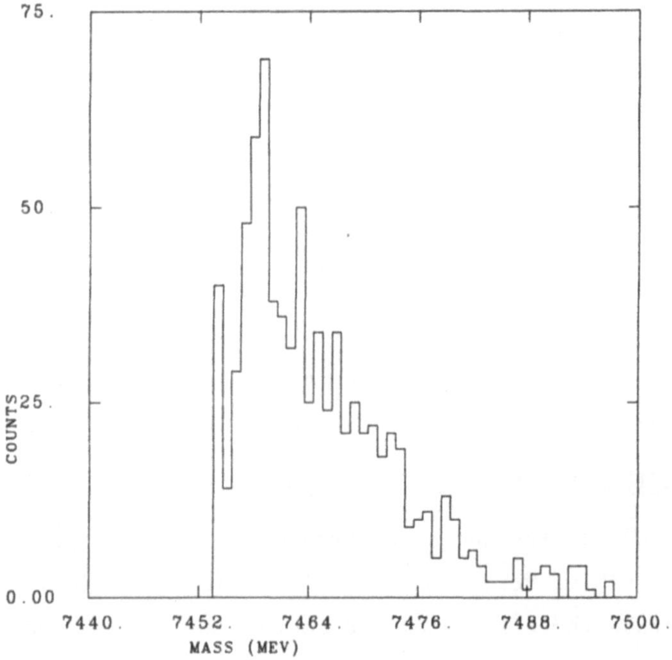

FIGURE 16 Two alpha invariant mass.

at these energies I show in Fig. 17 a photograph, taken by the Riverside group, of 1.4 GeV/nucleon La in the steamer chamber with the magnetic field turned off. The spatial extent of the delta ray cloud is quite large, and if one is interested in detecting low and high charge particles simultaneously the fact that delta rays are charge-one particles cannot be neglected. This problem is a major one for the field and has been the object of considerable study by the HISS group. The goal of our study was to develop and demonstrate a design to satisfy the requirements listed in Table III. This year Fred Bieser and Toshio Koybayshi have conducted extensive tests of our proposed solution outlined in Table IV. The 12-plane test chamber is schematically shown in Fig. 18. The design uses multiple measurements in the

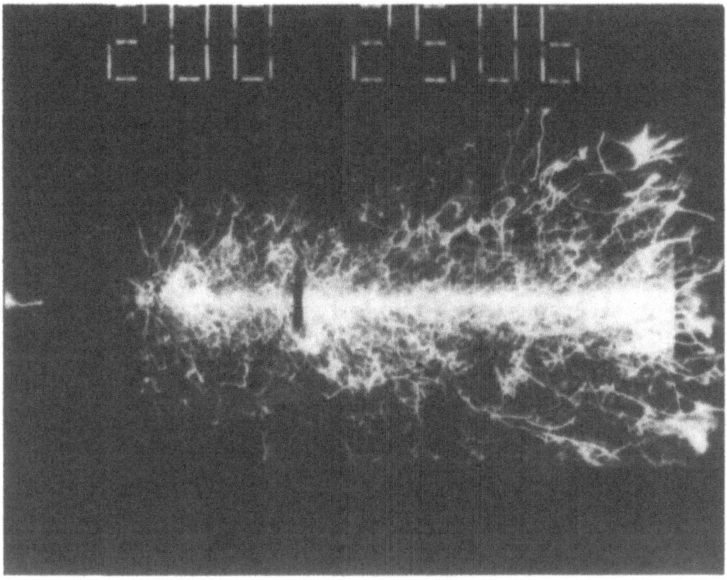

FIGURE 17 Lanthanum streamer chamber event.

S,T,U directions to achieve sufficient redundancy to identify multiple tracks. The drift cell configuration is designed to allow variation of the gain, keeping the drift field relatively constant (Fig. 19). The ground planes between the twelve modules capture the electrons produced outside of the sense planes and isolate the drift fields from one another. The variation of drift field when the gain is changed by 400 volts is shown in Fig. 20. We see a rather small variation allowing drift time calibrations to be relatively independent of operating voltage.

In Fig. 21 we have plotted the efficiency curves for the drift chamber obtained for the Z=1,5 beams. The Z=1 case shows the standard behavior: the chamber is plateaued at 100% efficiency and the single hit efficiency is almost 100% indicating that the chamber gives one hit per incident particle. As we increase the incident beam charge we see

TABLE III Requirements for a drift chamber for high energy heavy ions.

GENERAL REQUIREMENTS

- Good angular resolution

$$\frac{\Delta A}{A} = \sqrt{\left(\frac{\Delta R}{R}\right)^2 + \left(\frac{\Delta Z}{Z}\right)^2 + \left(\gamma^2 \frac{\Delta\beta}{\beta}\right)^2} \sim 0.2\% \text{ at } A = 100$$

- Good double track pair resolution

REQUIREMENTS UNIQUE TO HIGH ENERGY HEAVY IONS

- Large dynamic range

  i.e. $1 \leqslant Z \leqslant 92$ at $0.5 \sim 2.0$ GeV/nucleon

  Energy loss (αZ^2) = 1 to $\sim$ 9000 for Z=1 to 92

- δ-ray rejection + full efficiency for Z=1

  No. δ-rays α Z^2

  δ-rays are minimum ionizing particles!

- Thin chamber

  due to { large reaction cross section; large energy loss (α Z^2) }

TABLE IV Solutions.

TECHNIQUES

- Constant Fraction Discriminator (CFD)
  - to pick up the timing signal of core ionization in the presence of δ-rays
  - **chamber is still sensitive to δ-rays**
- Pulse height information from every wire
  - to find the cell hit by the heavy ion
  - to reject spurious hits from δ-rays
- Thin chamber
  - single gas volume
- Distributed planes
  - simpler track finding
- Good mechanical rigidity + good positioning
  - $\delta_{\text{position}} \sim 40\mu$ for all wires

STRUCTURE + ELECTRONICS

- Cell structure
  - 1 cm drift length with field shaping
- 12 planes : S T U S′T′U′ S′T′U′ S T U
  - 0°, ± 60°, 10 cm apart
- Modular plane on a thick base plate
  - Aluminum stiffner
  - $\delta_{\text{position}} \sim 40\mu$ from known line for all wires
  - Single gas volume : 40 (H) × 30 (V) × 120 (L) cm^3
- P-10 gas : Ar 90% + CH_4 10%
- Single hit TDC (limited by CFD)
- Current integrating ADC

TEST BEAMS

- Energy: ~1 GeV/nucleon
- Beams: H (Z=1), He (Z=2), Li (Z=3)
Be (Z=4), B (Z=5), Ne (Z=10)
Fe (Z=26), Xe (Z=54), U (Z=92)

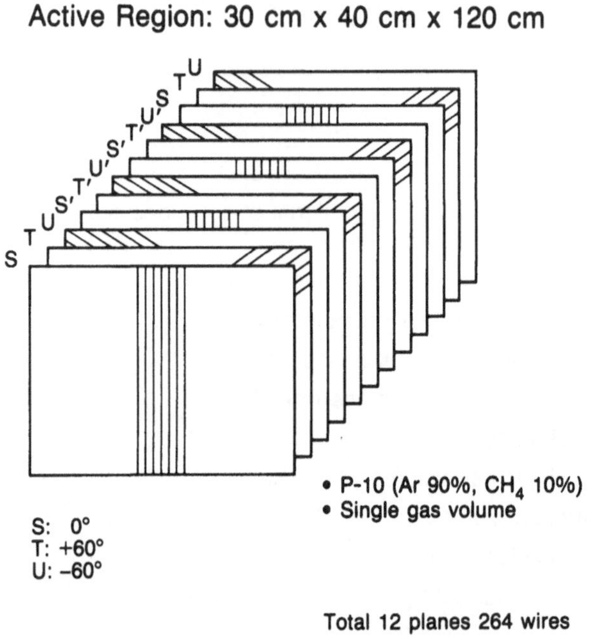

FIGURE 18 Schematic view of prototype drift chamber.

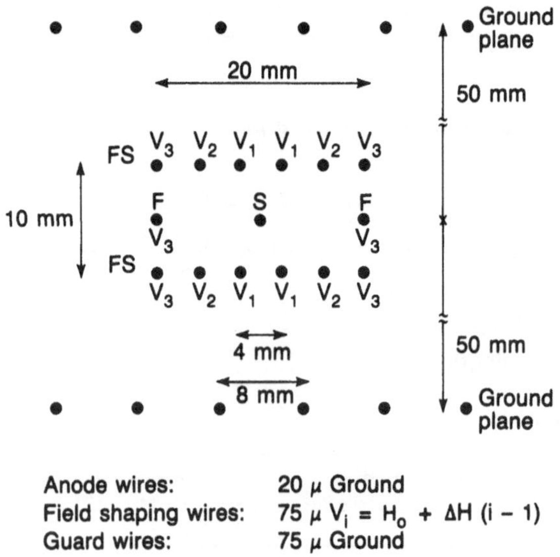

FIGURE 19 Cell geometry.

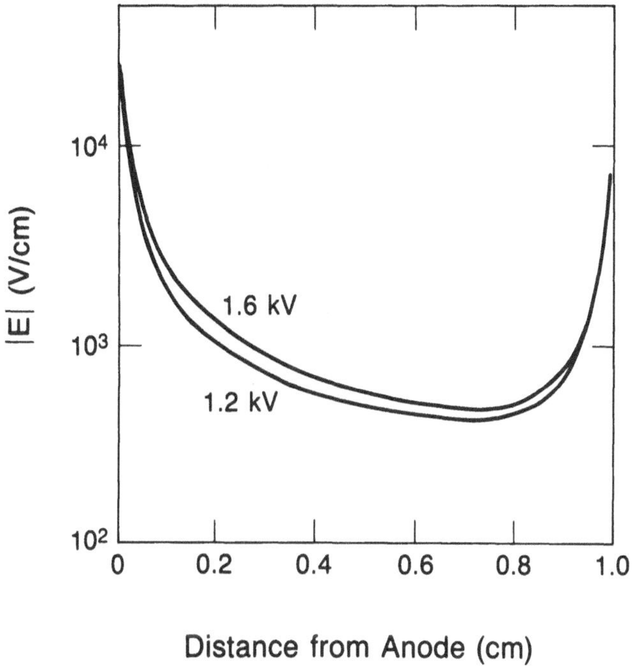

FIGURE 20 Electric field in midplane.

that the voltage at plateau decreases as expected due to the larger de/dx. However, the single hit efficiency drops off, and the higher charge cases never reach 100%. This is due to delta rays at large angles to the track depositing sufficient energy to trigger other wires in the chamber. It is clear that a chamber that is running at 1.5 kV in order to see protons will see single hits for a boron track only 50% of the time.

The fact that the single hit efficiency does not reach 100% means that in the region around Z=5 the delta rays can deposit as much energy as the primary track in a given cell. At higher charges this effect decreases and it again becomes possible to achieve 100% single hit efficiency over a very limited range of incident charge. These data alone tell us

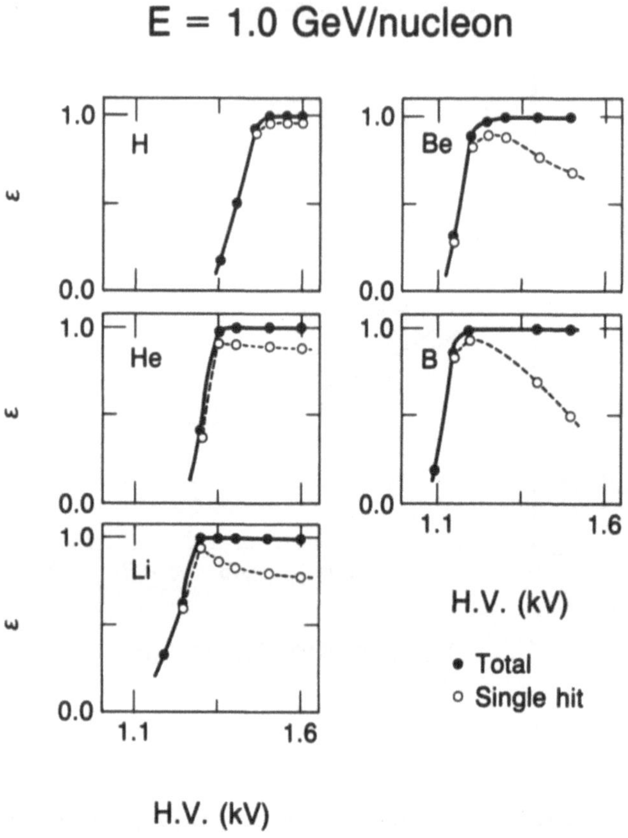

FIGURE 21 Efficiency curves for charges 1-5.

that in the drift chamber, with its constant fraction front end, the worst case for suppressing delta ray triggers in a single cell will be in the region of incident particle charges of Z=5.

In what follows I limit myself to discussing the response of a single plane of the drift chamber. The average number of hits per plane as a function of incident charge and voltage is shown in Fig. 22. The qualitative result from the earlier streamer chamber photograph is seen with a vengeance. The total number of wires in a plane is 20,

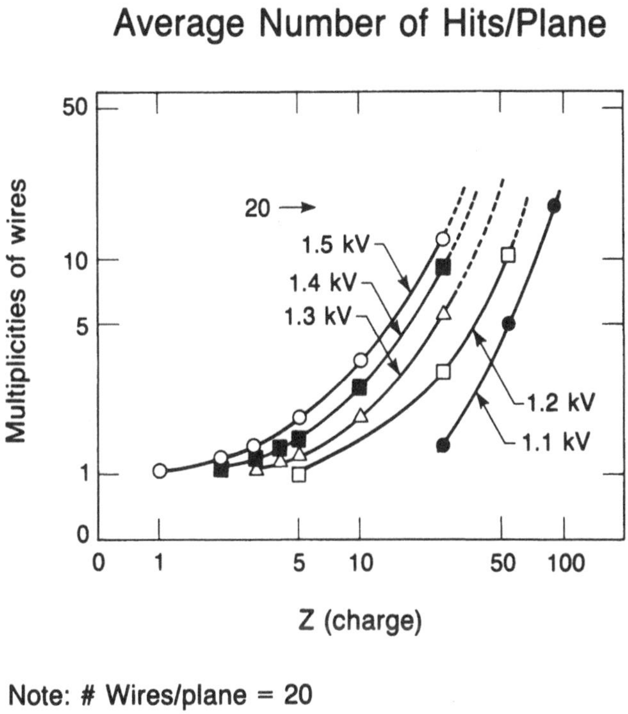

FIGURE 22 Multiplicity as a function of charge.

covering a distance of 20 cm on each side of the primary particle. Examination of these data makes it clear that there is no operating voltage covering a large variety of charges with single hit response. For the higher charges every wire in the chamber gives a signal regardless of the gain.

So TDC information is not enough; we need to examine the charge deposited as a function of position. Figure 23 shows the pulse height spectra as a function of cell number for Fe,Xe and U projectiles. As we hoped the cell with the heavy ion is clearly separated over this wide charge range by the ADC information. An interesting fact is that the distribution of energy around the track is not the same

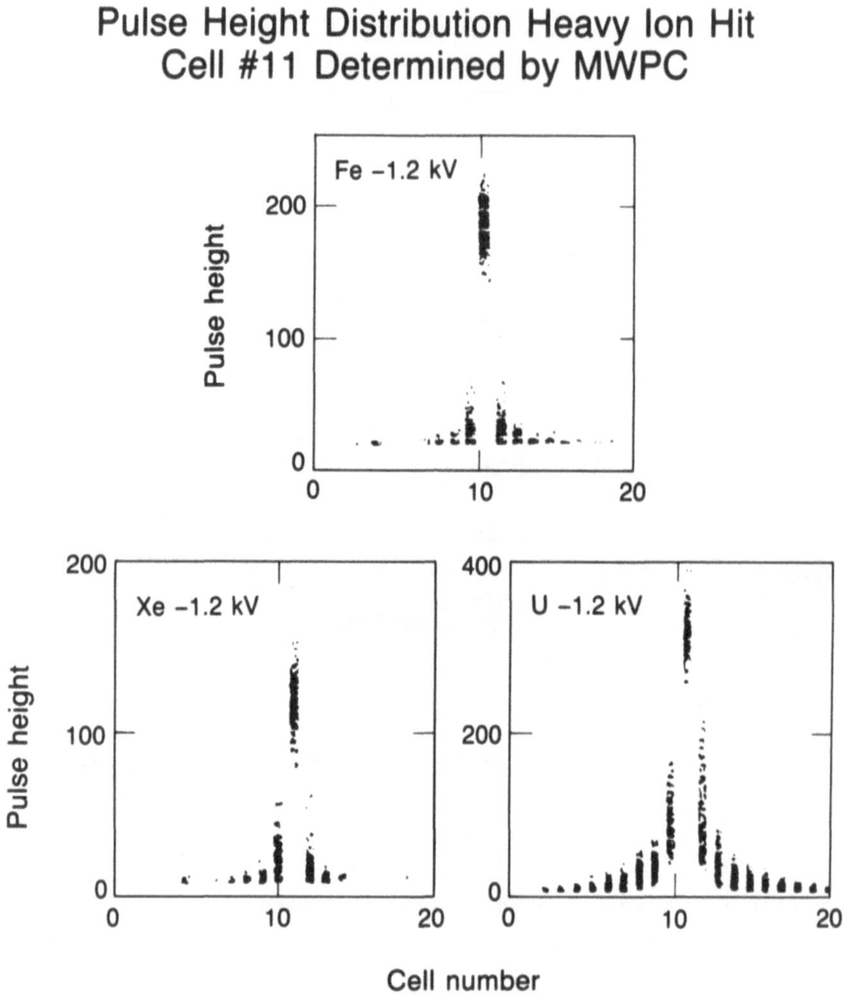

FIGURE 23 Pulse height distributions vs. cell number.

shape for the different incident particles. We have not calculated this yet, but feel that it is connected with the multiple scattering of the delta rays.

Figure 24 demonstrates the selection of the proper drift cell by choosing the largest pulse height in each plane. The plots for incident particles Fe, Xe, and U are arranged in sets of each S,T,U combination and show on the

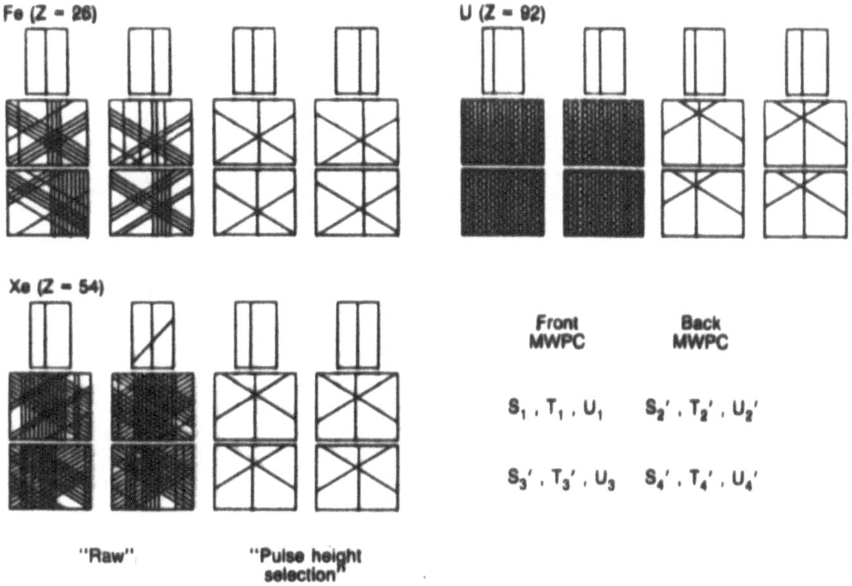

FIGURE 24 Typical wire patterns at H.V. = -1.2 kV.

left the uncut data and the ADC-selected wires on the right. Thus for a single chamber voltage of 1.2 kV we can use the ADC information to locate the proper cell; later we will see the accuracy of the TDC information within the selected cell.

At this point I would like to discuss the energy deposition in a little more detail. To illustrate the distribution of the central core ionization we plot in the upper half of Fig. 25 the distribution of pulse heights for all wires in a plane for a set of Xe events. In the lower half of Fig. 25 we see the "selected" and "rejected" spectra. The selection criterion was to take the largest pulse height seen in the entire plane. This gives the energy deposited in a two-centimeter-wide interval about the track. The rejected spectrum is what is left. Because these two spectra overlap, an absolute cut in pulse height will not always

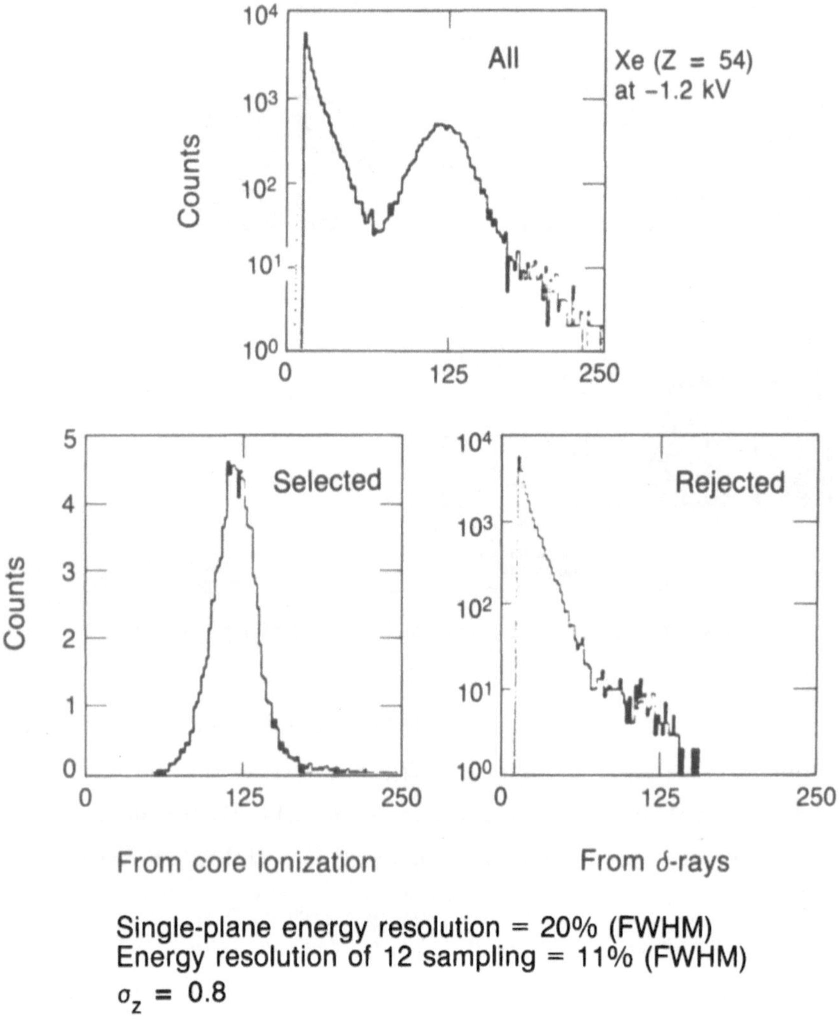

FIGURE 25 Single wire pulse height distributions.

select the correct cell due to the variations in both energy loss and possible variations of response depending on track position within a cell. Note that the pulse height information summed over the 12 planes available gives a rough charge estimate. This information should prove useful in trajectory reconstruction in conjunction with other more sensitive but less segmented charge detectors. In Fig. 26 we see the pulse height from the two-centimeter region about the track varies as the 1.5 power of the charge. This quantifies the shape change mentioned earlier. As the charge of the projectile increases we see that more of the energy is deposited at greater distances from the trajectory. Also for charges somewhere between 10 and 26 the spectrum of energy deposited in this region becomes gaussian. As a consistency check we sum all the energy seen in one plane in

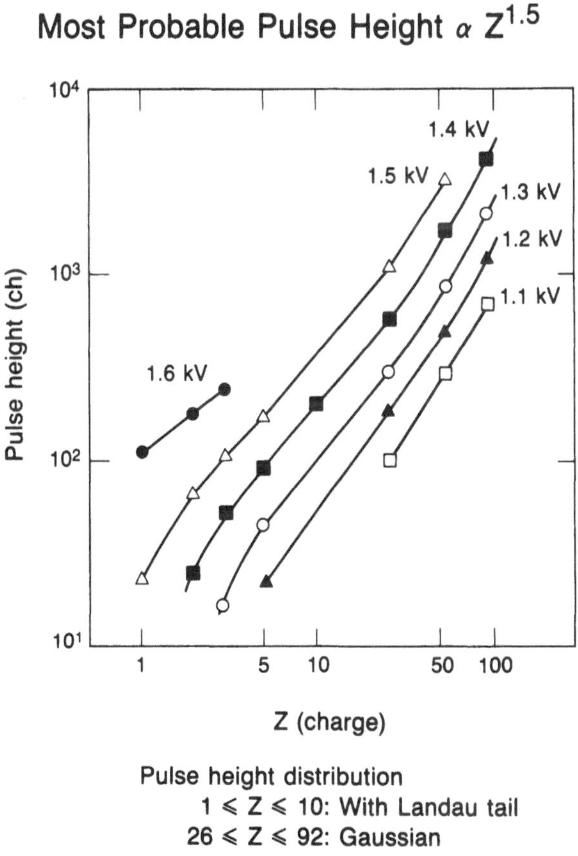

FIGURE 26 Most probable pulse height of one wire vs. particle charge.

Fig. 27. The Z-squared dependence is recovered along with the correct Landau shape for all charges.

Using the trajectory as determined by the multiwire we can examine the spatial linearity and resolution of the drift chamber. Figure 28 shows the TDC response of the drift chamber for charges ranging from 5 to 92 recorded at the same chamber voltage. Some nonlinearity near the anodes develops for the lower charges. We believe this is due to changes in the pulse shape from the radical change in the distribution of electron arrival time at the anode, and it indicates that calibrations will change slightly as a function of charge. The chamber resolution for tracks 0.5 cm

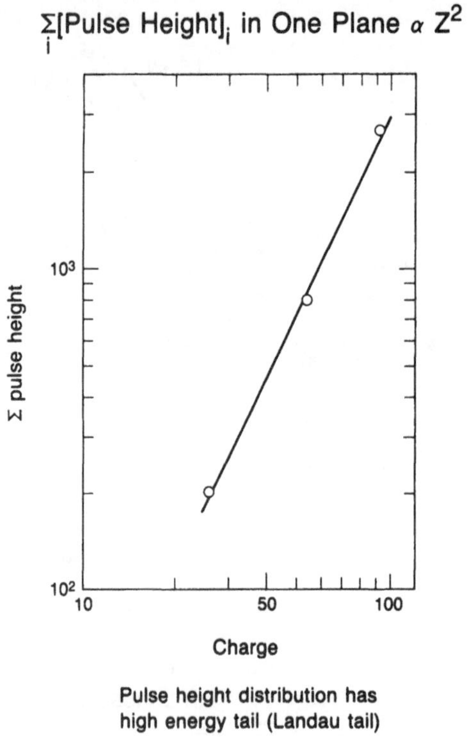

FIGURE 27 Sum of pulse heights from all wires in a plane.

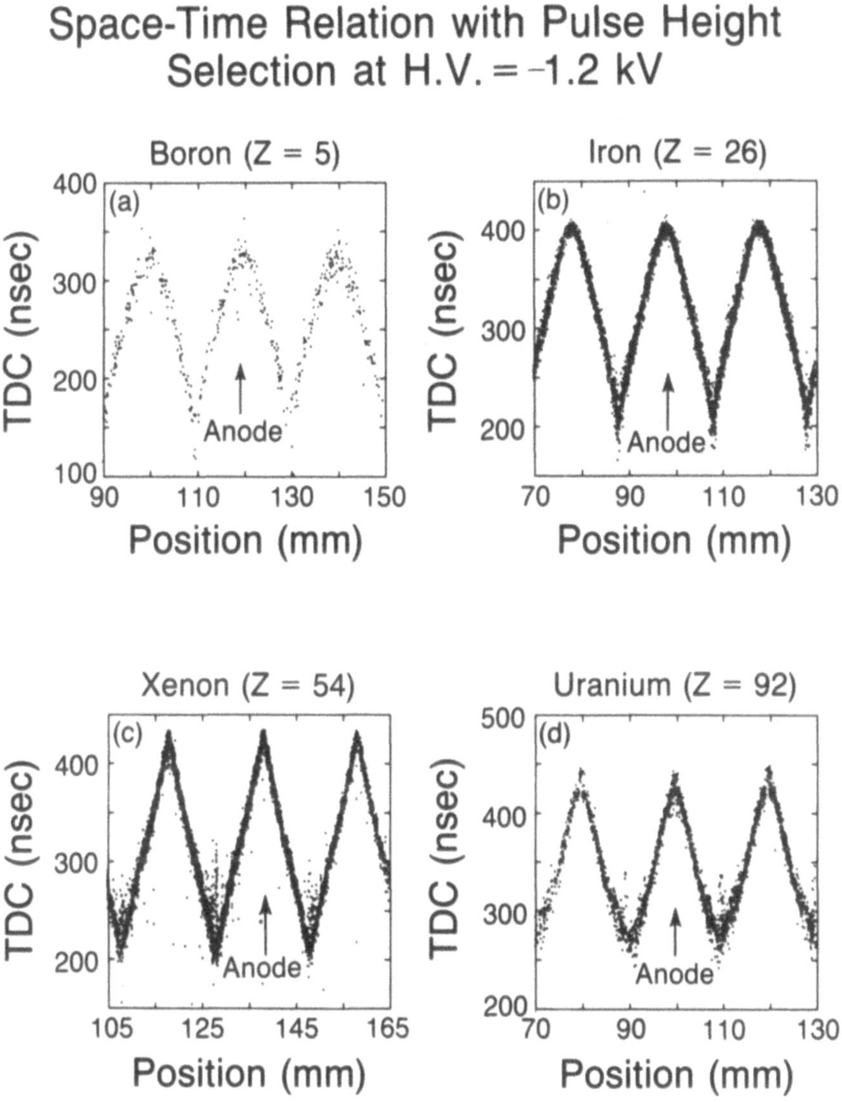

FIGURE 28 Space-time relation for various charges.

from the anode is shown as a function of chamber voltage and particle charge in Fig. 29. The optimum operating voltage varies with charge so that the dynamic range and accuracy requirements of the experiment must be considered when determining the running voltages. The plotted resolutions are a slight overestimate because the trajectory uncertainty of the MWPC's is not removed.

Another concern is particle pair separation. If there are other tracks near a heavily ionizing trajectory, the delta ray cloud can mask their signals. This cylinder of confusion is most probably a basic limitation of gaseous trajectory measurement devices applied to heavy ions. To see the possible magnitude of this limitation we use the pulse height information to show the minimum detectable charge as a function of distance from primary tracks of Fe,

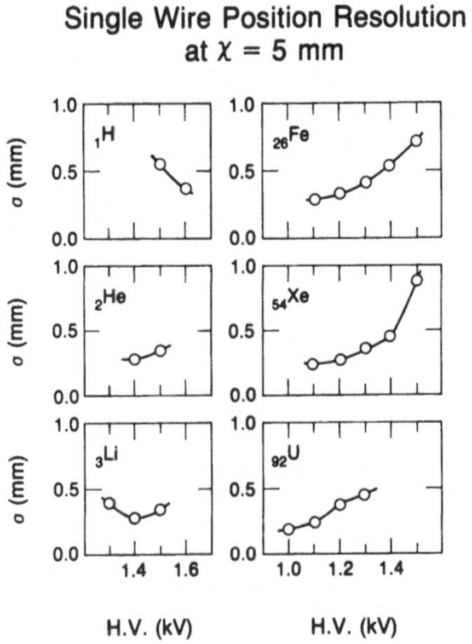

FIGURE 29 Single wire position resolution for various charges.

Xe, and U (Fig. 30). Some improvement may be made using peak sensing ADC's, utilizing redundancy and magnetic fields, but we don't expect too much.

The results presented here will form the basis for our design of the large area tracking chamber at HISS. These results and conclusions are summarized in Table V.

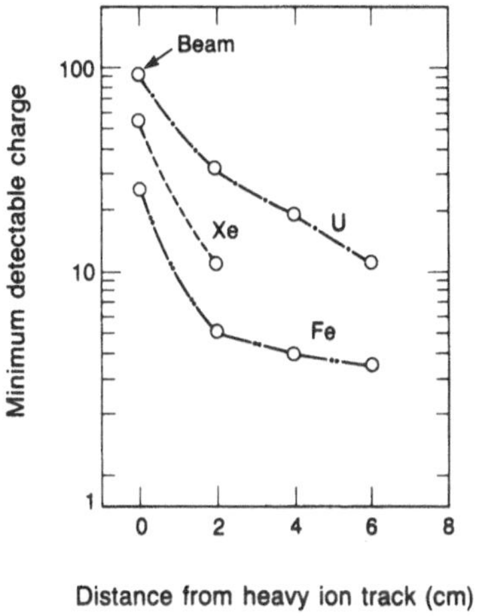

FIGURE 30 Estimate of double track pair resolution from pulse height information.

TABLE V Bevalac test results.

RESULTS

1. Efficiency
   - ~100%
2. Pulse Height Information
   - reject δ-ray signals, ~ 100%
   - visible energy loss in one cell, $\propto Z^{1.5}$
   - total energy loss (Σ pulse height), $\propto Z^2$
   - shape
     $1 \leqslant Z \leqslant 10$: asymmetric + Landau tail
     $26 \leqslant Z \leqslant 92$: symmetric (Gaussian)
3. Position Resolution + Time-Distance Relation
   - Multiple scattering limit at the optimum voltage for each charge
   - Parameterization of time-distance relation as a function of charge at a given voltage for actual experiment required
4. Double Track Pair Resolution
   - estimate from pulse height distribution
   - real 2 track finding algorithm being developed
5. Charge Resolution or Energy Loss Resolution
   - Using 12 sampling
     $26 \leqslant Z \leqslant 92$: averaging $\sigma_Z \sim 0.8$
     $1 \leqslant Z \leqslant 10$: truncated mean

CONCLUSIONS

1. Limited Dynamic Range
   - Requiring ~ 300 μ resolution

High Voltage	Range
−1.6 kV	$1 \leqslant Z \leqslant 12$
−1.4 kV	$2 \leqslant Z \leqslant 26$
−1.2 kV	$5 \leqslant Z \leqslant 92$
−1.1 kV	$8 \leqslant Z \leqslant 92$

2. Double Track Pair Resolution
   - Detection of Z=1 possible outside ± 20 cm from uranium track (worst case)

REPORT ON THE 1984 LBL WORKSHOP ON DETECTORS FOR RELATIVISTIC NUCLEAR COLLISIONS*

L.S. SCHROEDER
Nuclear Science Division
Lawrence Berkeley Laboratory
University of California
Berkeley, California 94720

In this talk I will be reviewing the highlights of the recent Workshop on Detectors for Relativistic Nuclear Collisions, held March 26-30, 1984, at the Lawrence Berkeley Laboratory. For a more complete report I refer you to the Proceedings of the Workshop.[1] The Workshop was also summarized at the Quark Matter '84 meeting in Helsinki this last June.[2] For additional information on detector requirements for high energy nuclear collisions, consult the talks of T. Ludlam, H. Specht, and A. Sandoval given at this Conference.

In the time available, I want to not only give you a feeling for the detector concepts discussed and in many cases developed at the Workshop, but also I hope to leave you with an overall impression of the tremendous strides that have been made in coming to grips with the very complex collisions that we will encounter at higher energies.

*This work was supported by the Director, Office of Energy Research, Division of Nuclear Physics of the Office of High Energy and Nuclear Physics of the U.S. Department of Energy under Contract No. DE-AC03-76SF00098.

1. INTRODUCTION

There were several reasons for having a dedicated Detector Workshop at this time, including: (i) at Brookhaven a transfer line between the Tandem Van de Graaff and the AGS will provide light ions ($A \leq 32$) at energies up to 15 GeV/nucleon beginning in late 1986; (ii) at the same time, an upgrade of the CERN PS injector by a GSI/LBL collaboration will provide ^{16}O ions in the SPS at energies from 15-225 GeV/nucleon beginning in late 1986; (iii) in late 1983, the Nuclear Science Advisory Committee of the Department of Energy and the National Science Foundation recommended[3] that the next major nuclear physics facility in the U.S. should be a relativistic nuclear collider to explore the domain of high energy and baryon densities and hopefully shed light on deconfined hadronic matter--the quark-gluon plasma (QGP); and finally (iv) several national laboratories (BNL, LBL, ORNL) in the U.S. have active programs to look at collider options.[4-6]

Since the new fixed target capabilities will be coming on-line within two years, we felt it was both timely and necessary to start looking at the detector requirements imposed by these much higher energies. For the more distant colliders, an early start on detectors is also called for. Earlier meetings had considered detector requirements,[7-9] but we wanted to go beyond these and in particular hoped to identify any detector elements thought to be on the critical path for investigating the QGP that require an R&D effort.

To provide a focus for the Workshop it was decided to concentrate on the detection of the QGP. However, be aware that the conclusions arrived at for detectors are really in the context of high energy and baryon density achievable in central high energy nucleus-nucleus collisions and apply to a potentially broader and richer range of phenomena than just the QGP. The individual working groups were roughly set up along lines of observables of the QGP.[10,11] Figure 1 is a representation of the evolution of a central nucleus-nucleus collision in the temperature (T) and relative baryon density (ρ/ρ_0) plane. Two trajectories, one emphasizing the baryon-rich plasma, the other the meson-rich (so-called "transparency") regime, are shown. In order for a detection scheme to be sensitive to the presence of the quark-gluon phase, measurements of direct radiation from the plasma itself (region 1) or from the rehadronized phase (region 2)

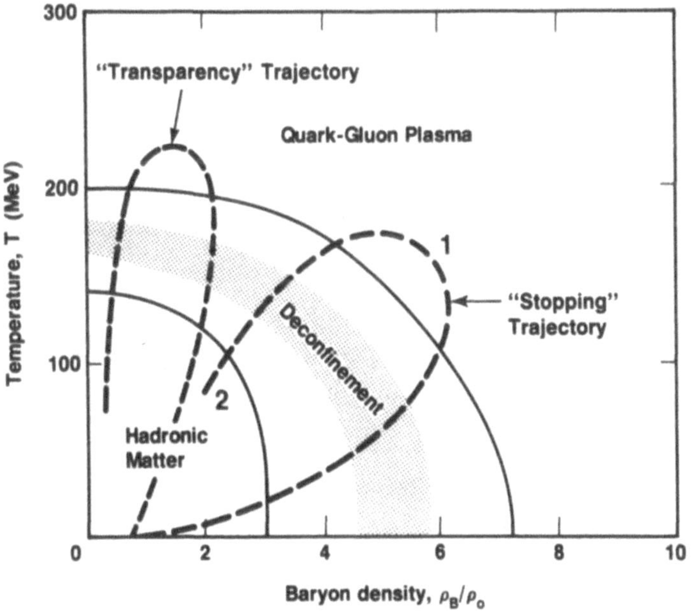

Experimental Signatures

Region 1: Direct radiation from plasma (photons, leptons, lepton pairs)

Region 2: Recondensed hadronic matter (π, K, p, $\bar{p}$,...) Possible free quarks, diquarks

FIGURE 1 Phase diagram with nuclear collision trajectories.

will be required. Figure 2 graphically illustrates some of the diagnostic tools that have been suggested for uncovering the QGP. Since the plasma is not thought to be stable, radiation from both regions could be present in each event. A device measuring hadrons in the final state would detect hadrons arising from ordinary hadronization processes found, for example, in nucleon-nucleon collisions and hadrons produced by quark and gluon recombination processes in the plasma phase. Theoretical guidance will certainly be required to differentiate between these sources.

At the Workshop we identified three areas of detector measurements which were used as the focus for individual groups. These were: (i) hadrons--generally concerning detection of strange particles and antibaryons, i.e., few particle measurements, often with reduced solid angle, (ii) event parameters--experiments with nearly 4π acceptance, measuring many of the particles in the final state, and (iii) penetrating probes--measurements of photons, leptons, lepton pairs, etc., which are thought to be suitable in probing the hot, compressed stage of the collision. In each category there was a working group for fixed target and collider geometries. Each group had conveners who organized and guided the efforts of the group.[12] The conveners are the ones who made the Workshop a success.

2. WHAT LIES AHEAD AT HIGHER ENERGIES?

As we increase the bombarding energy the most immediate consequence in a central nucleus-nucleus collision is that the multiplicity of final state particles will increase dramatically. At the Bevalac the GSI/LBL Plastic Ball experiment (see talk by H.-A. Gustafsson in this Proceed-

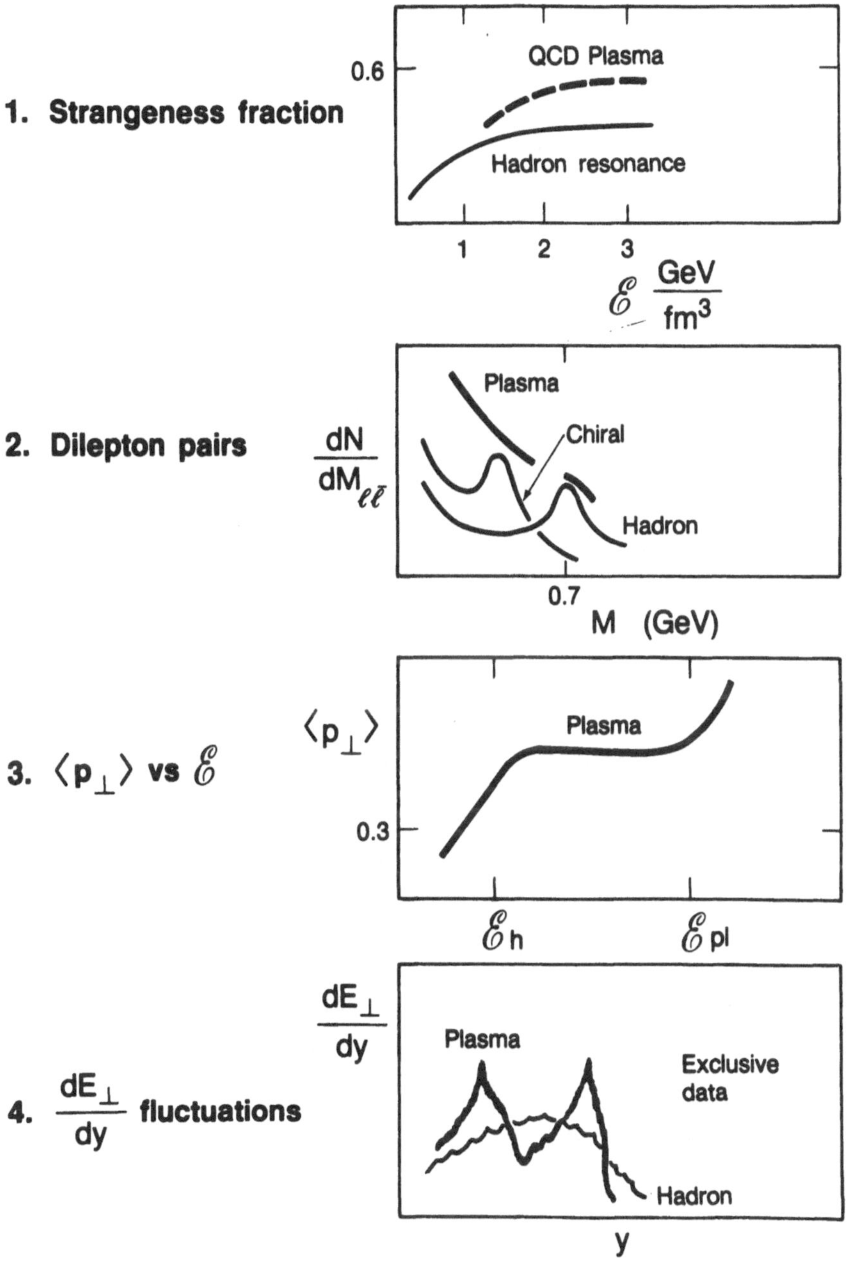

FIGURE 2 Some possible diagnostic tools for looking at the quark-gluon plasma.

ings) observes ~100-200 charged particles being produced in central Au + Au collision at 1.05 GeV/nucleon. In this region, the final states are dominated by nucleons and light fragments with pions making up ~10-20%. At 10 GeV/nucleon, corresponding to energies available at the AGS, we anticipate several hundred pions will be created, comparable to the number of nucleons. At much higher energies, such as the Si + Ag JACEE event[13] at ~4-5 TeV/nucleon (laboratory energy), we expect to see a thousand or more mesons (π,K) being created. These estimates are in line with results from recent event simulations based on extrapolations of pp and pA data. Many detector-related questions naturally arise when considering the high multiplicity final states resulting from central collisions, including: (i) to what extent will tracking of individual particles still be possible? (ii) will calorimetry replace tracking in high multiplicity environment? (iii) what are the appropriate central collision triggers? (iv) how to record and process all the information?, and (v) are there existing detector designs (e.g., UA1/UA2) that can do the job?

3. DETECTORS IN FIXED TARGET AND COLLIDER GEOMETRIES

A brief comment about some of the differences to be expected for experiments done in fixed target versus collider modes. As one increases the incident energy, the center of mass for fixed target experiments moves rapidly forward. This means that the large multiplicity of particles produced in these collisions will be tightly focused in the forward direction, putting severe demands on the granularity of detector elements if one wishes to preserve individual particle identification (ID) capability. At the same time the momenta of the outgoing particles will span a large dynamic range, again testing one's ability to provide particle ID over as large a range as possible. However, the large momentum boost experienced by unstable particles (e.g., Λ^0, $\overline{\Lambda}^0$) will make them more accessible to detection by increasing their apparent lifetime. In collider geometry, particle emission is over 4π, somewhat easing the demands on detector granularity. The energy range of particles being produced will generally be in the few hundred MeV to few GeV range at mid-rapidity (generally thought to be the region of most interest for the QGP). Here particle identification techniques are generally well understood. The most forward ($\theta_{cm} \approx 0$-$15°$) and backward ($\theta_{cm} \approx 165$-$180°$) sections of collider detectors correspond to large values of rapidity and will look quite similar to detectors at forward angles for fixed target work.

Now I want to review results from the individual working groups. Again, I can only cover some of the highlights and refer you to the Workshop Proceedings[1] for a more complete picture.

4. HADRONS-FIXED TARGET

This group considered fixed target experiments measuring single-particle or few-particle observables. Generally such experiments would be done with reduced solid angle coverage. Experiments considered were: (i) measurement of strange particles and antibaryons for flavor tasting of the QGP, (ii) Hanbury-Brown/Twiss (HBT) like-particle interferometry to obtain information on the space-time evolution of the particle source, and (iii) production of heavy flavors such as $\phi(1020) = \phi(s\overline{s})$, $\psi(3100) = \psi(c\overline{c})$, to be used as probes of the quark content of the QGP. A description of two of the detectors considered follows below. Other detectors such as streamer chambers (see talk by A. Sandoval in this Proceedings), solid state track detectors, and emulsion chambers, although discussed in the Workshop Proceedings, will not be covered here.

Considerable attention was paid to the question of how to trigger on central collisions. One possibility is the so-called "spectator veto" where one vetos events that produce near beam-rapidity particles in the projectile fragmentation region. Specifically, the spectator nucleons associated with the projectile will be emitted into a narrow cone of $\Delta\theta < 1°$ about the beam axis at 15 GeV/nucleon, and $\Delta\theta < 0.1°$ at 225 GeV/nucleon.

Figure 3 shows a relatively large solid angle spectrometer being designed by Gruhn et al.[14] for detecting strange baryons and antibaryons. Particular attention is given to Ω production since being an (sss) system it could be a very sensitive probe of the plasma phase of the collision. The main elements of the detector are a superconducting magnet (3-7 T) to sweep away low momentum particles from the forward cone, a multiplicity array for centrality tagging, and a downstream projectile calorimeter in which low energy deposition would be used to signal a central event. The heart of the apparatus is a micro-TPC (μTPC) that will be used to recognize and identify both the initial and final strange particle decays (e.g., $\Xi^- \rightarrow \Lambda^0\pi^-$ followed by $\Lambda^0 \rightarrow p\pi^-$). Three planes of drift chambers would be used in front of the μTPC to increase momentum resolution.

Small solid angle devices also received attention. They offer several advantages, including: (i) excellent particle identification; (ii) good momentum resolution,

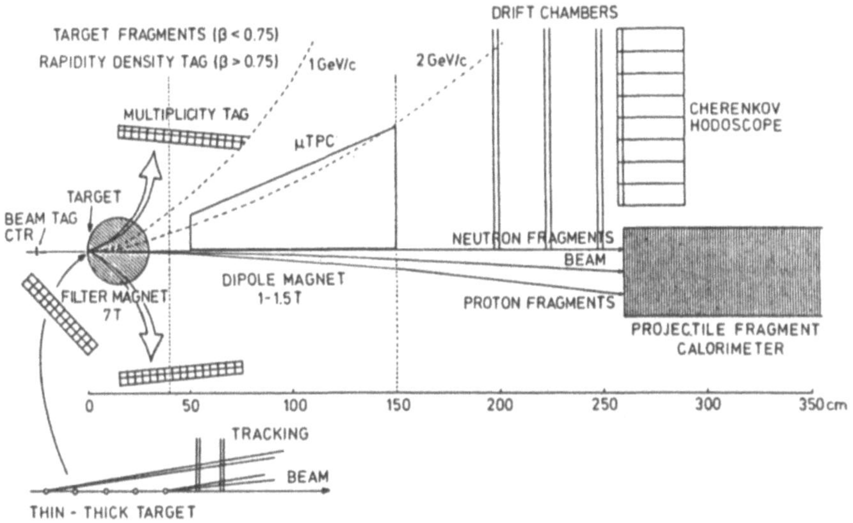

FIGURE 3 Proposed layout of an experiment to study strange baryon and antibaryon production with ^{16}O beams at the CERN SPS.

$\Delta p/p \lesssim 1\%$; and (iii) are capable of taking high beam intensity. Figure 4 shows an example of a small solid angle ($\Delta\Omega \approx$ 10 msr) spectrometer being considered for studying charged particle spectra ($\pi^{\pm}$, $K^{\pm}$, p, $\overline{p}$, Λ^{0} decays) in the central rapidity region ($5° \lesssim \Theta_{Lab} \lesssim 20°$ for ^{32}S + Au collisions at 15 GeV/nucleon). For 10^7 particles on target per sec up to 10 particles per event could be tracked and identified in the spectrometer. The main elements of the detector are the track segment projection chamber (TSPC), a large aperture vertical deflection magnet, and several arrays of threshold Cherenkov and aerogel counters. A lucite multisegmented Cherenkov multiplicity filter would be used to tag central events. The device would be employed to measure inclusive spectra of strange and nonstrange particles. Comparison of these spectra at several beam energies would provide information on the "nuclear temperature" of the participant zone,

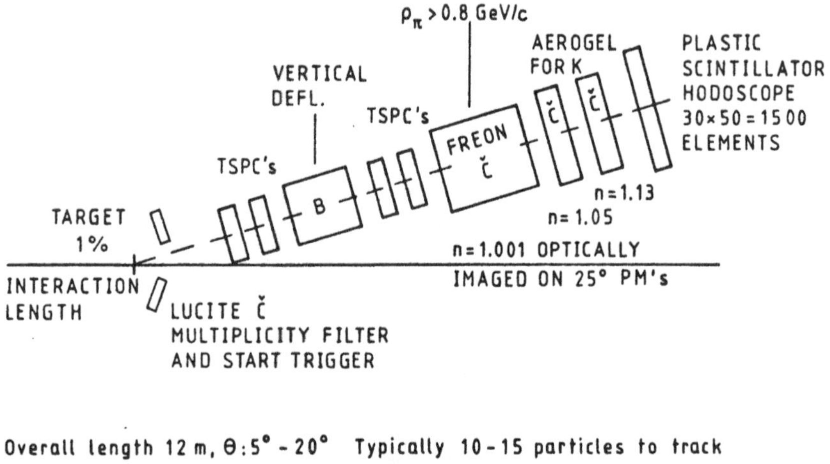

FIGURE 4 Single arm spectrometer being considered for possible use at the AGS (15 GeV/nucleon) with light ion beam ($A \leq 32$).

and the strange particle yields have of course been suggested as a probe of the plasma phase.[10,11,15]

5. EVENT PARAMETERS--FIXED TARGET

The group concentrated on detectors for fixed target work having nearly 4π solid angle (i.e., complete azimuthal coverage). Four types were discussed, of which I shall present only the results for the so-called "idealized" detector. Two others, a nonmagnetic spectrometer and a streamer chamber, are very close to the systems being described by A. Sandoval elsewhere in this Proceedings.

The detectors were designed for both 15 GeV/nucleon ^{32}S at the AGS and 225 GeV/nucleon ^{16}O beams at the SPS. The range of physics they wished to cover included: (i) measurement of negative particle (mostly π^-) and total multiplicity, (ii) pseudo-rapidity ($dn/d\eta$) and rapidity (dn/dy) distributions for charged particles, (iii) total $\pi^{\pm}$ and π^0

energy, (iv) E_T and p_T, (v) shape measurements (e.g., equilibration flow, jets), and (vi) particle correlations (e.g., HBT).

The idealized dipole detector conceived by this group is shown in Figure 5. It consists of a target surrounded by a cube of CCD detecting planes which act as vertex detector and multiplicity counter inside the uniform magnetic field region. Three-dimensional tracking chambers are spread throughout the volume of the magnet, with electromagnetic and hadronic calorimeters (located inside return yoke of dipole coils) and time-of-flight counters providing angular coverage (down to $\Theta_{Lab} \approx 20°$). Downstream of the magnet would be additional tracking chambers, possible RICH (Ring Image Cherenkov) detectors for particle identification and more time-of-flight and calorimetry.

As suggested earlier, the number of particles will be large. For example, the results of a HIJET calculation[16] for ^{16}O + Au at 15 GeV/nucleon for impact parameters b $\leq$

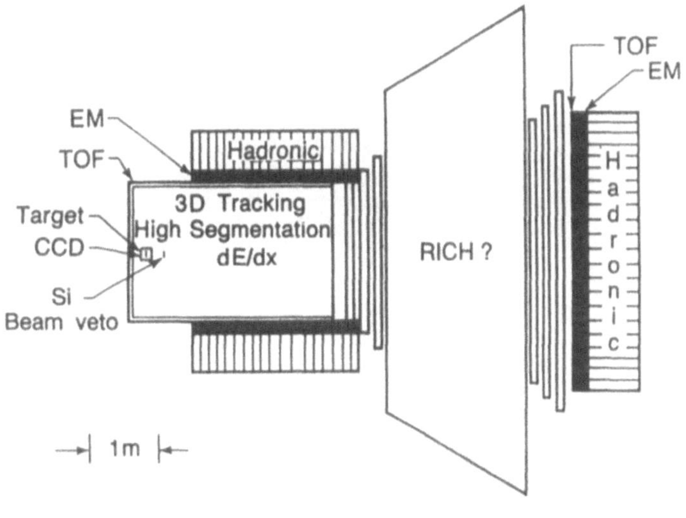

FIGURE 5 Idealized detector.

4 fm suggest several hundred particles. To see what this means for a detector like the idealized system of Figure 5, the group simulated what a typical central event at 15 GeV/nucleon would look like. Figure 6(a) shows such an event in the nonbend plane where pattern reconstruction is easier (tracks are straight lines). Figure 6(b) shows the same event in the bend plane (for a 0.5 T field). Clearly at 225 GeV/nucleon, due to the larger number of particles

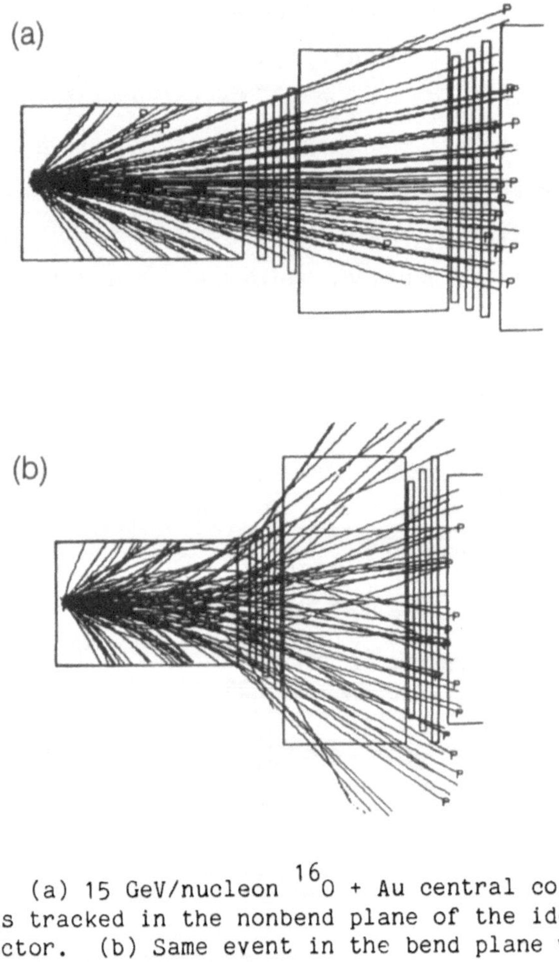

FIGURE 6 (a) 15 GeV/nucleon ^{16}O + Au central collisions tracked in the nonbend plane of the idealized detector. (b) Same event in the bend plane with a field of 0.5 T.

and their stronger forward focus, it will be even harder. The group suggests that a system like the European Hybrid Spectrometer which uses a series of dispersing dipoles spaced along the beam line would possibly be a better solution. They further concluded that heavy beam (Au + Au) fixed target experiments look very difficult at 15 GeV/nucleon and only calorimetry at 225 GeV/nucleon look feasible, even for ^{16}O.

6. PENETRATING PROBES--FIXED TARGET

Efforts were directed at systems designed to detect directly produced electromagnetic probes (photons, leptons, lepton pairs) in light ion collisions over the energy range of 15-225 GeV/nucleon. Since these probes do not interact strongly with matter, they should carry information on the hot, compressed stage of the collision.

Although several different detectors were identified by the group, I will only mention plans to measure high mass muon pairs using a combination (NA34) of a spectrometer being designed for proton-nucleus experiments and the existing NA3 spectrometer to be used as the downstream muon spectrometer. Its elements are shown in Figure 7. Strong filtering (≈10 interaction lengths) after the target is used to remove hadrons and limit decays of π's and K's into muons. In this way they hope to get around problems arising from high multiplicities, lepton identification, and combinatorics. Multiple scattering in the filter will of course impose limits on the mass resolution and p_T of the dimuon pair. The front-end of the detector consists of a vertex detector and hadron calorimeter. In addition, they will also incorporate electron identification in the front-

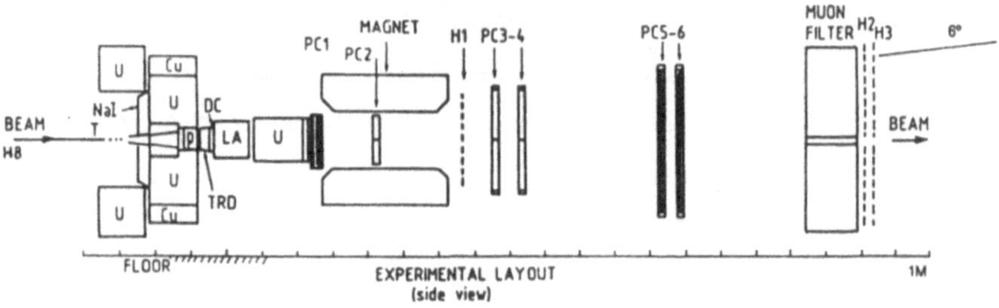

FIGURE 7 NA34 detector being proposed for light ion experiments at CERN.

end so that the low mass spectrum (m_{ee} < 1000 MeV) can be studied. However, at the time of the Workshop, studies indicated that due to the large multiplicities expected (they quote ~300 charged particles within a 5° cone for ^{16}O + ^{195}Pt at 225 GeV/nucleon), individual electron tracking and identification may not be possible. Furthermore, the 150 π^0's expected/central event will yield between 1 and 2 e-pairs via Dalitz decay alone--complicating the process. A possible future development is the addition of a pair spectrometer (placed in an opening in the calorimeter section) to measure soft photons and electrons. Figure 8 shows, as an example, previous results on "low mass" dimuon pairs taken by the NA3 component of the system. These data represent dimuon pairs with p_T > 2 GeV/c, a requirement that significantly increases the laboratory energy of the muons, thereby reducing the effects of multiple scattering. Please refer to the talk by H. Specht in this Proceedings for more complete information on the NA34 system.

7. PENETRATING PROBES--COLLIDER

This working group employed the same criteria for the physics and kinematic regimes of interest as described in Sec-

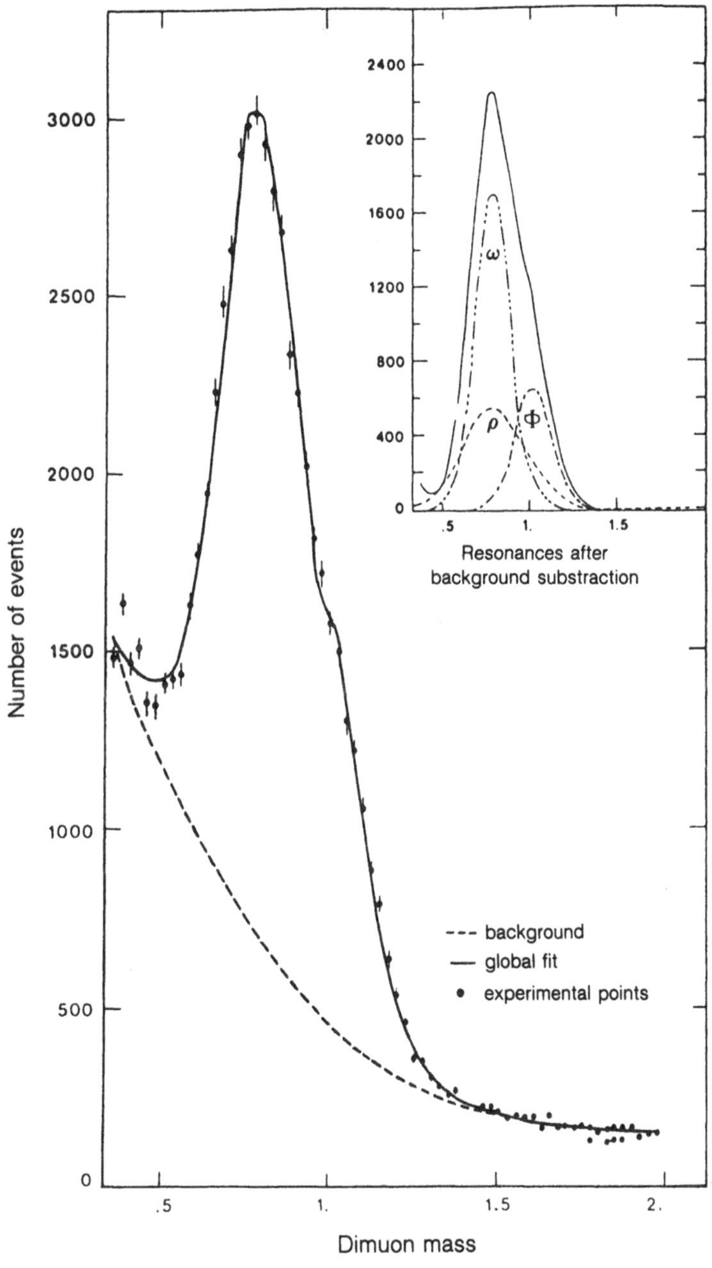

FIGURE 8 $\mu^+\mu^-$ mass spectrum obtained by NA 3 experiment in 150 GeV/c π^-p interactions.

tion 6, the difference being that they were concerned with experiments done in the colliding beam mode, where the particles of interest spread out over 4π. Figure 9 displays a picture of the geometry and kinematics in the collider frame of reference (positions of detector elements are shown). Note that the average momenta associated with different rapidity values are not large (<1-2 GeV/c) for values of y near mid-rapidity. Also note that the extended source present in the collider will require some close-in tracking capability in order to differentiate between multiple event vertices.

The group did not undertake the design of a specific apparatus; however, they did conclude that a device as shown in Figure 10 (asymmetric version of a detector discussed at the Bielefeld[7] meeting) had many of the features required to

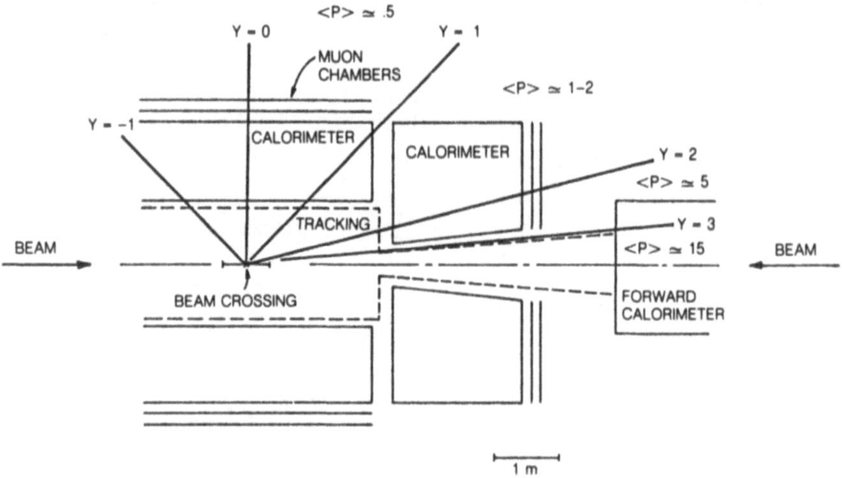

FIGURE 9 Schematic layout of the forward half of a collider detector showing the angular correspondence with pseudo-rapidity ($y = -\ln(\tan\Theta/2)$); this corresponds to the rapidity of particles whose $m \ll p_T$. The approximate average momentum, <p>, is given for each rapidity interval.

make it a useful dimuon detector. That is, it should be compact (to filter out hadrons) and be able to detect muons with $p_\mu > 5$ GeV/c (momentum needed for effective hadron filtering). Also if one is interested in studying the small pair mass and transverse momentum characteristics of the plasma, one is restricted to $|y| > 2$, where the geometry and kinematics are similar to fixed target experiments, consistent with the design in Figure 10. The practical limit for pair mass muons in the central region in a collider would be $M_{\mu\mu} > 3$ GeV--well above the mass region expected to be strongly populated by a thermalized quark-gluon plasma.

Using low mass electron pairs for probing the plasma will be difficult due to the high multiplicity expected. It was suggested that any electron detector should exploit the reduced numbers of particles expected at mid-rapidity. Dif-

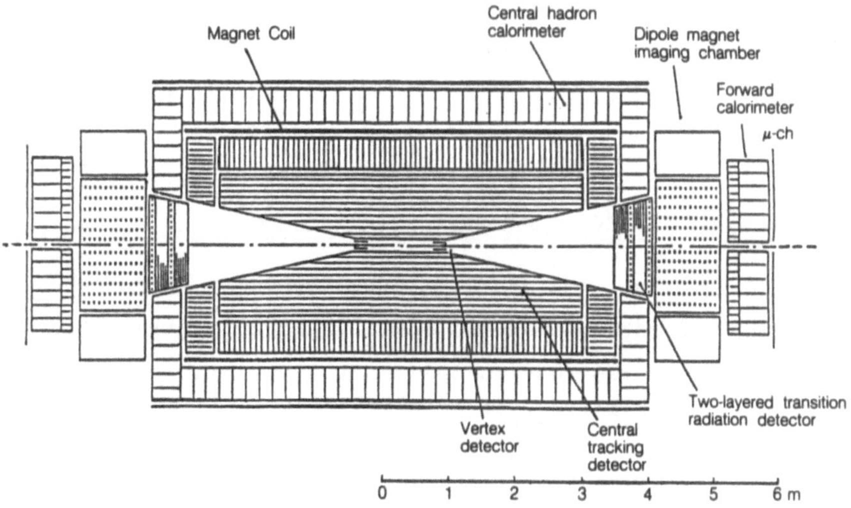

FIGURE 10 A 4π detector for high energy colliding beam operations derived from a workshop study (Ref. 7) for a fixed target device at the CERN SPS. To make this detector, two of the SPS fixed target designs were placed back-to-back.

ficulties arise due to tracking in this environment, and an electron/hadron rejection of $\gtrsim 10^5$ is needed to be able to identify direct electrons. A specialized detector with relatively small aperture at $\Theta_{c.m.} = 90°$ appears to be the best choice. The group noted that electron identification techniques are not easily incorporated in a large acceptance system, such as the one shown in Figure 10.

High p_T photons ($p_T \gtrsim 8$ GeV/c) are rare processes in hadron-hadron collisions and require high luminosities to be seen. However, they may be enhanced in central nucleus-nucleus collisions and should be looked for (luminosity will be a dominant issue). Shower counters, capable of resolving individual photons, will be required--all in a high multiplicity environment. The group concluded that much work remains in this area before experiments are undertaken.

One final physics point is worth noting. They discussed the possibility that jet-production[17] may be an interesting new signature for the plasma. Collider energies in excess of 50-75 GeV/nucleon will be required before such studies can even be started.

8. HADRONS AND EVENT PARAMETERS--COLLIDER

This group considered detectors that address a broad range of needs for both hadron and event parameter experiments. For hadron studies they considered detection of baryons, mesons, and their antiparticles with the idea of obtaining information on: (i) strange/nonstrange and baryon/antibaryon production ratios; (ii) relative yields of particles with $|s| = 0, 1, 2, 3$; (iii) $\pi\pi$, KK, pp intensity interferometry; and (iv) studies of multipion and multikaon systems.

Event parameter measurements include: (i) multiplicity and energy and momentum flow as functions of rapidity and azimuthal emission angle, (ii) correlations between dn/dy and $\langle p_T \rangle$, and (iii) fluctuations in these observables.

Also considered were the similarities and differences for collider detectors operating between 2 + 2 GeV/nucleon and 100 + 100 GeV/nucleon. One immediate comment is that if the plasma is characterized by a temperature, say T = 200 MeV, the average energies of particles emitted from such a thermal source will be relatively low (e.g., π's ≈ 500 MeV, K's ≈ 400 MeV). This means that the detector (particularly at mid-rapidity) will generally be sampling soft hadrons. This puts an added burden on calorimetry for these particles. In addition, the relatively low energy reduces the survival probability for those particles which are unstable (e.g., ~88% of the pions produced from a 200 MeV plasma will survive a 3 meter flight path and ~52% of the kaons).

The group's general aim was to define a relatively simple collider detector, one employing tracking over a limited solid angle while providing calorimetry over the remaining nearly 4π. In this scheme, the target/projectile fragmentation regions were not considered, since at collider energies most of the fragmentation products remain inside the beam pipe. Figure 11 shows the layout of a collider detector. It consists of the following elements: (i) a multiplicity barrel surrounding the interaction region capable of measuring $dn/d\eta$ with good accuracy, (ii) electromagnetic and hadronic calorimetry over at least the central two units of rapidity ($40° \lesssim \theta_{c.m.} \lesssim 140°$), and (iii) an open region into which one could place either a visual detector like a streamer chamber or a charged particle spectrometer with the ability to track up to 20 particles. The general philosophy

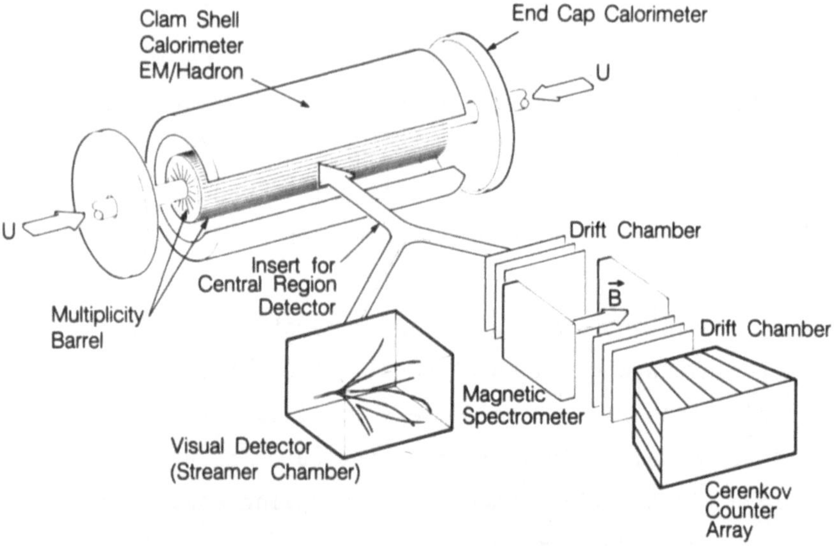

FIGURE 11 Possible collider detector.

was to use the calorimetry section for triggering (high p_T or E_T) and the streamer chamber or spectrometer arm for measuring single and few particle correlations. The capabilities of a streamer chamber are well known for relatively high multiplicity events and for observing decays of unstable particles like $\Lambda^0 \rightarrow p\pi^-$. The magnetic spectrometer arm would have to possess good tracking and particle identification capabilities. Note that the bending will be in the vertical plane, allowing equal acceptance for positive and negative particles. Such a device, concentrating on the physics of the central region, should have a large dynamic range of operation, i.e., work as well at a few GeV/nucleon as at tens of GeV/nucleon, since the plasma once achieved should be characterized by a single temperature (relatively independent of incident energy).

The group performed a simple Monte Carlo study of the magnetic spectrometer arm's acceptance for HBT experiments.

They assumed: (i) 10^5 pairs needed to provide a sensitivity of Δ (radius) $\approx$ 0.2 fm, Δ (lifetime) $\approx$ 0.7 fm/c, and Δ (intercept-degree of coherence) $\approx$ 0.1; (ii) luminosity($\mathcal{L}$) of 10^{24} $cm^{-2}sec^{-1}$ for uranium-uranium collisions; and (iii) rate $\propto \sigma\mathcal{L}(dn/dy)^2$. From the Monte Carlo they obtained 1.5×10^4 like sign π-pairs/day, so that about one week's run would provide detailed HBT information.

9. CONCLUSIONS

Several lessons were learned from the Workshop, including:

1. Experiments are do-able, but: (i) in high flux environment one will probably have to give up tracking all particles and go to calorimetry, (ii) selective tracking will be required (e.g., to distinguish between multiple interaction vertices in collider experiments).

2. R&D issues identified: (i) particle identification (thin TPC's, RICH, TRD (Transition Radiation Detectors), (ii) low mass detectors needed to reduce conversions of γ-rays ($\gamma \rightarrow e^+e^-$), secondary interactions, and production of δ-rays (particularly by high-Z projectiles) that can produce a soft background in detectors and decrease spatial resolution of drift/wire chambers, and (iv) use and development of high technology items (e.g., Li-strip detectors).

3. Need for input from high energy physics: (i) both their interest and participation, (ii) experience and expertise with large detectors such as UA1 allows us to obtain information on tracking algorithms, pattern recognition problems, and how to calibrate large-scale systems.

4. More theoretical work needed on question of signatures of the quark-gluon plasma (QGP)--theorist's R&D.

10. ACKNOWLEDGMENTS

I want to express my appreciation to the Organizing Committee of this Conference for inviting me to speak. Special thanks to the conveners at the LBL Detector Workshop; without their considerable efforts, the Workshop would not have been the success it was.

REFERENCES

1. Proceedings of the Workshop on Detectors for Relativistic Nuclear Collisions, edited by L.S. Schroeder, LBL-18225 (September 1984).
2. L.S. SCHROEDER, Detectors for High Energy Nucleus-Nucleus Collisions, LBL-18310 (August 1984), to be published in the Proceedings of the Quark Matter '84 Meeting, held in Helsinki, Finland, June 17-21, 1984.
3. A Long Range Plan for Nuclear Science, A report by the DOE/NSF Nuclear Science Advisory Committee, (December 1983).
4. RHIC and Quark Matter, Proposal for a Relativistic Heavy Ion Collider at Brookhaven National Laboratory, BNL-35237 (August 1984).
5. L.S. SCHROEDER, The Berkeley Mini-Collider, LBL-17990 (June 1984), to be published in the Proceedings of the Conference on the Intersections Between Particle and Nuclear Physics, held in Steamboat Springs, Colorado, May 23-30, 1984.
6. G.R. YOUNG. J.A. MARTIN, and J.B. BALL, Nucl. Phys. A418, 361 (1984).
7. See detector related sections of the Proceedings of the Bielefeld Workshop, Quark Matter Formation and Heavy Ion Collisions, edited by M. Jacob and H. Satz, (World Scientific, Singapore, 1982).
8. See detector related sections in the Proceedings of the Third International Conference on Ultra-Relativistic Nucleus-Nucleus Collisions (Brookhaven, September 1983), edited by T.W. Ludlam and H.E. Wegner, published in Nucl. Phys. A418 (1984).
9. See detector related sections in the Proceedings of the Workshop on SPS Fixed Target Physics in the Years 1984-1989, Vol. I and II, edited by J. Mannelli, CERN 83-02 (February 10, 1983).

10. M. GUYLASSY, Nucl. Phys., A418, 59 (1984).
11. L. McLERRAN, Signatures for New Phenomena, to be published in the Proceedings of the Quark Matter '84 Meeting, held in Helsinki, Finland, June 17-21, 1984.
12. The conveners were: Fixed Target: Hadrons--C. Gruhn (LBL), I. Otterlund (Lund); Event Parameters--K. Foley (BNL), and A. Poskanzer (LBL); Penetrating Probes--J. Carroll (UCLA), M. Goldberg (Syracuse); Collider: Hadrons and Event Parameters--W. Carithers (LBL), T.J.M. Symons (LBL), G. Young (ORNL); Penetrating Probes--T. Ludlam (BNL), L. Madansky (Johns Hopkins).
13. T.M. BURNETT et al., Phys. Rev. Lett., 50, 2062 (1983).
14. C. GRUHN et al., SPS Proposal, CERN/SPSC 84-13 (1984).
15. J. RAFELSKI and M. DANOS, GSI-83-6 (1983) and references therein.
16. T. LUDLAM, private communication.
17. T. LUDLAM, F. PAIGE, L. MADANSKY, Very High Energy Probes of the Quark-Gluon Plasma, Proceedings of the Workshop on Detectors for Relativistic Nuclear Collisions (Ref. 1 above), p. 115.

EXPERIMENTS AND DETECTORS FOR HIGH ENERGY HEAVY ION COLLIDERS

T. LUDLAM
Physics Department
Brookhaven National Laboratory
Upton, New York

1. INTRODUCTION: The Nature of the Experiments

In the previous talk, by Lee Schroeder,[1] you have seen an overview of the detector considerations for future experiments with high energy nuclear beams. By and large, these experiments are shaped by the expectations for new phenomena to be realized when nuclear matter is brought to extreme states of temperature and density. Lee has concentrated on the fixed-target experiments which are being prepared for the relatively near future at Brookhaven and CERN: experiments at collision energies which are expected to approach the limit of nuclear transparency and thus produce high energy density through compression and heating of baryon-rich nuclear matter.

In the present talk my focus will be on the problems and possibilities for experiments at the highest collision energies achievable in man-made accelerators; i.e. colliding beams of heavy nuclei at c.m. energies $\gtrsim$100 GeV/amu, well beyond the threshold of nuclear transparency. Here the final state consists of two hot, dense, baryon-rich fireballs flying away from each other at large rapidity (the fragmentation regions), and thermally-produced particles with near-zero net baryon number populating the central rapidity range. The matter produced at central rapidity (the lab frame for a collider) may reach extremely high temperatures and energy densities, and it is here that one expects to produce thermodynamic conditions similar to those which existed when the early universe condensed from a plasma of quarks and gluons to a gas of hadrons.

Figure 1 shows the layout of the Brookhaven accelerator complex, including the partially-completed ISABELLE colliding beams accelerator which is now being proposed as a dedicated Relativistic Heavy Ion Collider (RHIC).[2] For the sake of specificity the discussion here will assume a collider facility with the design parameters of RHIC: beam masses and energies up to gold (A≃200) at 100 GeV/amu with luminosity $L > 10^{26}$ cm^{-2} sec^{-1}.

We are familiar with many examples of large, expensive and extraordinarily complex detection systems for studying collisions of elementary particles in e^+e^-, pp and $\bar{p}p$ colliders. Detectors at a heavy ion collider, while they may also be complex and expensive, are likely to be very different in their design and function from their counterparts in high energy physics experiments. Machines such as LEP, the $\bar{p}p$ colliders, and the Superconducting Super Collider are undertaken to explore the realm of hard scattering processes to resolve phenomena at the shortest possible interaction distances. This leads to experiments detecting leptons and jets at high transverse momenta, which are rare events, and little or no concern for sensitivity to the "soft" particles which are the typical reaction products.

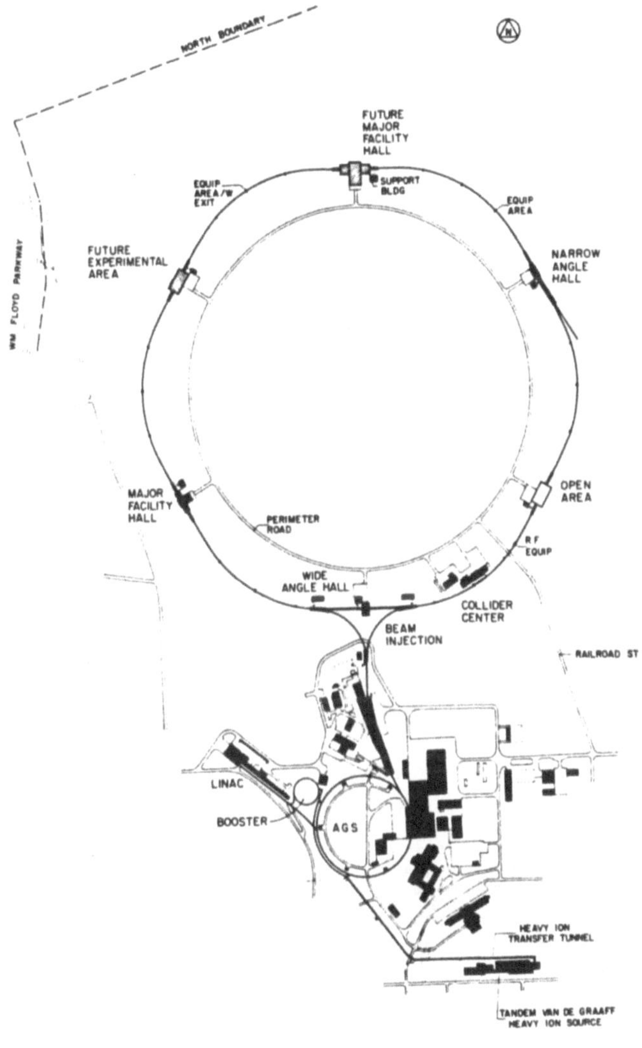

Fig. 1. Site map of the Brookhaven accelerator complex, showing the 30 GeV Alternating Gradient Synchrotron (AGS) proton accelerator with heavy ion injection from the Tandem Van de Graaff facility. The AGS would serve as injector to the (unfinished) colliding beams facility.

TABLE I. Experimental signals for new states of matter.

SIGNAL	COMMENTS	
INCLUSIVE PARTICLE SPECTRA PARTICLE INTERFEROMETRY	INDICATORS OF TEMPERATURE, SIZE AND DENSITY	GLOBAL EVENT PARAMETERS
MULTI-PARTICLE CORRELATIONS IN RAPIDITY; ENERGY FLOW	LONG RANGE CORRELATIONS AND MACROSCOPIC FLUCTUATIONS CHARACTERISTIC OF FIRST-ORDER PHASE TRANSITION	
LOCAL CHARGE CORRELATIONS	COLOR SCREENING EFFECTS IN PLASMA DIFFERENT FROM NORMAL PAIR PRODUCTION BY VACUUM POLARIZATION	INDICATORS OF A PHASE TRANSITION
PARTICLE FLAVOR RATIOS	CHEMICAL EQUILIBRIUM IN HOT PLASMA GIVES A LARGE NUMBER OF STRANGE PARTICLES AND ENHANCED $\bar{\Lambda}/\bar{p}$ RATIO	
STABLE MULTIQUARK STATES	6-QUARK AND HIGHER CONFIGURATIONS READILY ASSEMBLED IN THE PLASMA	
DIRECT PHOTON PRODUCTION ($m_T = p_T$) LEPTON PAIR PRODUCTION (VIRTUAL PHOTON: $m_T^2 = m_{PAIR}^2 + p_T^2$)	$m_T \lesssim 50$ MeV: COHERENT EMISSION FROM LOCAL CHARGE FLUCTUATIONS $50 \lesssim m_T \lesssim 500$ MeV: HADRONIC DECAYS; SOME COHERENT EFFECTS $500 \lesssim m_T \lesssim 3$ GeV: DIRECT EMISSION FROM PLASMA $m_T \gtrsim 3$ GeV: APPROACH TO EQUILIBRIUM; STRUCTURE FUNCTIONS OF QUARKS AND GLUONS CHANGE AND ARE COMPUTABLE IN PERTURBATIVE QCD	PENETRATING PROBES: DIRECT INFORMATION FROM THE PLASMA
HIGH-p_T JETS	MEASURES PROPAGATION OF QUARKS AND GLUONS THROUGH NUCLEAR MATTER; HADRONIZATION PROPERTIES REFLECT THE "REAL SEA" OF QUARK-GLUON PLASMA	

By contrast, the quark-matter physics of high energy nuclear collisions emphasizes the long distance aspects of QCD, and the signals of interest involve particles whose momenta are characteristic of the equilibrium thermodynamic conditions reached in the collision. This means that most experiments will be designed to look for signals carried by relatively soft particles. Table I summarizes the kinds of signals which are often discussed.[3] They emphasize measurement of detailed global patterns of energy flow (in which the importance of individual particle measurements is suppressed); the relative rate of heavy flavor production (strangeness and charm); short-range correlations among small numbers of particles, including resonance masses and particle interferometry (the latter is a completely new form of measurement made possible by the high multiplicity and large spatial extent of interesting events - these measurements require precise measurement of individual tracks, possibly over a limited solid angle); and the detection and reconstruction of relatively rare froms of particle emission which are thought to serve as "penetrating probes" of the initial stages of the nuclear collision (virtual photons, seen as lepton pairs, or hard-scattered quarks and gluons seen as jets). In general these types of measurements must be made on an event-by-event basis, with some kind of trigger selection on small impact parameter and large energy deposition. The final states thus selected will be very different from those encountered in high energy physics experiments: the particle multiplicity may range into the thousands. An educated guess at the average rapidity distribution of charged particles for three different collider energies is shown in Figure 2. Figure 3 relates the rapidity spectrum to intervals of laboratory angle in a possible

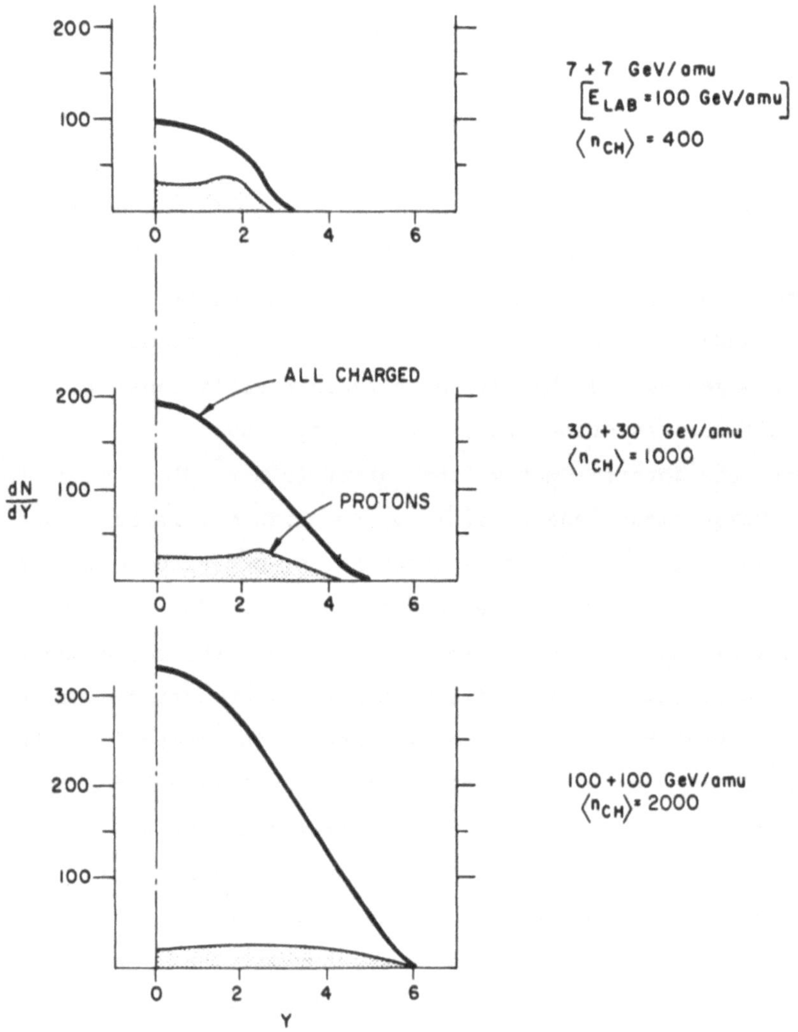

Fig. 2. Illustrating the expected charged particle spectra for Au + Au collisions at various collider energies. The shaded areas are the net proton densities extrapolated from the stopping power measurements in proton-nucleus collisions. $\langle n_{CH} \rangle$ is the mean multiplicity of charged particles.

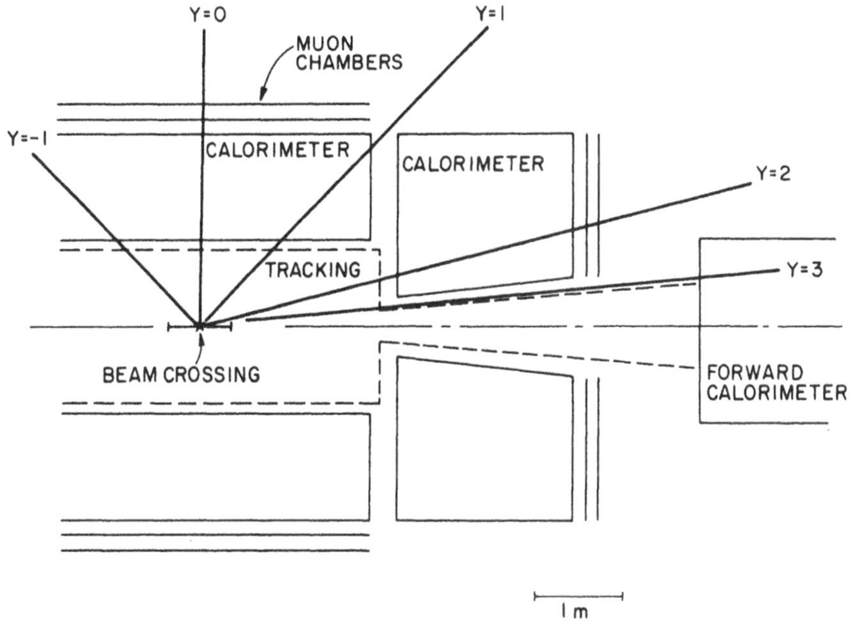

Fig. 3. Schematic layout showing the geometry and kinematics of a detector system for colliding beams experiments. The angular intervals corresponding to various rapidity (y) values are indicated. A full solid-angle detector would have the same coverage on either side of y = 0; this illustration shows coverage of the central region and one of the fragmentation regions.

layout of detectors.

In the following sections I will examine some specific detection problems for these experiments. The required detection technology for tracking, calorimetry, particle identification and fast trigger decisions has a great deal in common with components of high energy physics experiments. For general discussions of the various detector types and their elementary properties some suggested sources are given in Refs. 3-5.

2. THE PROBLEM OF TRACKING

The problem of tracking in these experiments is one of dealing with extraordinarily high multiplicities of particles: upwards of 1000 charged tracks in the central rapidity region (Fig. 2), corresponding to ~200 per steradian of solid angle. In point of fact, this density of tracks is not greater than those encountered in present experiments looking at high-p_T jets. The measurement of local track densities of this magnitude is not beyond the capability of present detector technology. The difficulty, of course, is that in nuclear collision events the detectors will see such densities uniformly over all angles.

It may be asked, with so many final-state particles per event, whether it is necessary to resolve individual tracks in order to determine the salient event parameters. Indeed it is true that many of the global properties of events can be obtained through local averages over many particles, e.g. with calorimeter measurements, in ways that would not be possible, because of large fluctuations, in elementary particle collisions.

Nonetheless, we have seen that there are cases where it is desirable to resolve single particles in high multiplicity nuclear collisions: the detection of strange particles and leptons, and the measurement of correlation functions for particle interferometry are some examples. It may be acceptable to make such measurements over a limited solid angle, sampling a small fraction of the total final-state particle count. In this case today's advanced, three-dimensional "imaging detectors" are up to the job. For example, the time projection chamber (TPC) being built for the ALEPH experiment at LEP is designed to handle track densities up to ~1000/steradian in jets which are localized to ~.01 steradian.[6] Other types of wire chambers, less expensive and with better rate capability than the TPC, are being refined to the point where they may be considered for such applications. Thus, the two-track resolution in the drift direction in the conventional wire chamber is typically on the order of 5 mm. Improvements in electronics, electrode geometry, and gas chemistry have reduced this by an order of magnitude in practical prototypes that are now being built.[7,8]

If it is necessary to cover large solid angles, recording hundreds of tracks in a single event, the difficulty lies not in the technical limitations of the detectors themselves, but in the data acquisition and processing. The system, consisting of detector and track reconstruction algorithm, must be designed so that the processing time remains approximately linear with the number of tracks up to very high track counts. The intrinsic limits have not been explored in real experiments, although studies of a particular case indicate that the level of ~1000 tracks per event can be reached with presently understood techniques.[7]

3. LEPTON MEASUREMENTS

Effective measurement of low-mass electron pairs in the high multiplicity environment of nucleus-nucleus interactions is widely recognized as an important and formidable challenge for the design of experiments. Detailed calculations of detector response and background rates have been carried out for specific cases of experiments planned for fixed target operation.[9,10,11]

The instrumentation must be capable of distinguishing electrons from a high flux of background hadrons: electron to hadron rejection at a level $\simeq 10^5$ is required. In the collider case this might be accomplished in a specialized detector of relatively small aperture of 90° in the c.m. with the tracks most favorably disposed in laboratory angle. For efficient detection of low-mass pairs a two-arm detector would be required (the arms back-to-back). Successive, independent Cerenkov measurements would provide good hadron rejection. The Cerenkov counters have to be well segmented. If ring-imaging Cerenkov detectors[12] are developed to their apparent potential, one has a very nice solution for this configuration: a multitrack detector with good position resolution which can be made effectively blind to the multitude of background hadronic tracks.

These technical requirements for electron/hadron discrimination notwithstanding, the most severe fundamental limitation to the study of electron pairs in the mass and p_T range $\lesssim 1$ GeV is the level of background electrons from secondary processes. Given the charged particle densities indicated in Fig. 2, we can expect in excess of 100 π° particles per unit of rapidity in the central region. Thus on the average there will be ≈ 1 e^+e^- pair per unit of rapidity from the decay $\pi^\circ \to e^+e^-\gamma$, and > 200 photons from $\pi^\circ \to \gamma\ \gamma$. If the detectors present only 2% of a radiation length of material thickness to induce pair conversions, the net result is > 10 electrons per unit rapidity in each event. This background can only be eliminated by correctly reconstructing all electrons and removing pairs with mass > 100 MeV. Design studies with detector systems optimized for this task (see Refs. 9, 10, 11) have rather consistently concluded that the limiting sensitivity for e^+e^- pairs in the mass range .5 - 1 GeV is

$$\frac{n(e^+e^-)}{n\ (\pi^\circ)} \geqslant 10^{-5}.$$

Specialized techniques for electron identification are not so easily incorporated in a large solid angle, general purpose detector such as that shown in Fig. 3. In the central rapidity region such a detector may achieve some e/π rejection through dE/dx measurements, if sensitivity is sufficiently good to work in the relativistic rise regime. This is a severe requirement for a detector which must also achieve efficient tracking at high multiplicities and remain relatively compact. Some additional hadron rejection may be gained with fine-grain segmentation in the first layers (≈20 rad.len.) of the calorimeter. Very soft hadrons ($\beta\gamma \lesssim 2$) may be rejected by time-of-flight, even in a compact geometry with ≈ 1 meter flight path, if the timing accuracy promised by Pestov counters can be realized in a practical way.[13] In the forward directions, where the laboratory momenta are high compact transition radiation detectors may be used for electron identification with good segmentation and hadron rejection as illustrated in Fig. 4. All of these techniques require efficient and precise tracking and momentum measurement.

Muons are identified by their ability to penetrate a thick absorber, and for high momentum muons this provides a straightforward means of reducing the high multiplicity hadronic background. Muons of momentum $P\mu > 5$ GeV/c are required for effective filtering of hadrons. Thus, in the collider geometry one may not be able to efficiently measure low-mass, low P_T muon pairs in the rapidity range near $y = 0$.

The main region of interest in the spectrum of transverse mass of muon pairs is (see Table I) $M_T \sim 1\text{-}10$ GeV. As with electrons, the direct muon signal is threatened by a potentially disastrous background from secondary processes originating with the enormous number of final-state hadrons. In this case the problem is with muons from pion decay. An experiment dedicated to the measurement of $\mu^+\mu^-$ pairs might take the approach of the Mark J experiment at PETRA,[15] shown in Fig. 5. Here an active absorber (calorimeter) immediately surrounds the interaction region, providing energy flow information and removing hadrons before they decay, followed by layers of magnetized iron interleaved with drift chambers to measure the trajectories of muons. For the nuclear beams case it will be critical to achieve rapid absorption of the mesons: An effective flight path of even 10 cm before absorption, a distance comparable to hadronic interaction lengths in dense materials, would still allow a finite probability for several fake pairs. Thus a properly designed experiment may require special beam pipes and carefully crafted compromises in the measurements of energy flow in the initial absorber in order to reduce the background to a manageable level. Such compromises will also affect the accessibility of the low end of the interesting mass range near $y = 0$.

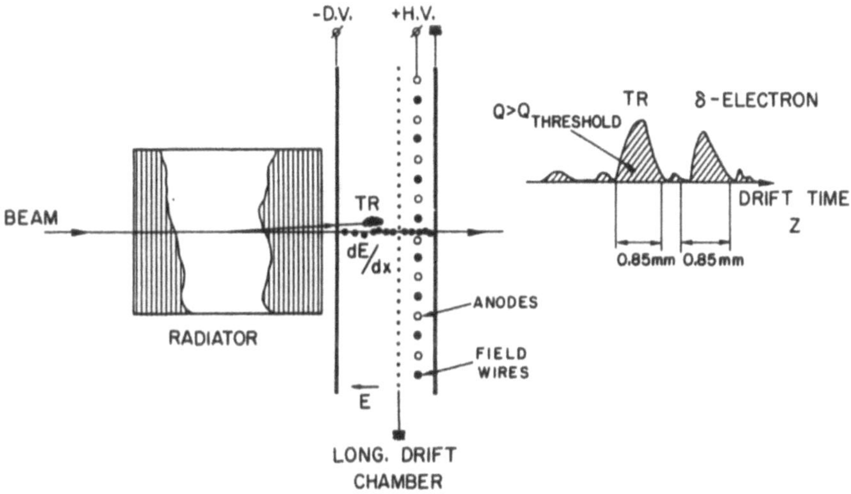

Fig. 4. Diagram showing the principle of transition radiation detection by "cluster counting" technique. The drift chamber photon detector and its high speed readout electronics are designed to resolve individual ionization electrons in the wire chamber. In this way the dE/dx signals from charged tracks (small clusters) can be distinguished from the signals due to absorption of x-ray quanta (large clusters), and the sensitivity to the TR signal can be significantly enhanced. (see Ref. 14).

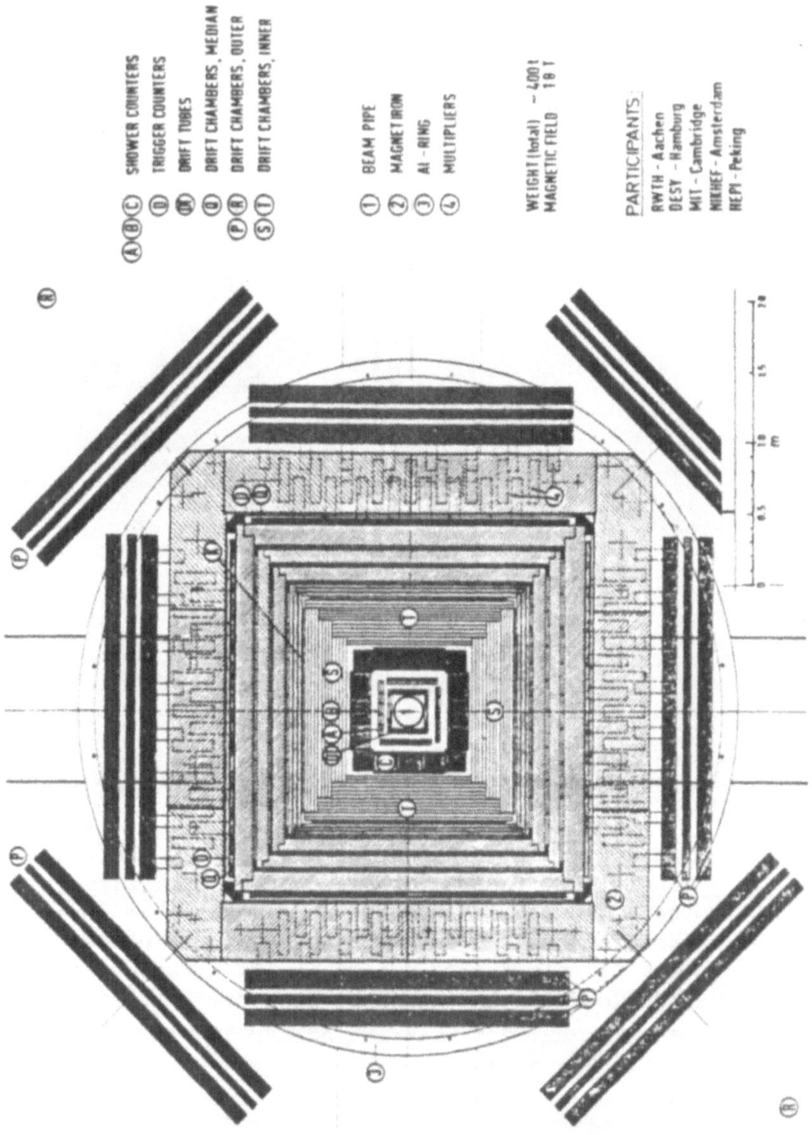

Fig. 5 View of Mark J detector for $\mu^+\mu^-$ measurements in colliding electron beams at PETRA. The beam direction is perpendicular to the plane of the drawing (Ref.15).

4. CALORIMETERS

Calorimeters measure the energy, position and direction of particles through total absorption of the incident energy in a dense material that has been laced with a matrix of active readout elements. The technology of these detectors has recently been brought to a high level of sophistication for many types of application in colliding beam experiments.[5] For nuclear beams experiments these devices will provide a powerful means of averaging over the single particle behavior to give precise patterns of energy flow, with signals developed on a time scale ($\lesssim$ 100 nanosec) which is useful for very fast trigger decisions. Calorimeters provide the only practical means of measuring the energy carried by neutral particles.

The limiting energy resolution of a calorimeter is determined by fluctuations intrinsic to the mechanism of shower development. The underlying phenomena are statistical processes whose effects grow in magnitude as $E^{1/2}$, where E is the deposited energy. Hence the limiting accuracy, expressed as a fraction of the total energy, improves with increasing energy deposit as $E^{-1/2}$. For sampling calorimeters, typical values for the fractional energy resolution (σ/E) and transverse shower size (X_T) are:[5]

Electromagnetic Showers (electrons and photons):

$$\frac{\sigma(E)}{E} \simeq \frac{.12}{\sqrt{E(\text{GeV})}} \; ; \; X_T \simeq 1 \text{ cm}$$

Hadronic Showers:

$$\frac{\sigma(E)}{E} \lesssim \frac{.5}{\sqrt{E(\text{GeV})}} \; ; \; X_T \simeq 10 \text{ cm}$$

The value of X_T gives the minimum physical size of individual read-out elements. The degree of angular segmentation is then determined by the distance of the calorimeter face from the interaction point. Figure 6 shows a calorimeter array consisting of four walls surrounding the intersection point of the colliding beams at the CERN ISR.[16] The calorimeter stacks, which here consist of interleaved layers of uranium, copper and plastic scintillator, are 1.2 meters thick. Each wall is subdivided into a checkerboard array of 20 cm × 20 cm "towers" which are read out individually. The tower size is determined by the dimensions of hadronic showers. For detection of electrons and photons the segmentation can be much finer: Fig. 7 shows, from the same experiment, an array of 4 × 4 cm^2 sodium iodide crystals which is placed on the inside face of one of the calorimeter walls.

In contrast to our problems with tracking, the performance of a calorimeter system such as this should be excellent for the high multiplicity nuclear beams case. In a collider with the parameters of RHIC, for instance, where Gold beams interact at 100 + 100 GeV/amu, the ~ 3000 final-state particles in a head-on collision carry away as much as 40 TeV of energy.

A calorimeter system covering the full solid angle and subdivided into 1000 cells would have enough energy deposited in each cell for an energy measurement at the level of ~10%. The angular distribution of the energy flow in each event is thus determined with extraordinary precision. The calorimeter need not be too deep, since the energy is carried by a large number of relatively low energy particles - an important difference from the design criteria for current high energy physics experiments. Also, a 4π calorimeter for

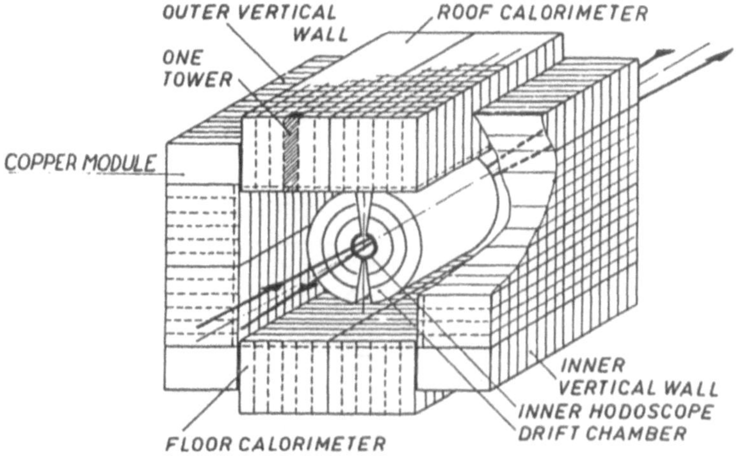

Fig. 6. The Detector Array of the Axial Field Spectrometer (CERN experiment R807/808) surrounding the ISR colliding beams (Ref. 16).

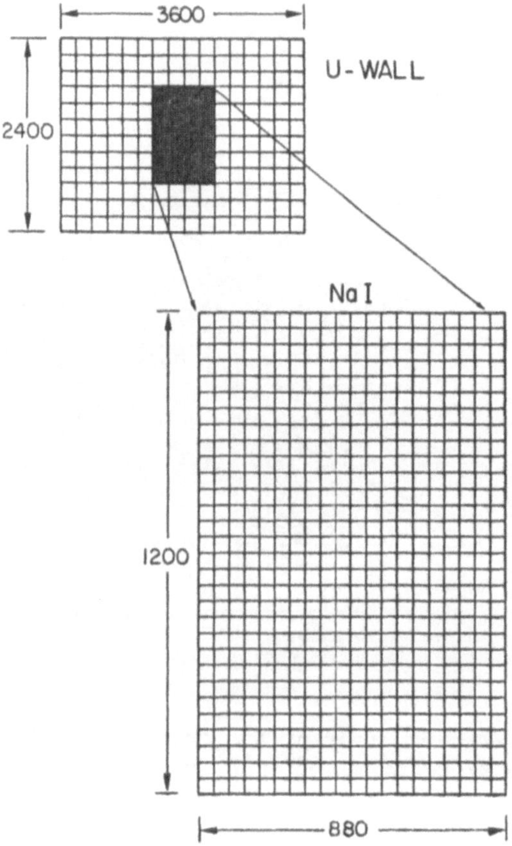

Fig. 7. An array of 600 sodium iodide crystals, each 4×4 cm^2, covering an area ~ 1 square meter on the face of a uranium calorimeter wall of the Axial Field Spectrometer.

nuclear beams experiments need not be rigorously "hermetic": The physics considerations which guide event selection do not hinge on cases where a large fraction of the total incident energy may be carried away by a single particle, and so the design can tolerate some cracks and other gaps in the coverage if necessary.

These considerations lead us to a detector concept unique to the physics of a high energy heavy ion collider:[17] a nearly - 4π calorimeter facility in which a number of small apertures are provided for special-purpose spectrometers. Events are selected according to precisely measured patterns of energy flow, while the several instrumented "ports" provide detailed measurements on individual particles in spectrometers whose apertures are small enough so that the tracking problem is manageable.

This work was supported by the U.S. Department of Energy under Contract DE-AC02-76-H00016.

REFERENCES

1. L. SCHROEDER, "Report on the LBL Workshop on Detectors for Relativistic Nuclear Collisions," these proceedings.

2. Brookhaven Proposal for a Relativistic Heavy Ion Collider, BNL 51801 (1984); see also: T. LUDLAM "RHIC and Quark Matter: A Proposed Heavy Ion Collider at Brookhaven National Laboratory", BNL 35237, to appear in Proc. of the Fourth International Conference on Ultra-Relativistic Nucleus-Nucleus Collisions, Helsinki, June 1984.

3. Phys. Today, October 1978, A special issue on "Detecting Particles."

4. C. W. FABJAN and H. G. FISCHER, "Particle Detectors", Rep. Prog. Phys., 43, 1003 (1980).

5. C. W. FABJAN and T. LUDLAM, "Calorimetry in High Energy Physics", Ann. Rev. Nucl. Part. Sci., 32, 335 (1982).

6. ALEPH collaboration, CERN proposal, LEPC/82-3Il (1982).

7. W. WILLIS, Nucl. Phys. A418, 425c (1984).

8. G. CHARPAK and F. SAULI, Ann. Rev. Nucl. and Part. Sci. 39, 285 (1984).

9. D. HENDRIE, "Dilepton Spectrometer Based on the Lampshade Magnet", Nucl. Phys. A418, 647c (1984).

10. H. GORDON et al., "Proposal for CERN Experiment NA34", CERN-SPSC/83-51 (1983).

11. H. SPECHT, "Detection of Photons and Leptons in High Energy Nuclear Collisions," These Proceedings.

12. B. BARRELET et al., Nucl. Instr. Meth. 200, 219 (1982).

13. Y. PESTOV and W. ATWOOD, Proc. Intl. Conf. on Instrumentation for Colliding Beam Physics, SLAC, 1982.

REFERENCES (cont'd)

14. T. LUDLAM et al., Nucl. Instr. Meth. 180, 143 (1981); C. W. FABJAN et al., Nucl. Instr. Meth. 216, 105 (1983).
15. Mark J. Collaboration, Phys. Reports 63, 337 (1980).
16. T. AKESSON et. al., Z Phys. C-25, 13 (1984).
17. W. WILLIS, private communication.

DETECTION OF PHOTONS AND LEPTONS IN HIGH-ENERGY NUCLEAR COLLISIONS

H.J. SPECHT
Physikalisches Institut der Universität Heidelberg
Heidelberg, Germany

Comment: The lecture delivered at this conference has had a strong overlap with several other lectures given by the author during 1984. To avoid the usual permutations, only an extended abstract is reproduced in the following. A full paper on the topic can be found elsewhere[1].

EXTENDED ABSTRACT

In the last few years, various signals have been suggested as probes of the transient plasma of locally unconfined quarks and gluons expected to occur in high energy nucleus-nucleus collisions. Among these, direct photons[2-9] and lepton pairs[8-12] are particularly attractive: the Stefan-Boltzmann behaviour of the thermal production rates ($\sim T^4$) favors the hottest stages of the collision, and the absence of final state interactions allows to directly view the interior. In terms of the transverse momenta or transverse masses $(M^2+p_T^2)^{1/2}$ involved, the region of 200 - 500 MeV/c should be the one most relevant for thermal production.

In high energy hadron collisions like pp or pN, the continuum region above a few GeV/c is reasonably well understood on the basis of single parton interactions: for inclusive real photon production the QCD Compton process[13], for e^+e^-- or $\mu^+\mu^-$-pair production the Drell-Yan mechanism[14,15]. The region below 1 GeV/c, on the other hand, is presumably dominated by non-perturbative QCD-processes and is at present not understood at all. In the case of real photons, a non-trivial signal exists for 10 MeV/c $\leq p_T \leq$ 100 MeV/c, $|x_F| \leq 0.01$, above what can be expected from the radiative decay of known hadrons[16,17)], while nothing is known for 100 MeV/c $\leq p_T \leq$ 1 GeV/c. In the case of lepton pairs, an "anomalous" continuum is known to exist for 200 MeV/c^2 $\leq M \leq$ 1 GeV/c^2 since several years[18-24], with a possible connection to the puzzling low-p_T inclusive electrons observed at the ISR[25-27]. It appears to be of great importance to not only follow the nuclear case, but also to achieve, experimentally and theoretically, more progress for the hadron case.

The forthcoming experiment of the NA34 collaboration consists, indeed, of two parts, one[28] devoted to pN collisions, the other[29] to AA collisions; the case of pA is, to some extent, incorporated in both. The overall lay-out, shown in Figure 1, contains five major components: a target-vertex detector system, a 4π-calorimeter with separate sections for electromagnetic and hadronic energy, a compact forward electron spectrometer, a forward muon spectrometer, and an external spectrometer at sideward angles. Smaller components exist, in addition, for the detection of real photons. A sixth major component, an electron spectrometer at larger angles, may be added in the future for the nuclear case.

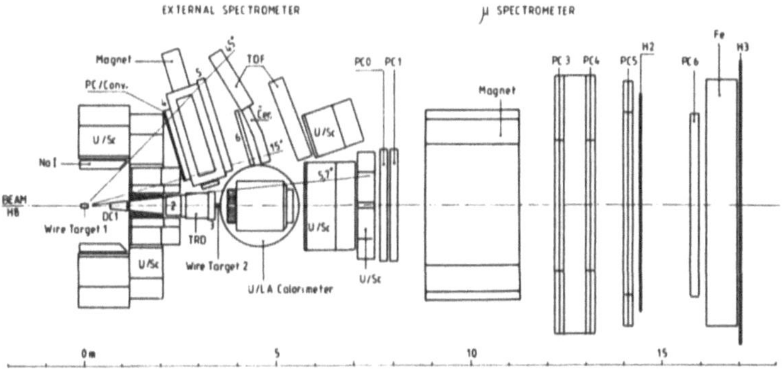

Figure 1: Top view of the experimental configuration of the NA34 collaboration[29]

The first part of the experiment[28] is devoted to a high-accuracy study of the production of electrons, muons and neutrinos. The main physical issues concern the problem of e/μ universality and the anomalous low-mass lepton pairs and low-p_T leptons. This is indeed the first experiment in the field which combines electron identification, muon identification, missing energy measurement (for neutrinos), and accurate vertex determination (for the direct identification of open charm decays).

The muon spectrometer uses major components of the previous NA3 dimuon experiment, but rearranged in a more compact form. The forward calorimeter also serves the purpose of the necessary hadron filter. Tracks in the muon spectrometer can be matched with upstream tracks inside the electron spectrometer to achieve an excellent mass resolution. The acceptance covers the mid- and forward rapidity part and the whole mass regime of interest.

The forward electron spectrometer utilizes special drift chambers with high multitrack capability and a calorimetrized weak field (dipole) magnet. Hadron rejection on a level $>10^5$ is achieved by combining a set of transition radiation detectors (TRD) with a high-granularity uranium/liquid argon calorimeter (U/LA). Photon conversions and low-mass Dalitz decays can be vetoed on the trigger level: The low field integral allows a straight-line interpolation between a downstream hit and the target, consisting of a beryllium wire with a diameter of 50 μm and a length of 20 mm (in the beam direction); a Si-pad detector at a distance of 15 cm, matched in accuracy to the overall system, then recognizes close pairs by double dE/dx at the respective intersection point. Several Si-strip detectors, also installed close to the target, serve for accurate vertex determination in connection with charm decays.

Both spectrometers become even more powerful in connection with the 4π -calorimeter. Apart from the forward part, this consists of the uranium/scintillator modules developed and used before in R807/808 at the ISR. For the first time therefore, the dependence of inclusive lepton- and lepton pair production on the hadronic environment, e.g. large transverse energy deposition, can be studied. It does not need to be stressed that pN and pA collisions appear equally attractive in this context.

The nuclear part of the experiment[29] emphasizes the

exploratory nature of the first phase. Calorimetry will play a decisive role, both in terms of self-interest (energy flow, rapidity structure) and trigger function (central collision-, high-E_T selection). The external spectrometer at wide angles, containing particle identification with silica aerogel Cerenkov detectors and time-of-flight hodoscopes and viewing the target through a narrow horizontal slit in the calorimeter, allows to study inclusive spectra (strangeness production, $\langle p_T \rangle$ versus E_T etc) as well as identical particle correlations, and to search for long-lived unusual charge or mass particles. The passive single wire target[28] is replaced by an active multiwire target to minimize secondary interactions and photon conversions, to detect the vertex with high accuracy, and to recognize and eliminate interactions from beam fragments.

The spectra of real photons are investigated with three different approaches. The first uses conversion of the photons in a special converter-wire-chamber arrangement immediately in front of the external spectrometer, thus covering an angular region of 15^0-45^0. This region is of particular interest because of maximum particle production in asymmetric nuclear collisions. The second method exploits a reuse of two arrays of 600 NaI crystals (each) from the R808 experiment. The third uses a small array of shielded BGO crystals, arranged as to give some directional sensitivity. This array will be movable to scan a wide rapidity region. All set-ups will also be used in connection with pN and pA collisions.

The muon spectrometer, situated behind the hadron filter

and thus not subject to charged particle loads in the hundreds, will be the major tool for lepton studies. It is served by a second active wire target immediately in front of the U/LA calorimeter to minimize the random combinatorial background from $\pi \to \mu\nu$ decays. Matching to upstream tracks becomes impossible, due to the prohibitive particle load in the (unusable) electron spectrometer, resulting in some loss of mass resolution (limited now by multiple scattering in the calorimeter). The acceptance of the instrument is remarkably good: for the running conditions foreseen with the 200 GeV/c ^{16}O-beam at CERN, $> 10^5$ events in the region of anomalous pairs can be accumulated over a running time of 600 h (assuming no new physics to occur).

As an approach complementary to the muon spectrometer, a new electron spectrometer is also under consideration. It is motivated by desires to extend the mass range of dileptons down to 100 MeV/c^2, to extend the rapidity coverage to larger angles, to obtain an improved mass resolution (complete ρ, ϕ separation), and to allow the first target to be used in connection with lepton pairs (more complete event information). The problem one is faced with in this case is twofold: a weak source of rather open electron pairs has to be detected in the presence of hundreds of charged particles; lack of complete detection efficiency for trivial low-mass pairs may create a prohibitive combinatorial background. Yet, extensive Monte Carlo simulations appear to be encouraging. The present concept consists of a ring image Cerenkov counter, a ring of BGO crystals, and a weak-field magnet together with a system of pad detectors, all arranged within 1 m around the target. The RICH, operated in a threshold mode, presents the main tool for hadron rejection. Momentum analysis occurs in the electromagnetic calorimeter; the magnetic field is chosen sufficiently low to keep the pairs together, but just high enough ($Bl \approx 0.01$Tm) to obtain a detectable deflection for charge determination. Further investigations are being pursued to find the optimal solution to this extremely challenging problem.

REFERENCES

1. H.J. SPECHT, in Quark Matter 1984, Proc. of the Fourth Intern. Conf. on Ultrarelativistic Nucleus-Nucleus Collisions, Helsinki 1984, to appear in Lectures Notes in Physics, Springer, Heidelberg
2. E.L. FEINBERG, Nuovo Cimento 34A, 391 (1976)
3. F. HALZEN and H.C. LIU, Phys. Rev. D25, 1842 (1982)
4. W.J. WILLIS, Workshop on Future Relativistic Heavy Ion Experiments, Ed. R. STOCK and R. BOCK, Darmstadt, GSI 81-6 (1981)
5. T. GOLDMAN et al., Phys. Rev. D20, 619 (1979)
6. J. KAPUSTA, Phys. Rev. C15, 1580 (1977)
7. J.D. BJORKEN and L. MCLERRAN, Fermilab-Pub-84/35-T (1984)
8. E.V. SHURYAK, Phys. Lett. 78B, 150 (1978); Sov. J. Nucl. Phys. 28, 408 (1978)
9. K. KAJANTIE and H.I. MIETTENEN, Z. Phys. C9, 341 (1981) and Z. Phys. C14, 356 (1982)
10. G. DOMOKOS and J.I. GOLDMAN, Phys. Rev. D23, 203 (1981) G. DOMOKOS, Phys. Rev. D28, 123 (1983)
11. S.A. CHIN, Phys. Lett. 119B, 51 (1982)
12. J. BADIER et al., CENS preprint, DPhPE 82-11 (1982). Note that eq.A.8 must be multiplied by 2.
13. T. FERBEL and W.R. MOLZON, Rev. Mod. Phys. 56, 181 (1984)
14. S.D. DRELL and T.M. YAN, Phys. Rev. Lett. 25, 316 (1970) and Ann. Phys. 66, 578 (1971)
15. R. STROYNOWSKI, Phys. Rep. 71, 1 (1981)
16. A.T. GOSHAW et al., Phys. Rev. Lett. 43, 1065 (1979) and Phys. Rev. D24, 2829 (1981)

17. P.V. CHLIAPNIKOV et al., preprint (1984) and Phys. Lett. 141B, 276 (1984)
18. K.J. ANDERSON et al., Phys. Rev. Lett. 36, 237 (1976), and G.G. HENRY, Ph.D. THESIS, Univ. of Chicago (1978), quoted in [15]
19. B. HABER et al., Phys. Rev. D22, 2107 (1980)
20. K. BUNNELL et al., Phys. Rev. Lett. 40, 136 (1978)
21. J. BALLAM et al., Phys. Rev. Lett. 41, 1207 (1978)
22. S. MIKAMO et al., Phys. Lett. 106B, 428 (1981)
23. D. BLOCKUS et al., Nucl. Phys. B201, 205 (1982)
24. M.R. ADAMS et al., Phys. Rev. D27, 1977 (1983)
25. L. BAUM et al., Phys. Lett. 60B, 485 (1976), M. BARONE et al., Nucl. Phys. B132, 29 (1978)
26. M. HEIDEN, Ph.D. THESIS, Univ. of Heidelberg (1982); CERN EP82-05 (1982)
27. T. AKESSON et al., subm. to Phys. Lett. (1984)
28. H. GORDON et al., Proposal P189 to the SPSC, CERN-SPSC/83-51 (1983); accepted as NA34
29. H. GORDON et al., Proposal P203 to the SPSC, CERN-SPSC/84-43 (1984); accepted as NA34/1
The present list of names: H. GORDON, T. LUDLAM, V. POLYCHRONAKOS, D.C. RAHM, I. STUMER, C. WOODY, T.AKESSON, H. ATHERTON, H. BREUKER, C.W. FABJAN, U. GOERLACH, S. KATSANEVAS, U. MJORNMARK, J. SCHUKRAFT, W.J. WILLIS, P. GLÄSSEL, A. PFEIFFER, J. SOLTANI, H.J. SPECHT, N.J. DiGIACOMO, P.L. MCGAUGHEY, W.E. SONDHEIM, J.W. SUNIER, S. ALMEHED, G. JARLSKOG, B. LÖRSTAD, L.A. HAMEL, C. LEROY, Y. SIROIS, G. BEAUDOIN, J.M. BEAULIEU, P. DEPOMMIER, H. JEREMIE, L. LESSARD, A. LOUNIS, S. MAYBUROV, A. SHMELEVA, V. CHERNIATIN, B. DOLGOSHEIN, Yu. GOLUBKOV, A. KALINOVSKY, V. KANTSEROV, P. NEVSKY, A. SUMARAKOV, V. SIDOROV, W. CLELAND, J. THOMPSON, A. GAIDOT, F. GIBRAT, G.W. LONDON, J.P. PANSART, D. BETTONI, M. GOLDBERG, N. HORWITZ, G.C. MONETI, L. OLSEN, O. BERNARY, S. DAGAN, D. LISSAUER, Y. OREN (Brookhaven, CERN, Heidelberg, Los Alamos, Lund, Montreal, Moscow, Novosibirsk, Pittsburgh, Saclay, Syracuse, Tel Aviv)

CONFERENCE PARTICIPANTS

Remy Anne, GANIL, France
Xenofon Aslanoglou, Florida State University, Tallahassee, Florida
R. L. Auble, Oak Ridge National Laboratory, Oak Ridge, Tennessee
Terry Awes, Oak Ridge National Laboratory, Oak Ridge, Tennessee
Cyrus Baktash, Oak Ridge National Laboratory, Oak Ridge, Tennessee
J. B. Ball, Oak Ridge National Laboratory, Oak Ridge, Tennessee
Dr. M. Baumgartner, Lawrence Berkeley Laboratory, Berkeley, California
F. D. Becchetti, University of Michigan, Ann Arbor, Michigan
Dr. F. Beck, Centre de Rescherches Nucleaires, France
James R. Beene, Oak Ridge National Laboratory, Oak Ridge, Tennessee
Curtis E. Bemis, Jr., Oak Ridge National Laboratory, Oak Ridge, Tennessee
Fred B. Bertrand, Oak Ridge National Laboratory, Oak Ridge, Tennessee
Carrol R. Bingham, University of Tennessee, Knoxville, Tennessee
Norbert Bischof, Physikalisches Institut der Universitat Erlangen, West Germany
G. Y. Bizard, Universite de Caen, France
J. L. Blankenship, Oak Ridge National Laboratory, Oak Ridge Tennessee
A. Breskin, Weizmann Institute of Science, Israel
Steven Bricker, Michigan State University, East Lansing, Michigan
Barry L. Burks, Oak Ridge National Laboratory, Oak Ridge, Tennessee
Dr. William Carithers, Jr., Lawrence Berkeley Laboratory, Berkeley, California
Ken Carter, Oak Ridge Associated Universities, Oak Ridge, Tennessee
Yuen-dat Chan, Lawrence Berkeley Laboratory, Berkeley, California
Thomas E. Chapuran, University of Pennsylvania, Philadelphia, Pennsylvania

Rachel Chechik, Weizmann Institute of Science, Israel
J. C. Cole, EG&G Idaho, Idaho Falls, Idaho
Tom Cormier, University of Rochester, Rochester, New York
Antonio Del Zoppo, INFN, Italy
L. C. Dennis, Florida State University Tallahassee, Florida
Dr. Richard M. Diamond, Lawrence Berkeley Laboratory, Berkeley, California
Jurgen Eberth, Universitat zu Koln, West Germany
Guenther Eichhorn, Space Astronomy Laboratory, Gainesville, Florida
Dr. Curt Ekstrom, NFL-Studsvik, Sweden
Y. A. Ellis-Akovali, Oak Ridge National Laboratory, Oak Ridge, Tennessee
Karl Erb, Oak Ridge National Laboratory, Oak Ridge, Tennessee
Claes Fahlander, Tandem Accerator Laboratory, Sweden
R. L. Ferguson, Oak Ridge National Laboratory, Oak Ridge, Tennessee
Malcolm M. Fowler, Los Alamos National Laboratory, Los Alamos, New Mexico
Anthony D. Frawley, Florida State University, Tallahassee, Florida
Dr. H. W. Fulbright, University of Rochester, Rochester, New York
Tony Gabriel, Oak Ridge National Laboratory, Oak Ridge, Tennessee
Daniel Gardes, Institut de Physique Nucleaire, Orsay, France
Jerry D. Garrett, Niels Bohr Institute, Denmark
Werner Gast, Kernforschungsanlage Julich, West Germany
Avigdor Gavron, Los Alamos National Laboratory, Los Alamos, New Mexico
Stuart B. Gazes, Lawrence Berkeley Laboratory, Berkeley, California
Donald S. Gemmell, Argonne National Laboratory, Argonne, Illinois
D. W. Gifford, University of Liverpool, England
Charles W. Glover, Oak Ridge National Laboratory, Oak Ridge, Tennessee
Yasuyuki Gono, RIKEN (IPCR), Japan
Douglas E. Greiner, Lawrence Berkeley Laboratory, Berkeley, California
Henry C. Griffin, University of Michigan, Ann Arbor, Michigan
Carl J. Gross, Florida State University, Tallahassee, Florida

E. E. Gross, Oak Ridge National Laboratory, Oak Ridge, Tennessee
Giuseppe Guarino, Instituto Nazionale di Fisica Nucleare, Italy
Dr. H. Gustafsson, Lawrence Berkeley Laboratory, Berkeley, California
Melvyn L. Halbert, Oak Ridge National Laboratory, Oak Ridge, Tennessee
James Hamill, Florida State University, Tallahassee, Florida
J. S. Hanspal, Florida State University, Tallahassee, Florida
Muhsin Harakeh, Kernfysisch Versneller Instituut, The Netherlands
Dr. J. C. Hardy, Chalk River Nuclear Laboratories, Canada
Alan Harmon, University of Virginia, Charlottsville, Virginia
Jack Harvey, Oak Ridge National Laboratory, Oak Ridge, Tennessee
Leigh H. Harwood, Michigan State University, East Lansing, Michigan
Kichiji Hatanaka, The Institute of Physical and Chemical Research, Japan
Jorma Hattula, University of Jyvaskyla, Finland
David C. Hendrie, Lawrence Berkeley Laboratory, Berkeley, California
Walter Henning, Argonne National Laboratory, Argonne, Illinois
D. C. Hensley, Oak Ridge National Laboratory, Oak Ridge, Tennessee
Lars Hildingsson, SUNY, Stony Brook, New York
Kari Honkanen, Washington University, St. Louis, Missouri
Dan J. Horen, Oak Ridge National Laboratory, Oak Ridge, Tennessee
Uwe J. Huttmeier, Florida State University, Tallahassee, Florida
Ileana Iori, University Milan, Italy
David Jenkins, Virginia Polytechnical Institute and State University, Blacksburg, Virginia
Noah Johnson, Oak Ridge National Laboratory, Oak Ridge, Tennessee
T. L. Khoo, Argonne National Laboratory, Argonne, Illinois
H. J. Kim, Oak Ridge National Laboratory, Oak Ridge, Tennessee
E. D. Klema, Tufts University, Medford, Massachusetts
Toshio Kobayashi, Lawrence Berkeley Laboratory, Berkeley, California

Kris Kwiatkowski, Indiana University, Bloomington, Indiana
Dr. M. Langevin, Institut de Physique Nucleaire, France
Arlene J. Larabee, University of Tennessee, Knoxville, Tennessee
Dr. Christoph Lauterbach, Sektion Physik der Universitat Munchen, West Germany
Jay A. LeVerne, University of Notre Dame, Notre Dame, Indiana
Don Leach, University of Washington, Seattle, Washington
I. Y. Lee, Oak Ridge National Laboratory, Oak Ridge, Tennessee
Yu-Sheng Liang, Shanghai Institute of Nuclear Research, China
Thomas Lindblad, Research Institute of Physics, Sweden
C. J. Lister, University of Michigan, Ann Arbor, Michigan
T. Ludlam, Brookhaven National Laboratory, Upton, New York
Dr. W. C. Ma, Vanderbilt University, Nashville, Tennessee
Charles F. Maguire, Vanderbilt University, Nashville, Tennessee
M. R. Maier, Michigan State University, East Lansing, Michigan
F. K. McGowan, Oak Ridge National Laboratory, Oak Ridge, Tennessee
Patrick L. McGraughey, Los Alamos National Laboratory, Los Alamos, New Mexico
Dr. G. K. Mehta, Indian Institute of Technology, India
Arialdo Moroni, INFN, Italy
P. J. Nolan, University of Liverpool, England
Jerry Nolen, Michigan State University, East Lansing, Michigan
Dr. Rainer Novotny, University of Giessen, West Germany
F. Obenshain, Oak Ridge National Laboratory, Oak Ridge, Tennessee
Kazuo Omata, University of Tokyo, Japan
Masumi Oshima, Oak Ridge National Laboratory, Oak Ridge, Tennessee
Stephen Padalino, Florida State University, Tallahassee, Florida
Ron Parker, Florida State University, Tallahassee, Florida
William Piel, SUNY-Stony Brook, Stony Brook, New York
Rodney B. Piercey, University of Florida, Gainesville, Florida
Franz Plasil, Oak Ridge National Laboratory, Oak Ridge, Tennessee
V. O. Rauch, Centre de Recherches Nucleaires, France

Russell L. Robinson, Oak Ridge National Laboratory, Oak Ridge, Tennessee
David P. Sanderson, Florida State University, Tallahassee, Florida
R. Sandoval, Gesellschaft fur Schwerionenforschung, West Germany
Dr. Hans-Jurgen Sann, GSI, West Germany
Dr. D. G. Sarantites, Washington University, St. Louis, Missouri
Royce O. Sayer, Oak Ridge National Laboratory, Oak Ridge, Tennessee
Dr. John F. Sharpey-Schafer, University of Liverpool, United Kingdom
Hans Rudolf Schmidt, Lawrence Berkeley Laboratory, Berkeley, California
M. Schmorak, Oak Ridge National Laboratory, Oak Ridge, Tennessee
L. S. Schroeder, Lawrence Berkeley Laboratory, Berkeley, California
Dr. Dieter Schull, GSI, West Germany
Yves Schutz, Oak Ridge National Laboratory, Oak Ridge, Tennessee
Alan Seamster, University of Washington, Seattle, Washington
Dan Shapira, Oak Ridge National Laboratory, Oak Ridge, Tennessee
B. Shivakumar, Oak Ridge National Laboratory, Oak Ridge, Tennessee
Dr. D. Sinclair, University of Oxford, England
Terrence Sjoreen, Oak Ridge National Laboratory, Oak Ridge, Tennessee
Dr. H. Specht, University of Heidelberg, West Germany
E. H. Spejewski, Oak Ridge National Laboratory, Oak Ridge, Tennessee
P. H. Stelson, Oak Ridge National Laboratory, Oak Ridge, Tennessee
Robin L. Stern, University of Michigan, Ann Arbor, Michigan
Robert Stokstad, Lawrence Berkeley Laboratory, Berkeley, California
Yasuharu Sigiyama, Japan Atomic Energy Research Institute, Japan
Dr. Jules Sunier, Los Alamos National Laboratory, Los Alamos, New Mexico
Samuel L. Tabor, Florida State University, Tallahassee, Florida
Stephen T. Thornton, University of Virginia, Charlottesville, Virginia

K. S. Toth, Oak Ridge National Laboratory, Oak Ridge, Tennessee
N. Tsoupas, Yale University, New Haven, Connecticut
Dr. P. J. Twin, Daresbury Laboratory, United Kingdom
J. Vanderplicht, Michigan State University, East Lansing, Michigan
J. H. van Dijh, Brookhaven National Laboratory, Upton, New York
Bob Van Nooyen, Delft University of Technology, The Netherlands
Paul Vincent, Brookhaven National Laboratory, Upton, New York
Michael F. Vineyard, Argonne National Laboratory, Naperville, Illinois
Petar Volkov, Institut de Physique Nucleaire, France
Dr. Shuxian Wen, Vanderbilt University, Nashville, Tennessee
Lars Westerberg, Tandem Accelerator Laboratory, Sweden
Gary D. Westfall, Michigan State University, East Lansing, Michigan
Kevin Wolf, Texas A&M University, College Station, Texas
Craig L. Woody, Brookhaven National Laboratory, Upton, New York
Glenn R. Young, Oak Ridge National Laboratory, Oak Ridge, Tennessee
Dr. J. Yurkon, Michigan State University, East Lansing, Michigan
Robert Zurmuhle, University of Pennsylvania, Philadelphia, Pennsylvania

AUTHOR INDEX